Foreword to the Consumers Union Edition

The word "crisis" attained prominence in the United States in the early 1970s, as the country awakened to massive energy problems. To many, the term implied a temporary maladjustment. After all, "They" were working at it, and sooner or later everything would be all right again. Now in the mid-1970s, a growing number of Americans are aware that the energy crisis is worsening and promises to be anything but temporary; and that "They," bombarded with information, data, interpretations, and recommendations from a variety of interests, do not really yet know how to make everything all right again. What's more, it is unlikely that anything can ever be all right again in the sense of a return to business-as-usual. It would appear that the United States—indeed, much of the world—is in for some wrenching changes in its energy-dependent ways of life.

What will these changes be? How can they come about in ways that do least harm to consumers, the environment, and the economy? Concerned citizens must address themselves to these questions as they help to make the momentous decisions confronting society. To assist in understanding and evaluating many of the choices facing us all, and the arguments supporting and opposing them, Consumers Union arranged with Ballinger Publishing Company and the Ford Foundation to make available to the readers of CONSUMER REPORTS this edition of *A Time to Choose: America's Energy Future,* the Final Report of the Energy Policy Project (EPP) of the Ford Foundation.

We consider this report to be, overall, an excellent discussion of energy issues and policies—essentially a consumer-oriented, environment-oriented work. It should be especially useful to nontechnical readers as an introduction to the general field of energy. Those who may want to delve further into particular subject areas will find helpful guidance in the appendixes and the bibliography.

One of the most interesting and valuable sections of this book, beginning on page 349, presents the comments of the EPP's Advisory Board about this Final Report of the EPP staff. The Advisory Board includes the heads of petroleum, power, and manufacturing companies, scientists, engineers, educators, and environmentalists. Their observations, often dissenting from the conclusions and recommendations of the report, provide a broad spectrum of opinion about various aspects of the energy crisis and how to deal with them.

As with all Consumers Union special publications not of our own making, this book represents the research and thinking of its authors and contributors. It should not be assumed that Consumers Union concurs with all their data, opinions, emphases, and recommendations. For example, even if the growth rate on energy use is slowed, as recommended in the report, society will still face grave problems of energy supply, environmental damage, inequitable effects on the poor, and stresses in international relations—problems that do not receive enough attention in the report. The book does not provide much guidance for the individual who wants to "do something," or for the state or local official who would like to take action at those levels of government.

In a number of places there are statements with which we, on the basis of our experience, would disagree. On page 123, for example, the reader is told that "frost-free refrigerators require as much as two-thirds more energy than standard models"; we have found this not to be so (CONSUMER REPORTS, November 1974). In the same paragraph the report states that consumers have no way to learn the energy requirements of air conditioners; actually, the appliance industry has a widely available listing of room air conditioners and their energy requirements. There are other points we believe to be inaccurate or, as in the case of the chapter on utilities, inadequately treated. And a woeful omission for this, a reference book, is an index.

All the same, *A Time to Choose* represents a major resource for better understanding of the vital issues confronting the United States, and we commend it to all thoughtful and concerned readers of CONSUMER REPORTS.

THE EDITORS OF CONSUMER REPORTS

This edition of *A Time to Choose: America's Energy Future* is a special publication of Consumers Union, the nonprofit organization that publishes CONSUMER REPORTS, the monthly magazine of test reports, product Ratings, and buying guidance. Established in 1936, Consumers Union is chartered under the Not-For-Profit Corporation Law of the State of New York.

The purposes of Consumers Union, as stated in its charter, are to provide consumers with information and counsel on consumer goods and services, to give information and assistance on all matters relating to the expenditure of the family income, and to initiate and to cooperate with individual and group efforts seeking to create and maintain decent living standards.

Consumers Union derives its income solely from the sale of CONSUMER REPORTS (magazine and TV) and other publications. Consumers Union accepts no advertising and is not beholden in any way to any commercial interest. Its Ratings and reports are solely for the information and use of the readers of its publications.

International Standard Book Number: 0-89043-100-0 (CU)

Library of Congress Catalog Card Number: 75-3617

A TIME TO CHOOSE

Final report by the Energy Policy Project of the Ford Foundation

A TIME TO CHOOSE

America's Energy Future

Final report
by the
Energy Policy Project
of the Ford Foundation

Ballinger Publishing Co.
Cambridge, Massachusetts
A subsidiary of
J.B. Lippincott Company

Published in the United States of America by Ballinger Publishing Company, Cambridge, Mass.

First Printing, 1974
Second Printing, 1974

Library of Congress Catalog Card Number: 74–14787

International Standard Book Number: 0–88410–023–5 (H.B.)
0–88410–024–3 (Pbk)

Printed in the United States of America

Library of Congress Cataloging in Publication Data

Ford Foundation. Energy Policy Project.
A time to choose: America's energy future.

1. Energy policy—United States. I. Title.
HD9502.U52F67 1974a 333.7'0973 74–14787
ISBN 0–88410–023–5 (HB)
0–88410–024–3 (Pbk).

Contents

Authors' Note

A word is in order as to where responsibility lies for the Energy Policy Project's conclusions. Our Advisory Board has played an important role in offering advice, and numerous other experts in industry, government and academia have also assisted us. And of course the Ford Foundation has played the key role, in sponsoring the Project. But this book was written solely by the Project's staff. Indeed, the privilege of studying the issues, weighing the conflicting views of our advisors and rendering our independent judgment was the essential attraction of the Project for those who joined the study team. The Ford Foundation has been faithful to its pledge to leave these judgments to us, and the views expressed here are those of the undersigned and no one else. When the staff disagreed, I, as Director of the Project, was the final arbiter of the issue. Throughout the course of the Project, we have striven for objectivity—recognizing that it is an often elusive goal in an area so value-laden as energy policy. Under the ground rules of the Project, we have stated our conclusions for all to see, and printed the views of the Advisory Board as well, in the hope that they will better enable citizens to make up their own minds. We hope at the very least our work will persuade the nation that it is indeed *A Time to Choose.*

S. David Freeman, *Director*
Pamela Baldwin
Monte Canfield Jr.
Steven Carhart
John Davidson
Joy Dunkerley
Charles Eddy
Katherine Gillman
Arjun Makhijani
Kenneth Saulter
David Sheridan
Robert Williams

Foreword

In December 1971 the Trustees of the Ford Foundation authorized the organization of the Energy Policy Project. In subsequent decisions the Trustees have approved supporting appropriations to a total of $4 million, which is being spent over a three-year period for a series of studies and reports by responsible authorities in a wide range of fields.

The present volume, though not the last to be published, is the Project's final report. I am glad to commend it to all who care about problems of energy and to express our hearty thanks to Mr. David Freeman and his colleagues. It is their Report, and it reflects their deep and informed conviction that we can and should have a national energy policy that serves the needs of all our people. Taken together with about twenty other volumes which will compose the Project's published output, this report fulfills the hopes with which we organized the Project.

This is also the time for us to thank the Project's Advisory Board. Our Trustees have had long experience with the study of complex issues of public policy, and they are well aware of the difficulty of conducting studies of this sort in a way that would at once assist interested citizens to understand hard issues while taking responsible account of the fact that the questions involved are genuinely difficult and surrounded by sharply contrasting interests and concerns. For this reason the Board made two decisions that have governed the work of this Project in general and the final report in particular. The first was that Mr. Freeman, as the Project's Director, should have both freedom and responsibility in the management of the Project and in preparing the final report. No other arrangement would have attracted a man of stature and integrity. The second decision was that both in reviewing special studies and in the preparation of the final report Mr. Freeman and his staff would have the continuing counsel of an Advisory Board. We were fortunate in recruiting an outstanding and varied group for membership in that Board and more fortunate still in the willingness of Gilbert White to serve as its Chairman. His services to the Project have been second in importance only to those of Mr. Freeman himself.

We assured the members of the Advisory Board that their own comments on the final report would be welcome, and that set of comments contains a great deal that is illuminating. We have been able to satisfy nearly all members of the Board of our fairness in this matter, but we have not thought it reasonable to overrule the all-but-unanimous view of the Advisory Board that no single advisor should be allowed unlimited space.

If energy questions were urgent and in some degree divisive in 1971, they have burned still more brightly, and engaged feelings more strongly than ever, in the last twelve months. The Project staff and many members of the Advisory Board brought long experience and strong prior convictions to the work of the Project, and the reader of the report and of the supplementary comments will see that beliefs reflecting the perspectives of the businessman, the conservationist, the professor, or the public servant are not readily abandoned even after repeated exchanges in committee meetings.

But careful readers will also discern, I think, something that I myself have observed in following the evolution of the Project and in watching the changes from one draft of this report to another: that men and women of intelligence and goodwill, when they have sensitive and honorable leadership, can help each other more than their initial differences would lead them to expect, and can work their way to agreements that may be at least as important as their differences. The report itself owes much to the advice and criticism offered along the way, and the remaining differences, though sometimes sharp, seem in their own way to underline the central message of the study—that it is truly *Time to Choose.* The measure of this agreement is accurately registered in the general statement of the Advisory Board on "Major Issues." There *is* an energy crisis. It did *not* come and go in 1973-74. It *will* last a long time. Conservation *is* as important as supply. We *do* need "an integrated national policy." This report constitutes a major contribution to the understanding of these quite fundamental propositions.

No one connected with this effort has ever supposed that a single private Project could be definitive. While important gaps have been at least partly filled by its special studies, the Project did not, and could not, cover all matters with equal care. In spite of the solidity of its central argument, this final report itself is, perhaps inevitably, somewhat uneven. I believe that the international energy scene, for example, and the political and economic role of the large energy corporation are subjects too complex and demanding to yield to the somewhat cursory treatment they receive in this report. Moreover, the report often reflects the tensions which inevitably exist when one is dealing with subjects on which convictions are strong while the available evidence is incomplete.

Yet it is important to remember that precisely because this is a *Time to Choose* we shall not always be able to wait for "all" the evidence before we act. For this reason among others I supported the recommendation of the Advisory Board that our initial plans for this final report should be modified to permit the inclusion of a statement of the Conclusions and Recommendations of Mr. Freeman and his colleagues. The Ford Foundation neither endorses nor rejects their judgments, but we do think it is right to have them clearly set forth. These are men and women whose study and analysis have made a major contribution to understanding of the fundamental fact that choices *must* be made. They have earned the right to say what choices they themselves would recommend, and their recommendations deserve attention.

For ourselves, we accept the shared view of the staff and the Advisory Board that this Project, with all its substantial achievements, should be viewed more as a beginning than as an end. The Foundation, through the work of its Office of Resources and the Environment, will maintain its own concern for this great range of subjects, and we will be alert for further opportunities for the encouragement of expert, disinterested, and relevant analysis. I close by repeating our grateful acknowledgment of indebtedness to all who have helped this Project try to meet that standard.

McGeorge Bundy
President, The Ford Foundation

A TIME TO CHOOSE

CHAPTER 1
Introducing energy policy

The energy crisis seems to have vanished as suddenly as it appeared. The gasoline lines have gone and auto companies are again advertising big luxurious cars. Americans once again take for granted a plentiful supply of energy.

Now that the threat seems to have lifted, why do we need a national energy policy? The fundamental fact remains that the United States has entered a new age of energy, and we have not yet adjusted our habits, expectations, and national policies to the new age. The Arab oil embargo, while it lasted, made us keenly aware that in twentieth century America, a fourth essential has been added to the age-old necessities of life. Besides food, clothing, and shelter, we must have energy. It is an integral part of the nation's life support system. And we can no longer expect to get it with so little trouble and expense as we did in the recent past.

The embargo made the American public conscious of the widening gap between energy consumption and domestic production, and of our unaccustomed but growing dependence on foreign supplies. Before it was over, energy prices, which had lagged behind other prices for a decade, caught up or passed them in one short, stunning burst.

Energy shortages not only focused attention on these problems, but also revealed the lack of a coherent national policy to deal with them. The country faced the reality that our energy budget was out of balance, and that we need to find a way to make more energy, or use less, or some combination of both. The public interest in energy policy naturally waned with the end of shortages, but the conditions that made the nation vulnerable to abrupt withdrawal of foreign supplies are likely to persist. So too are the environmental problems of energy. If the indifference and neglect that helped to create the energy gap continue, the United States could drift into a serious, long lasting energy-environment crisis.

The objective of the Energy Policy Project has been to explore the range of energy choices open to the United States, and to identify policies that match the choices. The range of choices is broad, both because the national resource base is diverse, and because there is plenty of room for improving our efficiency in the use of energy. The trend of recent years has been toward rapid growth in energy use—rapid enough to double consumption every fifteen years. But trend, as Lewis Mumford has said, is not destiny. If we have learned one thing from our work in the Project, it is that our nation's energy future is not compelled to follow a single narrow path.

As a way of cutting energy-related problems down to manageable size, the Project has given special attention to the possibilities for saving energy. We believe that the scope of potential energy savings and the benefits of slower energy growth have not yet received their just due in the national energy debate. In this book we hope to demonstrate that slower energy growth than we have recently experienced can work without undermining our standard of living, and can also exert a powerful positive influence on environmental and other problems closely intertwined with energy.

At the same time, the Project has not neglected the question of energy supply. The United States still possesses a large storehouse of energy resources in the ground. Indeed, the resource base is large enough to support continued energy growth at much the same rates as those of the past, at least until the end of this century. The problems, as we shall later describe in detail, are to find and produce these resources at a rapid rate without doing serious injury to other social goals.

It is our judgment that while real, workable choices exist, none of them is easy or automatic. An energy future based on the pattern of the past will require at least as much positive action by lawmakers, administrators, industry leaders and citizens as one that rests upon a more conservative energy growth rate. To manage rapid growth today without disruptive shortages demands, among many other things, skill (and luck) in juggling foreign policy entanglements, grave damage to the environment, and soaring costs. Government must inevitably participate in making and carrying out these hard choices.

To slow our growth in energy use will also require a national effort. It means using energy more efficiently so that a slowdown in this sector will not seriously impair economic growth and job opportunities. The Project's engineering and economic studies convince us that saving energy is possible without either disruptive social change or coercive government action. But this will require skill, forethought, and consistency at every level of government, as well as ground rules that make it reasonable for energy producers and consumers to live with slower growth.

We believe that the bundle of policies and actions that add up to slower growth offers a more sensible response to the nation's present and future energy situation than attempts to continue growing at our accustomed pace. But whatever course the nation chooses, some choice is better than none.

Drift is surely the worst of the alternatives before us. No one can foresee everything the future holds, and plans must change as new circumstances arise. But a sense of direction for energy policy is essential because many decisions must mesh consistently together, and because it takes a long time to make things happen in the energy world. For example, it takes a minimum of three years to build an oil refinery; it takes three to five years to locate a new offshore oil field and bring it into production; and it may take as long as ten years to plan and build a nuclear power plant. Fundamental to any such plans are decisions about the size of the energy supply the country needs.

Basic changes in patterns of energy use occur slowly too. The nation's energy-consuming stock of cars, trains, buses, houses, stores, and factory machinery cannot change overnight. While much can be done in a short time to tighten energy use in existing plants, buildings, and transportation systems, it will take years, even decades, to fully replace old stock with more efficient new equipment.

The process of making energy decisions is also time consuming. A change in the tax laws affecting oil companies, for example, or the application of new clean air standards affecting coal, may occupy voters and their representatives for several years. Many decisions require collaborative action by local, state, and federal governments, and lack of coordination often produces stalemate and delay.

Finally, research and development in the energy field is inherently ponderous. Twenty-five years elapsed between the infancy of atomic power and the startup of the first commercial nuclear plant. By 1973, half the electric power plants under construction were nuclear, but nuclear power that year still supplied only about one percent of the nation's energy.

Energy policy has momentum, like a mammoth supertanker carrying a quarter-million tons of crude oil, which cannot stop in less than twenty minutes and three nautical miles. Just as the decisions to develop atomic power in the 1940s are shaping energy supply in the 1970s, so too, decisions taken today will have a vital effect on energy use and energy production in the year 2000.

Intelligent decisions require a clear understanding of the goals of energy policy. One obvious goal is to ensure adequate supply. Even the recent mild shortages have demonstrated that energy is indispensable for satisfying basic human needs—feeding people, keeping them warm, providing jobs and getting them to work. If the flow of energy stops, our high-energy civilization stops too.

But the goal of adequate supply is deceptively simple. A single-purpose program will not be satisfactory because there are other important social goals related to energy that must also be pursued. Among these goals are safeguarding the environment; prudently managing foreign affairs, without undue pressure from energy problems; keeping the real social and economic costs of energy as low as possible; taking care that changing energy policies do not place extra hardships on the poor or unfair burdens on the regions that supply energy for the rest of the nation.

In the past, conflicts between these goals either did not exist or were ignored. Nonhuman, nonanimal energy seemed an unmixed blessing. It freed people from brute labor, gave them comfort and mobility, and lighted the dark. So long as the country had energy to burn, and so long as the environmental damage from producing and using it was little noticed and still less understood, the simple growth ethic—"more is better"—went without serious challenge.

Energy consumption and production rose steadily together for about a century. But within the last decade, while growth in demand accelerated, domestic production leveled off. By 1973, annual energy use in the United States had reached 75 quadrillion Btu's[a]—30 percent of the world's energy consumption, for 6 percent of its population—but production from native sources was only 62 quadrillion Btu's.

From 1950 to 1973, U.S. energy consumption increased at an average annual growth rate of 3.5 percent, while domestic production rose more slowly at just under 3 percent. In the eight years after 1965, consumption raced ahead at 4.5 percent a year; but since 1970, growth in domestic production has been at a virtual standstill.[b]

Practically all the increase in energy consumption after 1970 came from oil imports. Oil is the mainstay of the U.S. energy economy, furnishing 46 percent of total energy consumption in 1973.[c] In the same year, 35 percent of the oil used in this country was imported. About 17 percent of total U.S. energy supplies originated abroad. Not until late 1973, when the flow of foreign oil was suddenly interrupted, did we begin to realize the implications of the change from our secure historical position of energy self-sufficiency.

Actually, the United States still has a vast and varied

---

[a] Btu (British thermal unit) will be used throughout this book as the common unit of measure for all forms of energy. It is the amount of energy needed to raise the temperature of one pound of water by one Fahrenheit degree. To put it in more meaningful terms, it takes about 150 million Btu's to heat the average house for a year; 100 million Btu's to run the average car. Conversion rates are approximately as follows:

1 42-gallon barrel of oil	= 5.8 million Btu's
1 cubic foot of natural gas	= 1,031 Btu's
1 kilowatt hour of electricity	= 3,413 Btu's
1 ton of coal	= 25 million Btu's

Following U.S. practice, we define quadrillion as 10^{15}.

[b] For a detailed discussion of the energy gap and how it grew, see the Energy Policy Project's Preliminary Report, *Exploring Energy Choices* (Ford Foundation, P.O. Box 1919, New York, N.Y. 10001).

[c] Natural gas was the second most important fuel, providing 31 percent; coal accounted for 18 percent, hydropower 4 percent, and the atom 1 percent.

supply of energy resources—indeed, these are large enough so that if we wish to accept the consequences, we could in the future again become self-sufficient. Its physical heritage is one of the two factors that gives America its broad choice of possible energy futures. The other is our national habit of extravagance in energy use. It is that very extravagance that gives us room to use energy more efficiently, to conserve without cutting to the bone.

The rate of energy growth in the United States is now slower than in most other nations, developed and underdeveloped. But this country had a long head start. The absolute level of energy consumption in the United States is very high: six times as high per capita as the world average, and far beyond that of most other affluent countries. In 1970 the Swiss, for example, used about one-third as much energy per person, and the West Germans less than one-half.

These comparisons are not meant to imply that the United States could or should cut energy use to European levels, nor indeed cut back consumption in absolute terms at all. What the Project has explored is the effect of slower growth, not a reduction in the level of energy use. The opportunities for eliminating energy waste we have identified, and the energy experience of other developed countries suggest that a pleasant, comfortable, civilized life could be enjoyed with slower rates of energy growth in this country.

Much of this book will be concerned with policies that are tied to one or another rate of energy growth. A general observation is in order here. When we speak of agreeing upon a direction for energy growth, we do not mean to imply that a centralized single agency or "energy czar" should give orders to the country. What we do mean is that the energy decisions that government and industry must make, long in advance of the time they will have their effect, must be shaped by a governing conception of how fast energy needs will grow.

The United States has basically a private enterprise, market economy. Many decisions concerning energy use are made by countless individuals on the basis of price. When the market works well, it automatically implements some of our choices for us. If energy resources become scarcer and more costly, prices rise, and energy consumers tend to look for ways to use less energy or to use it more efficiently. At the same time, higher prices encourage producers to search for more supply.

The market is a very important means of carrying out energy decisions—billions of decisions by millions of energy purveyors and energy consumers. But it is misleading to speak of the market as though it has an autonomous life of its own. Political decisions, or the lack of them, crucially affect the way the market works. And some of the fundamental decisions affecting energy can be made only by government.

Regulation to protect the environment is a classic example of how political decisions affect the market. Air and water used to be regarded as "free goods," and the illusion was that industry could dump poisonous wastes without cost to anybody. In fact, as we have learned, everyone pays those costs in the form of environmental degradation. But some pay more than others. The East Kentucky farmer whose land is ruined with mud slides from the strip mine atop the mountain pays more than the Chattanooga machinist who heats his house cheaply with TVA electricity made from the coal from the same strip mine.

Effective strip mine reclamation laws would add to the cost of coal, and therefore to the electricity generated from it. The market would then more truly reflect the full cost of the coal to society. But politics must make it happen. Kentucky must pass an effective law and then enforce it or, the Congress must enact uniform national strip mine controls to be enforced by federal agents.

Because the marketplace fails to take account of foreign policy concerns, government may, to protect the national interest, take actions that interfere with the price system. Oil production in the United States has a higher value to the nation than production in Libya, for example, because it is more secure. Yet the market fails to reflect this higher value of domestic oil.

Middle Eastern oil is artificially high in price at present, but it is very cheap to produce; it is within the power of Middle East producers to undercut the price of synthetic oil made in America.

This would, however, mean that synthetic oil plants might not be built at all, or that they might go broke. The government must, therefore, decide whether the synthetic oil is worth more to America than the value that the market may assign to it. To ensure a reliable supply of synthetic oil, should the government protect domestic producers against potential competition by tariffs or purchase guarantees, thus raising prices to consumers?

Government may decide to intervene in the market for reasons of social equity. In time of shortage, when a necessity of life such as energy becomes scarce and expensive, the market mechanism deals harshly with the poor. It may take a long time for the market to right itself—time for producers to bring forth more supplies, time for consumers to buy thriftier cars or insulate their houses.

Meanwhile, the affluent can pay energy prices that include windfall profits. But what of retired people, barely making it on Social Security, who suddenly must pay twice as much for propane to heat their houses? What about the low income commuter who drives 40 miles to work from his home in a rural county, and faces gasoline bills that rise by one-third in four months? Energy is not simply another item that a consumer can shop around for or do without. It is a necessity that consumes 15 percent of a poor family's income. Soaring prices hurt, and hurt badly.

Therefore, a political decision must be made: to restrain windfall profits by price controls; to tax some of them away and spend the proceeds in ways that will benefit the poor; or to do nothing.

The federal government, by reason of its ownership of half the nation's most accessible remaining energy resources, wields enormous influence over energy supply. With its holdings of oil and gas in the ocean depths, and coal and uranium under western lands, it is the greatest single owner of fuels in the ground. Decisions about how to lease this land, and how fast, are greatly affected by judgments about how much energy the country will need, by environmental concerns, by the political influence of energy companies, and indirectly, by fuel prices.

Other government decisions are required. The Atomic Energy Commission must decide whether to license new nuclear plants and where to locate them. Local governments write building codes governing energy use in houses, stores, and office buildings. State governments regulate the prices and oversee the expansion plans of gas and electric utilities, which account for more than one-third of the energy market.

Among the most important government decisions are those determining the size and direction of federal spending for energy research and development. Federal R & D money approached $2 billion for fiscal 1975, almost three times as much as two years earlier, before energy assumed greater importance in the eyes of government decision makers. The way this money is allocated will have a decisive impact on the future of competing sources of energy and conservation efforts.

In the past, government R & D funding has been confined to new sources of supply, primarily atomic energy. Almost nothing has gone into energy conservation. The Atomic Energy Commission's five-year plan for federal R & D for 1975–1979 allocated only 7 percent of the funding toward improving the efficiency of energy use. It will be crucial to assign energy conservation opportunities a much higher priority in the future in order to achieve the benefits that slower growth in energy would bring.

Our sense of direction—or lack of it—about energy growth has a powerful influence on many of these decisions. In the past it was taken for granted that growth would continue at prevailing rates. A decision to examine the implications of growth need not imply government coercion, or even a shift toward more government decisions and fewer private ones. Rather, it means that federal energy decisions which are often in conflict could be coordinated and made more consistent through the guidance of a top level council on energy policy. We believe that such a council would be an effective aid to policy making. But making sense out of federal energy policy does not constitute a move toward socialism.

How can citizens make their views known? In the same way they take part in other political decisions. By voting, writing to their Congressmen, making political contributions, writing letters to the newspapers, attending meetings, and generally raising their voices. Citizens can attend public utility commission hearings on gas and electricity prices; they can take part in utility siting decisions wherever possible; and they can file written comments on the environmental impact statements that federal law now requires for major government actions.

To be sure, these methods of citizen participation are sometimes unsatisfactory. Taking part in utility hearings can be a frustrating, often futile, experience. Environmental impact statements simply describe the problems, but have no legal force to stop a government action—though comments from citizens can be effective in spotlighting the need for design improvements (as happened in the case of the Trans-Alaska pipeline). They can also cause delay, which in turn may be a source of frustration for those citizens who want development to proceed.

Citizens may feel relatively powerless as individuals,

but they have a latent force which can affect energy policy—as the energy companies recognize by their spending of millions of dollars on advertising campaigns to influence public opinion. Certainly, the sudden rise in congressional interest in energy during the Arab oil embargo was a response to the highly vocal interest ordinary citizens were then expressing.

It remains to be seen whether citizens, in the absence of shortages, will sustain their interest in energy. In the past, so long as the energy industry delivered the goods, energy prices were relatively low, and energy had not yet threatened the environment in an unmistakable way, most citizens were content to let industry make the major decisions.

But the fact is that the private interests of energy companies and the broader public interest do not always coincide. What is good for business is not always good for the rest of the country. Growth and environmental quality sometimes conflict, and growth is the foremost goal of most businesses, including the energy business. The "reasonable" and "balanced" concern for the environment that oil and electric power companies advocate in their ad campaigns usually means tipping the scales toward growth when there is a collision of interests.

Although experts can provide information and guidance, they cannot be the arbiters of these conflicting interests. The final decisions on energy related problems are based on value judgments. In case of conflict, the public, usually through its elected representatives, has the right and the responsibility to determine which values are more important.

A case in point is air quality standards. Industry, especially the auto industry, with support from oil companies, has attacked the Clean Air Act's auto emission standards as unreasonable and the prime cause of gasoline shortages (because antipollution devices use extra fuel). Congress and the Environmental Protection Agency (EPA) have so far taken these protests with a grain of salt, although the EPA did postpone application of tougher standards for a year. So far, the public has not accepted industry's argument that clean air is not worth the moderate penalty in fuel efficiency exacted in the present design of control devices.

On the other hand, Congress did resolve the five-year Alaska pipeline dispute in accord with the oil industry's position. Under pressure of fuel shortages, Congress endorsed the pipeline in 1973, and declared by edict that all requirements of the National Environmental Policy Act had been met.

Struggles of this kind, where two important goals —supply and environment—are in combat, can neither be left to laissez-faire market decisions nor settled through expertise. Resolutions emerge through a messy process of lobbying, legal maneuvers, propaganda, and conflict of regional interests. In a word, politics; and energy policy is a part of it.

We do not mean to suggest that citizens have an easy job in contending with the energy industry in the political arena when the industry's interests are at stake. As discussed in Chapter 9, oil companies have a long and impressive record of political success at all levels of government. Their political efforts are well financed, well organized, well supported by strategically placed leaders in Congress; and they have staying power. Citizens lack most of these advantages. The advantage they do have is that of numbers, and votes; but it is undeniably difficult for consumers to organize and use their power of greater numbers on a lasting basis.

It is this Project's conclusion that the size and shape of most energy problems are determined in large part by how fast energy consumption grows. Some problems, of course, such as high prices and their impact on the poor, must be faced whatever the policy adopted on conservation. But slower growth makes many energy-related problems less formidable.

It is, of course, a mistake to regard energy conservation as an end in itself; that puts the cart before the horse. Conservation is worthwhile as a means to alleviate shortages, preserve the environment, stretch out the supply of finite resources and protect the independence of U.S. foreign policy.

By the same token, energy growth is valuable only because it brings us more useful goods and services, warms our houses, and makes possible our vacation trips. More mundanely, energy gets us to work and keeps the office machinery and industrial plant going. If we could continue to enjoy these things in much the same way, with slower energy growth through greater efficiency, that achievement is worth considerable effort and perhaps some small sacrifice.

Energy futures

To assist our analysis of energy choices, the Energy Policy Project has constructed three different versions of possible energy futures for the United States through the year 2000. Figure 1 offers a comparative sketch of these three possible paths to the future. The three alternate futures, or scenarios, are based upon differing assumptions about growth in energy use. In many ways they are quite dissimilar, but each scenario is consistent with what we know about physical resources and economic effects.

The scenarios are not offered as predictions. Instead, they are illustrative, to help test and compare the consequences of different policy choices. In reality, there are infinite energy futures open to the nation, and it is not likely that the real energy future will closely resemble any of our scenarios. Our purpose is to spotlight three possibilities among the many, in order to think more clearly about the implications of different rates of energy growth. What are their effects on the economy, the environment, foreign policy, social equity, life styles? What policies would be likely to bring about each one? What resources are needed to make each of them work?

Certain common characteristics have been built into all three scenarios. They all include enough energy to provide the population with warmth in winter and air conditioning in summer; several "basic" appliances that would seem the height of luxury to most people in the world; and cars for most families, as well as other means of transportation.

We recognize that fundamental reforms are needed in America to bring "the good life" to all citizens. But we have included enough energy in our three versions of the future to allow it to happen. Where people live, how they get to work, how much they drive the family car, and the kind of car they own, differ from one scenario to the next. But none of them skimps on amenities. All the scenarios provide for energy growth over today's levels, and household comforts take a major share of such growth in each. All three scenarios include enough energy for more material prosperity than the country now enjoys.

A most important similarity among the scenarios is that all are based on full employment and steady growth in gross national product and personal incomes. The lower energy growth scenarios provide major savings in energy with small differences in the GNP from historical growth

Figure 1–**Scenarios: energy use in 1985 and 2000**

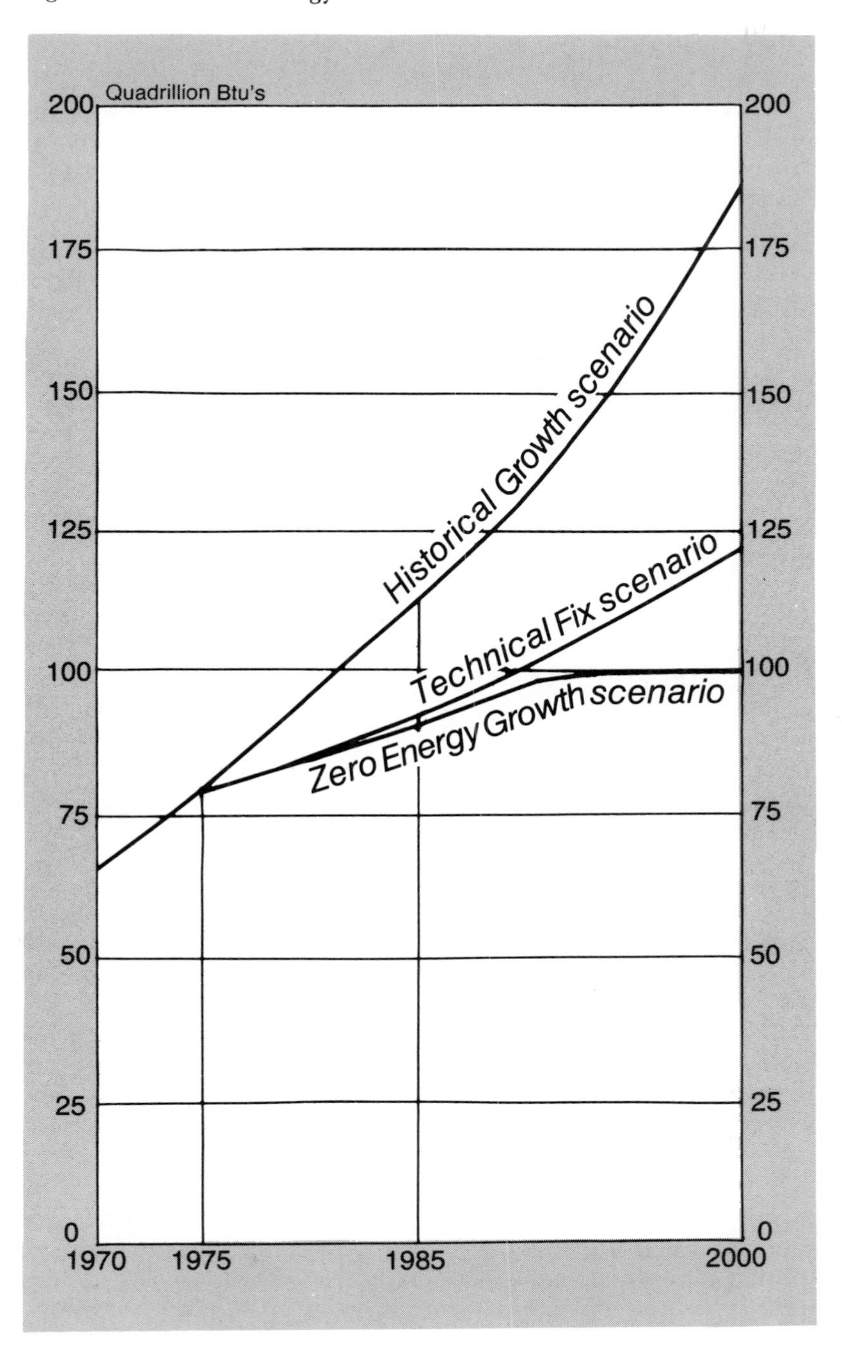

Source: Energy Policy Project

trends. Employment opportunities are, if anything, better. In all three scenarios, the real GNP for the year 2000 (discounting the effect of inflation) is more than twice what it is today. Of course, we recognize that the chief economic ills of the early 1970s are inflation and high unemployment, and we offer no cures for them here. But in the energy futures we are exploring, a lack of energy will not be a cause for these problems.

In this introductory chapter we intend simply to raise the curtain on our scenario studies; the following three chapters describe them in detail. We should state at the outset that the assumptions underlying the scenarios are not arbitrary. We have tested and supported their plausibility by economic and technical studies.

The energy growth rates for each scenario were built up from component parts. Rather than mechanically projecting into the future an overall energy growth rate, the Project examined trends for each energy using sector and subsector. All the scenarios use the Census Bureau's projections for population and households.[d] The projections for sectors such as transportation, home appliances, and industrial use were developed from various sources on the basis of historical growth, and were modified for the other scenarios. The growth rates built up from component parts were then cross-checked by a macroeconomic modeling effort to simulate the impacts on the economy generally of these levels of energy consumption. Appendixes A and F set forth the results of this work.

Our first scenario, *Historical Growth,* assumes that energy use in the United States would continue to grow till the end of the century at about 3.4 percent annually, the average rate of the years from 1950 to 1970. It assumes that no deliberate effort would be made to alter our habitual patterns of energy use. Instead, we assume that a vigorous national effort would be directed toward enlarging energy supply to keep up with rising demand. By 2000 we assume energy use would amount to about 187 quadrillion Btu's annually.

Present price and productivity trends cast some doubt on the likelihood that historical growth trends will persist. Even so, there are two persuasive reasons for exploring a *Historical Growth* scenario. First, it is the one

---

[d] For population, we used the Census Bureau's Series E, which projects a population of 236 million in 1985, and 265 million in 2000. See Appendix A for sources for other trends.

assumed by many government and industry leaders, and it has been the basis for important government and industry planning. Second, any analysis of future energy policy must examine the consequences of a continuation of historical growth. No one can be sure it will not take place; if the future is like the past, new uses of energy may appear that cannot be foreseen.

The second scenario, *Technical Fix,* differs little from *Historical Growth* in its mix of goods and services. The rate of economic growth is very slightly slower so that by 2000 the real GNP is nearly 4 percent less than in *Historical Growth* (but still more than twice as high as in 1973). This scenario reflects a conscious national effort to use energy more efficiently through engineering know-how—that is, by putting to use the practical, economical, energy-saving technology that is either available now or soon will be.

The Project's work indicates that if we were to apply these techniques consistently, an energy growth rate of 1.9 percent annually would be adequate to satisfy our national needs. This is little more than half the rate of *Historical Growth. Technical Fix* would use about 124 quadrillion Btu's a year in 2000—one-third less than *Historical Growth,* a saving four-fifths as large as our current total consumption. Yet the effect on the way people live and work—on material possessions, jobs, comfort, travel convenience—would be, our research tells us, quite moderate. *Technical Fix* is leaner and trimmer, but basically on the same track as *Historical Growth.*

Our *Zero Energy Growth* (*ZEG*) scenario represents a modest departure from that track. It would not require austerity, nor would it preclude economic growth. The real GNP in this scenario is approximately the same as in *Technical Fix,* and it actually provides more jobs. It includes all the energy-saving devices of *Technical Fix,* plus extra emphasis on efficiency. Its main difference lies in a small but distinct redirection of economic growth, away from energy-intensive industries toward economic activities that require less energy. An energy excise tax, by making energy more expensive, would encourage the shift.

Compared with the other energy futures, a *ZEG* future would have less emphasis on making things and more on offering services—better bus systems, more parks, better health care. About 2 percent of GNP would be diverted through the higher energy taxes to these public purposes—purposes designed to enhance the quality of life, as defined in the scenario.

The *ZEG* scenario assumes a modest rise in total energy use by 2000 but a declining rate of growth, which slows to zero before 1990. Total energy use would reach a level of 100 quadrillion Btu's a year, and remain on that lofty plateau.

Where would the energy come from to supply these various growth patterns? We must keep in mind the long lead time required for enlarging energy supply. The next few years are likely to be tight under any option, and brand new sources of energy are a decade or more away. Looking to the late 1970s and beyond, however, the supply problem can be solved if we decide on a policy now—and implement it.

The Project's research, as well as the results of independent studies we have commissioned, lead us to conclude that it is physically possible, even from domestic sources alone, to fuel the *Historical Growth* rate, during the later years of the century. It would not be easy, and we have serious doubts that it is desirable. But it could be done. It would mean very aggressive development of all available energy sources—oil and gas onshore and offshore, coal, shale, nuclear power. Increased reliance on imported oil could somewhat relieve the pressure on domestic energy sources.

Under the *Historical Growth* scenario there would be little scope to pick and choose among sources of supply, no matter what economic, foreign policy, or environmental problems they might raise. For example, no matter how we juggle the mix of sources, coal and nuclear power would have to be the mainstays of energy supply by the year 2000. Together they would furnish more energy than all sources combined provided in 1973.

Supply options are more flexible in the *Technical Fix* scenario. The slower growth in energy consumption permits more flexibility and a more relaxed pace of development. The nation could halt growth in at least one of the major domestic sources of energy—nuclear power, offshore oil and gas, or coal and shale from the Rocky Mountain region—and still demand less from the other supply sources than *Historical Growth* requires.

Zero Energy Growth would allow still more choice in supply from conventional sources. After 1985 this scenario could also permit use of cleaner, renewable, but smaller scale energy sources such as windpower, rooftop solar power, and recycled waste to meet a larger share of the total energy demand. Still, it should be remembered that even in this scenario the national energy appetite would be very large. Even if there were no further annual growth in energy use after the 1980s, the nation would still need to find enough supplies every year to meet an energy demand one-third larger than that of 1973.

The next three chapters describe the scenarios in detail, and illuminate the breadth of the energy choice that lies before us. Energy growth and economic growth can be uncoupled; they are not Siamese twins. From the early 1870s to 1950, GNP per capita rose sixfold, while energy use per capita little more than doubled.[e] Moreover, this happened at a time when the economy was rapidly shifting from agriculture to more energy-intensive industry. Economic growth far outpaced energy growth because the efficiency of energy production and use was dramatically improving.

From 1950 to 1973, energy and economic growth very closely coincided. Overall efficiency improvements in energy had come to a halt. If progress were to resume in getting the most useful work out of each barrel of oil and ton of coal, economic growth could again surge ahead of energy growth, and the nation would find it easier to avoid a crisis in energy and energy-related public concerns.

---

[e] Gross national product per capita, according to Simon Kuznets's estimates, was $223 (1929 prices) in the five-year period, 1869–1873. It was $1,233 in 1950. Energy use per capita was 99.0 million Btu's in 1870 (including wood, the most important source of nonhuman, nonanimal energy at that time). In 1950, energy use per capita was 214.4 million Btu's.

CHAPTER 2

The Historical Growth scenario

If historical growth is to be America's future then the first question that government and industrial planners must ask is simply: how much energy must be provided? The amount of energy used in a year depends upon a host of factors, including the price of energy, the level of overall economic activity, and the introduction of new technologies and processes. But since lead times for adding to supply are five to ten years, some estimate of future demand is needed to guide decision making.

Long term average growth rates are commonly the basis of such industry and government decisions. During the period between 1950 and 1970, the United States experienced a growth rate in energy consumption of 3.4 percent. During the 1965–1973 period, when growth in economic activity was particularly intense, energy consumption grew at an average rate of 4.5 percent; but recent price increases and shortages in energy suggest a slower growth rate in the future. Some supply sectors, such as electric utilities, have grown steadily at rates higher than the overall average, and their future planning has been on the basis of a 6 to 7 percent growth rate.

The *Historical Growth* scenario examines the consequences of continuing growth in energy consumption for the remainder of the century at the 1950–1970 average rate of 3.4 percent per year. Industry and government planners, more afraid of being blamed for energy shortages than for energy surpluses and waste, typically plan supply expansions based on a continuation of past trends. They assume that demand will materialize, stimulated, if necessary, through advertising, subsidies and promotional pricing. Accordingly, we selected the 3.4 percent figure for analysis because it is in line with many recent government and industry forecasts.

The absolute size of our current energy consumption is now quite large, and continuation of historical growth means the addition of larger and larger absolute amounts of supply capacity each year. It means adding each year new energy production equivalent to 1.3 million barrels of oil per day in 1975, 2.0 in 1985, and 3.3 in 2000. By 1985 this would be equivalent to adding one new Alaska pipeline each year. As a consequence, the social and economic impact of such development—not to mention the environmental impact of energy production and use—affect more and more people, who rightfully demand a say in how these activities are carried out. An examination in detail of this scenario is required to help make that political debate better informed.

What do we need the energy for?

In the *Historical Growth* scenario, energy use roughly doubles between now and the year 2000 in our homes, in commerce, and in transportation, while industrial energy use more than triples. (See Table 1 and Figure 2.) It is useful to identify what this energy would buy in human satisfaction.[a] In common with all our scenarios, it would furnish enough energy by 2000 to heat and centrally air

[a] A complete, detailed account of the requirements for energy in the *Historical Growth* scenario, by sectors, appears in Appendix A. This *Historical Growth* analysis of energy requirements is used as a basis for calculating the requirements in the *Technical Fix* and *Zero Energy Growth* scenarios.

Table 1–**Energy consumption in the Historical Growth scenario**
(Quadrillion Btu's)

	1973			*1985*			*2000*		
	Fuel	*Elec-tricity*	*Total*	*Fuel*	*Elec-tricity*	*Total*	*Fuel*	*Elec-tricity*	*Total*
Residential	9.4	2.3	16.3	9.7	4.5	22.9	10.0	7.5	30.1
Commercial	6.2	1.4	10.4	8.4	2.3	15.1	9.5	4.4	21.3
Industrial	21.4	2.7	29.5	34.5	6.1	52.1	55.8	15.2	96.9
Transportation	18.8	—	18.8	26.0	—	26.0	38.4[a]	—[a]	38.4
	55.8	6.4	75.0	78.6	12.9	116.1	113.7	27.1	186.7

Note: "Electricity" here is the electrical energy directly consumed in each sector. The totals for each sector are greater than the sum of fuel and electricity because they include in addition the heat wasted in producing electricity. For example, in 1973, two units of energy were wasted for every unit of electricity consumed, so that for the residential sector the total consumption is $9.4 + 2.3 + (2 \times 2.3) = 16.3$.

[a] If particular emphasis is placed on electric power (see Table 4) then some transportation energy would be provided by electricity. If one-half of urban auto traffic and one-half of railroad traffic were electrified in the year 2000, then 6.0 quadrillion Btu's of direct fuel requirements for transportation would be replaced by 2.1 quadrillion Btu's of electricity, with total fuel requirements remaining essentially unchanged.

condition all the 100 million households in the country. There would be enough energy for water heating, a cooking stove, a freezer, a dishwasher, and a big frost-free refrigerator in every home. Electricity would account for essentially all the energy growth in the residential sector, with more than one-third of the homes in 2000 being all electric.

In accord with present trends, a smaller share of the population (265 million by 2000) would live in the country or in densely packed center cities. More would live in suburbs. A greater share of homes, even in the suburbs, would be multifamily units—apartments or townhouses. More people would live in factory-made mobile homes. And a larger proportion of Americans would live near coasts, and in warmer, sunnier parts of the nation.

Along with increasing suburbanization would go greater dependence on cars. *Historical Growth* would fuel 138 million large, powerful automobiles (a little better than one car for every two people) with worse than mediocre fuel economy—11.4 miles per gallon. This compares with 89 million cars (averaging 2.3 people per car) in 1970, getting 13.6 miles per gallon.

People would travel by car about 15 percent more in 2000 than they do today; but growth in air travel is a more conspicuous feature of this scenario. By 2000 people would

Figure 2–**Energy consumption in the Historical Growth scenario**

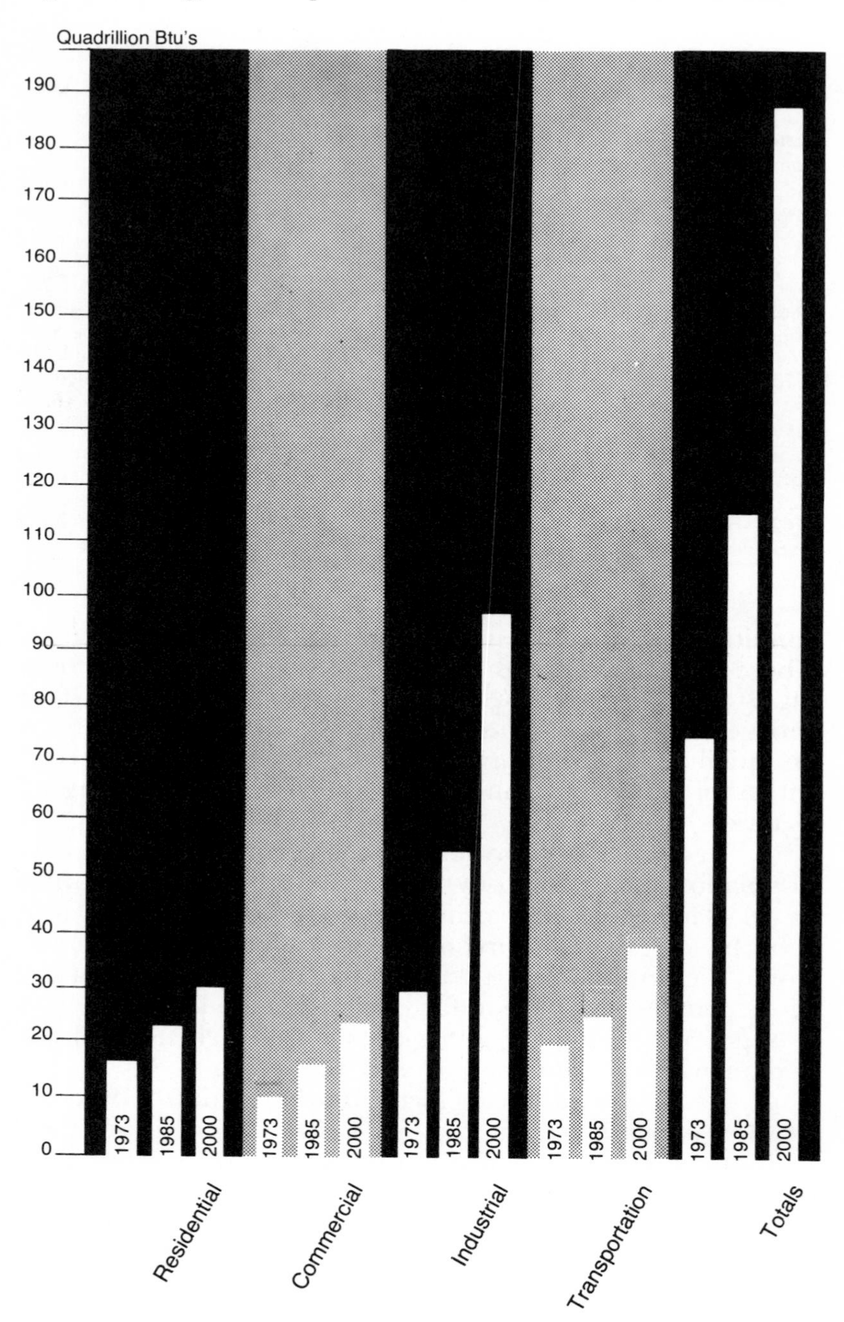

Source: Energy Policy Project

travel by air more than five times as much as in 1970, and air freight traffic would be up eighteenfold.

In all our scenarios, industry and commerce are assumed to be prosperously growing, with little unemployment. *Historical Growth* includes an annual increase in gross national product of 3.6 percent until 1985, and 3.3 percent thereafter until 2000 (reflecting the slowdown in population growth). In accordance with well established present trends, there would be slightly more employment in service jobs—teaching, selling in stores, insurance and real estate, government jobs, health care, public utilities and so on—than today.

Industrial energy would account for more than half of all energy consumption in 2000, compared to 40 percent today, reflecting the high value society would put on material goods. In projecting trends of industrial energy use, we gave special attention to five key, highly energy intensive manufacturing activities: aluminum, steel, paper, plastics and cement. Three of these industries—steel, cement and paper—follow historical growth trends that are slower than the growth in GNP. But growth in aluminum and plastics is much faster. A very large proportion of their growth is directly due to packaging. If the trends of today and the recent past continue, by the end of the century much more aluminum and plastics would be used to wrap or contain things than is used for *all* purposes today.

In addition to all the specific energy uses we have counted up, projected out, and provided for in this scenario, we have also made generous allowance for uses that are as yet unknown. About 10 percent of residential and commercial energy for the year 2000 is set apart for this purpose.

Altogether, *Historical Growth* would furnish an energy use per person that is twice as high in 2000 as in 1973. An important caveat is that not all this energy would be available to consumers. It takes energy to make energy. Consumption as high as this scenario assumes is likely to require large amounts of energy supply from marginal sources, such as oil shale or coal gasification, that demand large amounts of energy in their own production. While today about a quarter of our energy consumption is accounted for in the extraction and processing of fuels, this fraction is expected to grow to more than one third by 2000.

The *Historical Growth* scenario, to sum up, assumes

that high in society's scale of values is the idea that "more is better." More big cars traveling more miles, more convenience foods, more packaging, more goods that may wear out fast but are easily replaced. The flavor of life in affluent American suburbs today suggests the character of this future.

The economics of historical energy growth

Through the use of an econometric model,[1] the Project has explored the relationships between energy prices, energy growth, and economic prosperity. (See Appendix F for details.) The rate of growth in energy consumption is dependent both upon the prices for various forms of energy and their impact on consumer demand. It is also powerfully influenced by government policies, which may either promote greater energy consumption or, on the other hand, may encourage and assist energy conservation. (See Chapter 3, *The Technical Fix scenario.*)

The most plausible circumstances for a continuation of historical rates of growth in energy consumption would be unexpected good fortune in keeping energy prices down, combined with government policies that promote consumption. According to the estimates of the economic model, crude oil prices in the range of $7 to $8 a barrel (1971 prices) between 1985 and 2000 would be consistent with historical rates of energy growth. Other fuel prices would also need to remain stable. In addition, productivity improvements in electric power generation would have to resume their historical rate of improvement of 5 percent per year rather than the 2 percent which has prevailed since 1968, a rather unlikely development in view of current difficulties with new power plants. This would result in resumption of the long term trend toward falling electricity prices.

While there is some evidence that these prices may be low enough to sustain *Historical Growth,* they are also high enough that many of the energy conserving technologies that the *Technical Fix* scenario proposes would be economically attractive. Historical rates of energy growth have occurred in a period of slowly falling real energy prices. Because energy represented a small fraction of most budgets, consumers had little incentive to explore economically attractive conservation measures.

There is no precise way to estimate how the economy

as a whole will adjust over time to the recent runup in world oil prices. It is possible that prices for oil (and other energy sources) would have to fall well back from current levels, to the $4 to $6 per barrel range, in order for historical growth rates to continue. In that case, unless the producers were subsidized, our supply research[2] suggests that marginal supplies needed to meet the 3.4 percent growth rate might be uneconomic.

At the same time, we must recognize the possibility that energy growth could resume at historical rates in a few years, even at higher prices, if unexpected developments take place in the rest of the economy. Although the *Historical Growth* energy requirements calculations in Appendix A include generous allowances for presently unknown applications for energy, it is always possible that new processes or products may be developed that are even more energy-intensive than we have projected, and that offer economic or consumption benefits which are so compelling that the higher prices will not dampen demand.

One factor that will contribute to high growth in the face of rising prices is the greater emphasis on electricity and synthetic fossil fuels, for which energy extraction and conversion losses are especially large.

Where will the energy come from?

Planners preparing for historical growth in energy consumption face this fundamental question: where will the energy come from?

In answering this question, it seems simple and obvious to begin with estimates of available energy resources. Yet in fact there is nothing simple about resource estimates. They depend as much upon economic and technical developments as they do upon geological facts, which themselves are uncertain. Appendix D discusses the factors that must be taken into account in estimating reserves and resources, and defines those terms.

As Table 2 shows, oil and gas appear to be more limited than other resources, raising the possibility that in a few decades our oil and gas may "run out." From an economic point of view, we would never actually exhaust such a resource entirely; as we draw on lower and lower grade resources, their increasing prices would make other resources more attractive.

The resource estimates and other Project research[3]

Table 2–**Major U.S. energy resources**
(Quadrillion Btu's)

	Reserves[a]	*Additional Recoverable Resources*[b]	*Remaining Resource Base*[c]
Petroleum	410–530	1100–2200	16,000
Natural gas	440–570	1100–2200	7,300
Oil shale	900–3400[d]	7,600[d]	150,000[e]
Uranium			
Thermal reactors[f]	310[h]	630[h]	8,000[i]
Breeder reactors[g]	22,000[h]	44,000[h]	560,000[i]
Coal	5,000	(unspecified)	78,000

[a] *Reserves* are economically recoverable resources in identified deposits, with the extent of the resource measured or inferred on the basis of geological evidence.

[b] *Additional Recoverable Resources* are resources judged economically recoverable on the basis of broad geological evidence, but for which insufficient, detailed knowledge is available to classify the resources as reserves.

[c] *The Remaining Resource Base* is the total amount of the resource estimated to be left in the ground, usually in deposits having some minimum grade or better.

[d] In deposits having 30 gallons of oil or more per ton of shale.

[e] In deposits having 10 gallons of oil or more per ton of shale.

[f] With thermal reactors, a metric ton of uranium oxide (yellowcake) yields one trillion Btu's.

[g] With breeder reactors, a metric ton of uranium oxide yields 70 trillion Btu's.

[h] For uranium oxide up to $10/lb.

[i] For uranium oxide up to $20/lb.

These resource estimates are best understood in relation to present and projected demands for various fuels, as shown in Tables 3 and 4.

Sources: See Appendix D.

suggest that production of oil and gas in quantities sufficient to sustain historical growth could be supported from domestic resources through the remainder of the century. Such a production program would not use more than half the estimated recoverable resources by 2000. Fuels from the remainder of the resource base that are not economically recoverable today could be used to support a declining production rate after 2000.

Such an analysis ignores the many constraints on discovering and producing that much oil and gas at the contemplated rate of growth. These constraints include: environmental concerns that will slow or prevent the exploration of much of the resource base, the long lead time for discovery, the pace at which the federal domain will be made available, and the bottlenecks of manpower, drilling equipment, and the like. The amount of oil and gas in the ground will not be a limiting factor in supplying historical energy growth for the rest of this century. A more likely prospect is that oil and gas production will clash with other social values, and those constraints will slow the pace at which resources can be discovered and brought to market.

The coal resource is unquestionably very large. Our research suggests that the resource base contains enough coal that can be extracted without increase in real costs to sustain sizable growth in production for many decades. Uranium resources are more limited, if used in the present nuclear reactors. With the breeder reactor, the uranium resource base becomes larger than coal. If oil and gas production peak over the next two decades, a combination of synthetic oil and gas from coal, oil shale, and electricity generated from coal and nuclear sources could provide for growth in energy requirements, if cleaner and cheaper sources are not meanwhile developed.

Not only is the domestic resource base large enough to support historical growth for the rest of this century and beyond, but many of these resources, such as western and midwestern surface minable coal, and onshore Alaskan oil and gas, are low *cost* resources (in narrow economic terms).

Currently, the high world oil prices set the pace for the high prices of domestic energy sources, particularly the fossil fuels. However, if the United States decided to impose a ceiling on domestic energy prices lower than world oil prices, or if world prices dropped, resumption of historical energy growth would be a distinct possibility. But producing these resources at the pace needed would require a great number of positive governmental actions, and would mean compromising many environmental goals that we have set for ourselves.

Alternative supply cases

If we take a conservative view of the likely fruits of energy research and development, there are three major sources of future supplies for the rest of the century: domestic fossil fuels, including synthetic oil and gas; nuclear power; and oil imports.[b] The relative importance of these various sources depends upon such factors as environmental acceptability, relative price, and government policy concerning reliance on imports.

To illustrate the breadth of supply options, we have evaluated three supply cases: Domestic Oil and Gas, High

---

[b] See Energy Supply Notes, Appendix C.

Nuclear, and High Imports. These alternative supply options for the *Historical Growth* scenario are summarized in Tables 3 and 4 and depicted in Figure 3. Before describing how these supply cases differ from one another, let us first consider their similarities.

A basic feature of all supply options under *Historical Growth* is that the supply mix shifts away from oil and gas. Today gases and liquids make up more than three-quarters of our energy supply. But in the year 2000 they would account for only about half the total supply in the *Historical Growth* scenario. In contrast, an even greater role is expected for coal and nuclear power, whose share of the

Table 3–**Historical Growth energy supplies**
(Quadrillion Btu's)

	Actual	*Domestic Oil and Gas*		*High Nuclear*		*High Imports*	
	1973	*1985*	*2000*	*1985*	*2000*	*1985*	*2000*
Domestic oil	22	32	40	32	34	27	27
Shale oil	0	2	10	2	10	1	2
Synthetic liquids from coal	0	1	5	1	5	0	3
Imported oil	12	10	9	10	9	22	32
Nuclear	1	10	40	12	50	10	40
Coal (except synthetics)	13	25	33	23	33	20	38
Domestic gas	23	29	37	29	31	26	27
Synthetic gas from coal	0	1	3	1	3	1	3
Imported gas	1	1	0	1	2	4	5
Hydro	3	3	4	3	4	3	4
Geothermal	0	1	2	1	2	1	2
Other	0	0	1	0	1	0	1
Conversion losses from coal synthetics	0	1	3	1	4	0	3
Total	75	116	187	116	188	115	187

Table 4–**Fuels for central station electric power, HG scenario**
(Quadrillion Btu's)

	Actual	*Domestic Oil and Gas*		*High Nuclear*		*High Imports*	
	1973	*1985*	*2000*	*1985*	*2000*	*1985*	*2000*
Coal	8.7	16	20	15	20	12	16
Nuclear	0.9	10	40	12	50	10	40
Oil and Gas	7.3	7	7	6	3	11	11
Hydro	2.9	3	4	3	4	3	4
Geothermal	—	1	2	1	2	1	2
Other	—	0	1	0	1	0	1
	19.8	37	74	37	80	37	74

energy supply increases from 20 to 50 percent between now and 2000. Roughly two-thirds of the growth in energy between now and 2000 in the *Historical Growth* scenario is due to coal and nuclear power.

Because electricity is the principal form in which coal and nuclear energy can be utilized, the emphasis on these fuels means a continuing trend toward greater electrification; electric power generation accounts for roughly 40 percent of total energy use in 2000, compared with 25 percent today. This will require sustained growth of coal production and development of new mines. In addition, current obstacles to the development of nuclear power must be removed.

Domestic Oil and Gas case

The domestic Oil and Gas case hinges on policies that favor rapid exploitation of the fossil fuel resource base. Sufficient financial incentives to industry and resolution of environmental concerns are essential to spur this. Given these policies, domestic oil and gas production could be rapidly increased through extensive offshore development of all major prospects and use of advanced recovery techniques for existing wells. Toward the end of the century, oil and gas supplies could be supplemented with synthetics from coal and shale. This allows a gradual decline in requirements for imported oil, even though overall oil and gas consumption are rapidly rising—nearly doubling by 2000.

High Nuclear case

This case is an examination of the potential of nuclear power to assume an especially important role in the energy economy. For the period till 1985, the principal effect of adopting optimistic high estimates of nuclear growth[c] is that the pace of coal development can slacken, thus relieving somewhat air pollution and strip mining problems. For the longer term—up until 2000—the high nuclear option would help to fill the gap created by lower petroleum supplies.

---

[c] 265,000 Mw(e) by 1985 and 1,100,000 Mw(e) by 2000.

Figure 3–**Energy supply for Historical Growth**

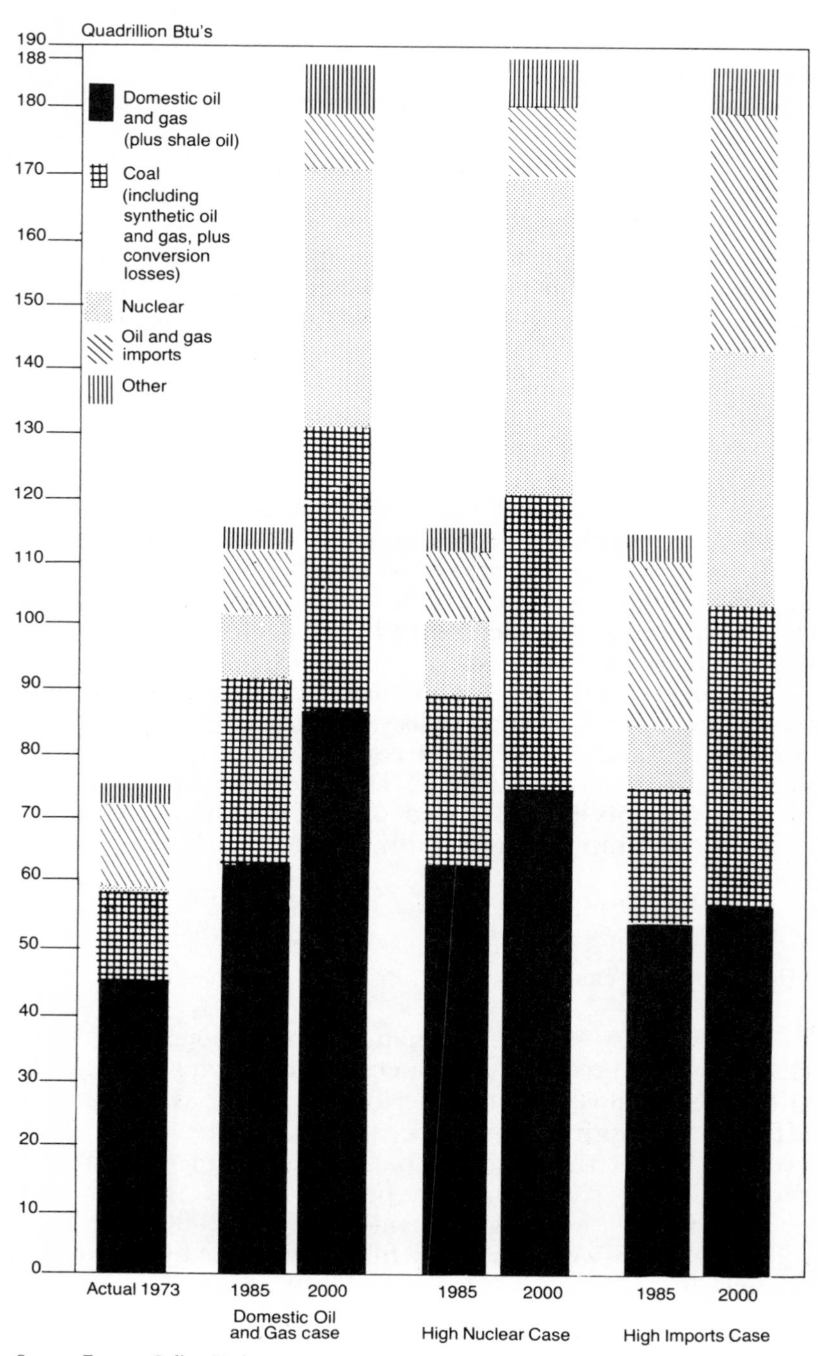

Source: Energy Policy Project

However, the ability of nuclear power to substitute for fossil fuels is limited both by technology and by price. Because nuclear power, for the next several decades at least, can be used primarily as electricity, this case depends in part on the degree to which technology will allow electricity to substitute for liquid fuels, especially in transportation. It is also extremely dependent on the ability of utilities and their suppliers steadily to improve the competitive position of electricity prices by raising productivity.

This case would mean increasing the electric utilities' share of total energy consumption from the 25 percent figure of 1973 to about 43 percent in 2000 (compared to 40 percent in the Domestic Oil and Gas case). This development is in line with the ongoing trend, but it may approach the limits to which electricity can be substituted for other fuels in this century.

It must be realized that under the High Nuclear option the pressure to develop oil and gas supplies continues. Oil consumption would still be up 70 percent in 2000 over today's level, compared to 90 percent in the Domestic Oil and Gas supply case. If we assume that extraordinary growth in domestic oil and gas production will not be forthcoming, some substitutes will be required in the form of synthetic oil and gas, imports, or coal. This scenario includes all of these, plus some contributions from unconventional sources.

High Imports case

Most current energy supply planning is based on the premise that the United States cannot count on large supplies of imported energy, particularly oil. For the near to intermediate term, however, the limits on increases in imported supplies are largely political and, therefore, subject to change. Furthermore, recent large increases in the world market price of oil have stimulated a new wave of oil exploration around the world. While such efforts have thus far been only moderately successful, it is quite possible that major new discoveries will permit new growth in imports to the United States. Offshore regions that have been little explored could produce large amounts of oil at prices near current world levels.

The principal consequence of such discoveries—if they are accompanied by a domestic policy decision to permit large scale imports with appropriate safeguards such as stockpiling—would be to slow down the development of marginal domestic sources such as synthetic oil and gas.

Implications of energy supply in Historical Growth

From these three supply cases flow several important consequences:

• If "self-sufficiency" is taken literally to mean no oil and gas imports, it is not possible by 1985 in any of the cases. If, starting immediately, an aggressive domestic liquids and gas development policy is pursued, it may be possible to approach "self-sufficiency" in the 1990s, but our best judgment is that imports of several million barrels per day would still be needed. Approaching "self-sufficiency" would most likely occur if public policy favored development of new energy sources over environmental concerns, and if oil prices are high.

• The United States's need for imports during the next decade or so from nations that have proved unreliable sources means that we must develop some form of stockpiling or insurance to protect against future energy blackmail. This poses obvious problems. It is difficult to accomplish with high world prices for oil and with the long lead times required to develop excess domestic producing capacity. But if we fail to face the consequences of a future oil embargo, we run the grave risk of a severe energy crisis.

• The relative roles of electric power versus oil and gas will depend both upon their relative prices and upon public policy decisions concerning their environmental acceptability. Another key factor is the number of new oil and gas discoveries that may result from massive development efforts. To increase substantially the production of conventional oil and gas, the offshore areas must be exploited.

• The main obstacles to rapid expansion of nuclear power are the limited uranium enrichment capacity, the shortage of skilled labor to meet construction schedules, and the poor reliability of operating plants, which now encounter unexpected down times of 20 to 40 percent. These problems must also be solved quickly to meet the forecasts for high nuclear growth. In the long run, the breeder reactor,[d] if it passes the tests of safety and economics, will greatly extend the energy potential of the uranium resource base.

• Coal, like nuclear power, has a very significant role in a *Historical Growth* scenario. Its major use will be to generate electric power, either directly or after limited treatment. Crucial to this use is the progress of various air pollution control technologies for using coal in power plants. A supplemental use of coal is to convert it to clean liquids and gas, which may be used directly for heating and transportation, in support of an oil and gas economy. Oil from shale may also be an important energy source toward the end of the century.

But synthetics both from coal and from shale are not likely to provide more than about 10 percent of energy by 2000. The large volumes of water required for synthetics production and conversion of coal into electricity may determine where such plants are located. For coal synthetics, water scarcity may dictate that coal mined in the arid West be shipped to plants in the East or Midwest where water is more plentiful. For oil shale, which cannot readily be transported, water supply may set a limit on development.

• In the next 25 years, regardless of which supply option is followed, the losses experienced in converting raw energy to usable form become increasingly important. The energy processing requirements (as shown in Appendix A) represent an expanding share of overall energy consumption. Growth in energy consumption by individuals and industry requires an even higher growth in fuel production because of rising losses in production. The losses will increase as natural oil and gas are replaced by synthetics and electric power, which must undergo inefficient conversion processes. As more marginal sources of fuel are used, more energy will also be required to extract fuel from the ground, to build the plants to convert it to usable form, and to provide the water supply and other required support facilities—and the net energy available to consumers will be reduced.

A projected 3.4 percent growth in overall energy production, therefore, will mean a lower rate of growth in

---

[d] The breeder reactor is a nuclear power plant designed to efficiently convert natural uranium (U-238), which is not a fuel, into plutonium (Pu-239), which is. By so doing, it produces more fuel than it consumes.

energy available at the point of consumption. Energy losses from inefficient conversion help explain the continuing growth in overall energy demand in the face of reduced population growth and saturation in the consumption of energy intensive goods and services. These losses are a challenge for research and development to develop greater efficiency in the production and conversion of energy.

- A strong federal energy R&D effort is essential to support this scenario. It would emphasize energy supply rather than consumption technologies, continuing the present federal priority. Each of the supply cases assumes that new energy sources will be developed and be in widespread commercial use before 2000. Most obvious are the production of liquids and gas from coal and shale, which would receive major support. Also needed are improved technology for oil and gas production in deeper water offshore, and more productive coal mining techniques. Less obvious today, but just as crucial, is the need for fundamentally new, potentially lower cost supply technologies such as the breeder, fusion power, geothermal, and central station solar energy to be available after 2000 to support a continuation of *Historical Growth*. If the nuclear-electric alternative turns out to be more advantageous, the breeder reactor would become essential to long term growth.

Experience with nuclear fission and the fossil fuels has shown that several decades are required before a fundamentally new technology is mature enough to supply a major amount of the nation's energy. Near term innovations would be supported to work on environmental and safety problems in the use of existing technologies such as nuclear power, oil and gas production, and coal combustion.

The institutional framework for supply development

Producing the domestic energy supplies necessary to satisfy this scenario would require a concerted effort by government and industry, with the support of the public.

The most fundamental policy question about energy supply involves the relationship between government and industry. There are two policy approaches that represent the extremes of government-industry relationships. One approach commonly discussed is for industry, working through the marketplace, to assume the primary responsibility for supplying energy. The government's role would be essentially supportive of private energy development efforts. Uncertainty about government's economic and regulatory policies would be minimized. Alternatively, the federal government could assume a strong leadership role as an active participant. Leadership could be exercised either by government entities (for example, government owned energy companies) or through detailed government planning and regulation of energy companies. The industry would have a quasi-utility status. Between these extremes lie various options in which government and industry share the leadership and planning functions.

The choice will be difficult. To meet the supply requirements for the *Historical Growth* scenario will strain any system. The question is not only which approach is most likely to work, but also which approach will gain the confidence of consumers, producers, and the other parties involved.

Market oriented approach

The traditional U.S. approach is to look to private industry to solve the technical and financial problems associated with a major economic development. The government provides a "favorable economic climate." In this case, a favorable economic climate involves continued tax incentives for the energy industry—especially for oil and gas, but also for coal and the utilities as well. It means the removal of price controls on natural gas and no new price controls on other fuels. It means making the federal domain available as necessary for development. It may also be necessary to provide some protection to investors in high cost oil and gas sources against the possibility of a sudden decline in world oil prices.

Expectations concerning government policy are probably as important to industry as the actual prices and tax levels. Industry's perceptions of the government's long term direction are as significant as the measures enacted along the way, for long term capital investments must be made on the basis of future expectation. The regulation of natural gas prices at the wellhead and the depletion allowance are two cases in point. Retention of the depletion allowance and removal of wellhead price control would be symbols of the government's intent to favor the oil and gas industries and, therefore, to encourage increased activity by industry.

The issue of protection of the U.S. oil industry and its cost to the economy is complicated by the uncertainty over future world oil prices. At one extreme, the government could move to an absolute prohibition on imported energy. The cost to the U.S. economy would be very high if all energy prices rose to the cost of the marginal U.S. barrel, while world prices were lower. For example, with a U.S. oil consumption level of eight billion barrels in 1985, a $2 per barrel difference—by no means inconceivable—would cost consumers $16 billion each year.

A variation would be for the U.S. government to guarantee a market for higher cost domestic fuels against the risk of lower priced imports, but permit the imports to take place. If the world prices remained at today's levels, government spending for such guarantees might be small; but if world oil prices should drop, such a scheme could also turn out to cost billions of dollars of public funds annually. The total costs to the society of domestic self-sufficiency must include secondary impacts on air, water, and land, many of which are not quantifiable. These costs could dwarf the direct and visible dollar costs.

Another area in which positive federal action would be needed is financing and capital formation. If this scenario is to succeed, the energy industries will require some $1.7 (in constant 1970 dollars) trillion between 1975 and 2000. (See Appendix B on capital requirements for the three scenarios.) This immense sum represents over 25 percent of the projected total national investment for plant and capital equipment in 1985 and about 30 percent if the trend continues to 2000. It compares with the energy industry's 21 percent of plant and capital equipment investment during recent years. Our studies[4] suggest that if utility regulations and fuels pricing policy permit high enough rates and profit margins, the capital can be attracted. But it will not occur automatically, and there is the possibility, especially for the utilities, that regulatory delays and political resistance to rate increases will make financing the *Historical Growth* option very difficult.

But these very rate increases, if they did take place, might inhibit growth in demand for energy. There are indications that energy growth at historical rates might not occur without substantial governmental subsidies. Federal subsidies, including funding R&D, tax incentives, and financial guarantees, are scarcely a novelty in the energy field. Among the new proposals, the chairman of the Michigan Public Service Commission has suggested that the federal government insure utility borrowing to finance power plants. And in New York, the state government has already purchased two Consolidated Edison generating plants (which the state financed at lower interest rates than the utility) rather than allow the utility to approach bankruptcy.

Schemes such as these, if adopted on a widespread scale, amount to taking a step back from recent moves (mostly in connection with environmental protection) designed to ensure that energy consumers pay the full economic and social costs of energy. Energy consumers would be subsidized by taxpayers, and consumption would be higher than it otherwise would be. Over the long term, taxes would be higher to compensate for artificially lower energy prices. These developments would constitute an unnecessary waste of resources, but they must be recognized as a likely feature of a *Historical Growth* energy future.

The net effect of a highly favorable economic climate for energy supply could amount to a "self-fulfilling prophecy" for energy growth. Once the investments were made to produce energy to satisfy a historical growth rate, it is almost inevitable that five to ten years later, when the power plants and coal gasification plants are in production, prices will be set to ensure that the energy is sold, with losses covered through some form of subsidy.

In addition to favorable economics, the federal government would need to provide industry with clearer policy signals on environmental, safety and other problems. At the extreme, the government could simply abandon the regulations developed during the late 1960s and early 1970s; but this would lead to unacceptable consequences in terms of occupational health and safety, and environmental degradation.

Another approach that industry generally favors would be sympathetic enforcement of the regulations, attempting to maintain their spirit but accepting less than full compliance and delaying implementation when this would seriously affect energy supply. Under this approach the philosophy would be "go ahead with the development now and correct the problems as they are found," rather than a philosophy of "do not start until there is assurance that the development will not cause unacceptable consequences."

Still, as more energy is consumed, tighter emission

controls are required to keep ambient air quality at acceptable levels. Thus the *Historical Growth* scenario would appear to call for more stringent controls in order to fulfill it without sacrificing air quality. We will need to develop an improved control technology that allows greater concentrations of energy use in heavily populated areas without making pollution worse—or else we will have to disperse the polluting activities into areas which are now relatively clean. Locating power plants, refineries, and energy intensive industries in rural or undeveloped areas could facilitate significant growth without tighter emission controls, but it would be at the cost of environments that are now unspoiled.

To implement the *Historical Growth* scenario, siting procedures must favor development without delay. Experience with the multitude of separate federal, state, and local approvals for a major facility has shown that the process is extremely time consuming and uncertain. Federal laws might be enacted that would preempt state and local authorities, as well as streamline procedures to authorize the acquisition of sites and construction of power plants, refineries, pipelines, transmission lines, and similar facilities in a single, unified procedure. A "one-stop" siting policy would mean that supply considerations, air and water pollution, health and safety, and land use would be dealt with in a single federal procedure. This would be a reversal of the present situation in which state and local interests have the power to delay indefinitely or halt completely objectionable developments.

Rapid growth of nuclear power is essential to the *Historical Growth* scenario. For energy growth to take place at historical rates, the public must be persuaded that nuclear power is safe enough to use on a large scale. Greatly enlarged coal production is also essential. Thus the nation must also accede to a heavy commitment to coal development—before air pollution control technology, coal mine health and safety improvements, and strip mine reclamation are adequately implemented. Unsolved environmental problems such as small particle air pollution, reclamation of surfaced mined arid lands, and uncertain nuclear risks will have to be accepted, with the hope they will prove solvable within a reasonable time.

In this scenario, the federal domain will provide a major source of future energy supplies. Primary emphasis would have to be given to rapid development of the resources, as opposed to optimizing revenues to the taxpayer or insuring environmental protection. Thus promising offshore oil and gas, coal geothermal, and shale lands would have to be leased as necessary. (See Chapter 11 for a detailed discussion.)

The key to managing federal lands in this scenario is to tie federal actions to development. Action making available oil and gas lands, geothermal, and shale should favor early production. But even under this scenario, further federal coal leasing does not appear warranted before 1985, since vast quantities are already under lease but are not being produced.

Government oriented approach

Under a rapid supply development scenario, the market oriented approach may prove inadequate in the face of enormous problems. Some technical bottlenecks will be inescapable. Construction delays, shortages of skilled labor, lack of heavy equipment, and inadequate water supply may well pose serious constraints, in addition to financing difficulties and environmental regulations.

To achieve the necessary supply growth, the federal government might need to identify bottlenecks, establish priorities for items in short supply, and take positive action to bring supply and demand into balance. In addition to actions directly influencing the energy suppliers, special training programs and loans to manufacturers of major components may well be needed. A new federal agency —an energy supply expediting office—could be established to continuously monitor every link in the supply chain and be given the authority and funding to help industry remove bottlenecks.

Public desire for a stronger voice in the decision making process affecting energy could be another reason for a more positive governmental role. This role could range from more stringent implementation of existing regulations and programs to nationalization of certain activities.

The government already has a number of regulatory powers that could be used in conjunction with its funding of research and development and public land management to take an aggressive part in shaping what industry does. The multitude of existing government actions required to expand most sources of energy is frequently the source of conflicting signals to industry and frustration to the public.

Unifying the federal actions into a coherent program could give the government a powerful tool for shaping future energy supply growth. If the country wants to be sure that energy will be available for historical growth—rather than adopting government programs that favor industry against environmentalists and consumer interests—it could well decide on a relationship similar to the sharing of roles in the electric power industry where, in some cases, government is both a regulator and a producer.

To coordinate government policies for rapid supply development, an Energy Policy Council could provide the guidance to existing agencies. The Council would be authorized to guide leasing of federal land, energy R&D, and other activities to encourage development. It would also be authorized to give policy guidance to the regulatory and siting agencies so that they could provide the financial incentives and regulatory approvals necessary for construction to move ahead on schedule.

A second step, previously mentioned, is the passage of federal siting laws so that construction can proceed despite state and local opposition. A third step is federal utility rate regulation, carried out on a regional basis, where the existing state regulatory system may be unable to sustain historical growth in an area of rapidly increasing capacity. (Utility reforms are suggested in Chapter 10.)

Another crucial federal responsibility is to put together a truly competent management team to make available government owned fuel resources, in a manner that serves the needs of the industry and the people in the region as well. (The planning process needed to deal with federal resources is described in Chapter 11.)

While these mechanisms could provide government incentives for industry to develop energy supplies, they would do little to satisfy a possible public desire for greater effective control over the energy industry. Federal chartering of the large integrated energy companies, which would open up corporate decision making to considerable public scrutiny has been proposed in the Senate. The federal charter would be a flexible device enabling the government to control various selected aspects of corporate activity. The appointment by the government of public interest members as corporate directors would be one means of public supervision. Corporations under federal charter also would be legally required to disclose information on reserves and costs.

The federal chartering concept is not entirely dependent on rapid growth in energy. Yet if the nation is to depend on a major expansion of oil production, the oil industry would have great leverage on public policy, thus providing all the more reason for measures to assert some degree of direct public control over the industry. (See Chapter 9.)

Another form of federal involvement would be the creation of a federal "yardstick" corporation. Following the example of the Tennessee Valley Authority in the electric industry, it has been suggested that a federal oil and gas corporation should be established that would explore, develop, and produce oil and gas on federal lands. In theory, this federal corporation would provide a benchmark for costs and prices, as well as a competitive spur to private enterprise to expand the search for oil and gas.

A fundamental question arising from this approach is whether the federal corporation would be designed to act as a giant oil company or as a political-social institution in which the production of low cost oil and gas was a secondary objective. The TVA experience provides some guidance. Like many government agencies, TVA in its early years was vigorous and gratified its supporters. But more recently it has been criticized as unresponsive to changing social needs, especially to environmental concerns. Nevertheless, if the United States is determined to sustain rapid growth in energy, and yet fears the concentration of power in the energy industry implicit in such a decision, the establishment of a federal yardstick corporation is worth serious consideration. (See Chapter 9.)

Another means of controlling oil company profits and operations, as well as assuring the necessary growth in production, is a utility type of regulation. A utility is responsible for making the investments necessary to supply service to its customers; and government in turn guarantees the utility a profit high enough to attract capital.

In the past, utility type regulation has been applied to companies with a natural monopoly. To expand the concept to the oil industry would represent a judgment that the industry is not behaving competitively and is unlikely to do so in the future. Such regulations could be applied most plausibly to the integrated companies whose refining and pipeline operations (and in the future, synthetic oil production) are capital intensive and not substantially different from natural gas pipelines and electric power plants, which are presently regulated.

Again, if the nation wants its energy and has little

faith in the private energy industry, the expansion of public utility regulation to the oil business may be desirable. The FPC already regulates the wellhead price of natural gas, and the extension of such regulation to oil is possible, although it would be bitterly opposed by the industry. The oil industry would probably claim that such a move would be self-defeating to a nation wanting more oil. Even so, such regulation could be confined to the major integrated companies and thus might indirectly encourage more activity by the independents. Yet if regulation provided ample earnings—as historically it has—there is no reason to suppose that the oil and gas industry would not continue to grow just as the electric power industry has grown under regulation for many decades.

A final—drastic—approach is for government to nationalize one or all of the energy industries. There are numerous precedents abroad for nationalized oil, gas, coal and electric industries. Energy is a vital industry to the modern industrial economy, and one way for the public to assert the national interest is for government to own and operate the energy sector of the economy. But in the United States, nationalization is a last resort. A fundamental defect is the absence of an independent body to monitor performance and curb the temptation to sacrifice economic efficiency to political considerations. When government takes over industry, there are few, if any, checks and balances on the operation of the nationalized sector.

In the United States, federal intervention so far has been limited to regulation and a yardstick presence. Yet if historical growth is the nation's choice, and the energy industry repeatedly fails to satisfy the public's energy appetite within reasonable profit limits, nationalization is an option that may loom larger and seem more attractive.

Reaching a consensus and moving forward

If new energy supplies on the scale required by this scenario are to be developed and made available, it is essential that a basic consensus be reached concerning the actions necessary. Any development program that does not satisfy the following needs is likely to encounter serious difficulties in gaining public acceptance.

- Reliable supplies to consumers at prices that do not include "unfair" profits.

• Adequate profits to companies and investors at reasonable risk.

• Realistic precautions to protect the environment that are satisfactory to the public.

• Energy development in sparsely settled regions that is satisfactory to most local residents and provides them a substantial share of economic benefits.

There is no program now in effect for carrying out energy development that simultaneously meets these needs. Nor does the political process in the United States seem likely to develop such a program. Yet one fact is clear: none of the supply development projections in this chapter is feasible in the absence of a broad public consensus to take the environmental and safety risks inherent in going ahead with such rapid development. Projects on the scale these energy supply projections require will not be acceptable politically if a sizable segment of society feels that such developments are too adverse to its interests and values.

Summary and conclusions

In this chapter, we have examined projections of the nation's energy future—often used as a basis for planning by government and industry—that call for a continuation of growth in energy consumption at rates approximately the same as in the past.

We have identified the quantities required for historical growth in energy and what this implies as to the way people will live in the future. It implies an America that is essentially a larger version of today: homes will be no better insulated that they are today; all households will have a full complement of major appliances and will also include presently unknown household energy consuming devices; travel by automobile will continue to increase and cars will be larger and less economical; air travel will greatly increase; improvements in energy use in industry will continue much the same as in the past; and government policies that directly affect energy consumption will continue to pay minimal attention to energy efficiency.

Our analysis of the response of energy demand to price increases shows that such a continuation of past growth patterns is also contingent upon low energy prices, and even (in the case of electricity) falling rates in the future. The prices that seem to be implicit in such projections are not realistic in view of current expectations about future energy prices, especially electricity prices. Thus growth in demand is likely to be realized only if the price of energy to the consumer is set by subsidy at a level considerably below its true social cost of production.

On the basis of information currently available, we must conclude that the continued growth of energy consumption at rates approaching those of the past is unlikely without a large scale government commitment. We do not believe that those government policies and actions necessary to make it happen are desirable.

CHAPTER 3
The Technical Fix scenario

In the past, energy was used lavishly in the United States because fuel was cheap and plentiful. Government policies also effectively promoted energy use through promotional pricing, tax advantages, and other forms of subsidies. In addition, some of the indirect costs of energy consumption were not included in the price.

These basic forces that fostered historical rates of growth in energy consumption are now changing. The combined effects of an international oil cartel and higher costs incurred in developing domestic oil and gas are leading to higher prices for these fuels; and economies of scale in electric power generation seem to have reached a dead end. The social and environmental costs of energy production and use are beginning to be reflected in energy prices. Promotional rate structures in the utility industry may also be expected to change.

Recent shortages have opened our eyes to the many opportunities for saving energy that formerly went unnoticed. Higher prices not only make these savings more attractive, but fear of shortages, pollution, and foreign policy concerns also combine to encourage the nation to embark upon a more frugal energy future.

The *Technical Fix* scenario is an attempt to anticipate the results if long term energy prices and government policies were to encourage greater efficiency in energy consumption. We find that there can be significant reductions in the growth rate of energy consumption without seriously affecting improvement in the standard of living expected between now and the end of the century.

Energy consumption

In the *Technical Fix* scenario annual energy consumption increases at an average rate of 1.9 percent per year between now and the year 2000—from 75 quadrillion Btu's in 1973 to 124 quadrillion in 2000. Future consumption patterns are summarized in Table 5 and Figure 4. (Detailed calculations of the energy budget in the scenario are given in Appendix A.)

This energy budget can provide essentially the same level of energy services (miles of travel, quality of housing and levels of heating and cooling, manufacturing output, etc.) as the *Historical Growth* scenario, if the nation adopts specific energy saving technologies, such as better insulation and better auto fuel economy, to perform these functions.

The energy saving technologies considered here are presently known, and the energy savings included are

Table 5–**Energy consumption in the Technical Fix scenario** (Quadrillion Btu's)

	1973			*1985*			*2000*		
	Fuel	*Elect.*	*Total*	*Fuel*	*Elect.*	*Total*	*Fuel*	*Elect.*	*Total*
Residential	9.4	2.3	16.3	9.3	3.0	18.2	7.7	4.3	19.3
Commercial	6.2	1.4	10.4	8.5	1.8	13.8	10.6	2.3	16.9
Industrial	21.4	2.7	29.5	30.7	3.2	40.0	50.2	4.8	63.1
Transportation	18.8	—	18.8	19.6	—	19.6	24.7	—	24.7
Totals	55.8	6.4	75.0	68.1	8.0	91.3	93.2	11.4	124.0

Note: Totals for each sector include waste heat produced at the power plants in generating electricity. See Footnote to Table 1.

Figure 4–**Energy consumption in the Technical Fix scenario**

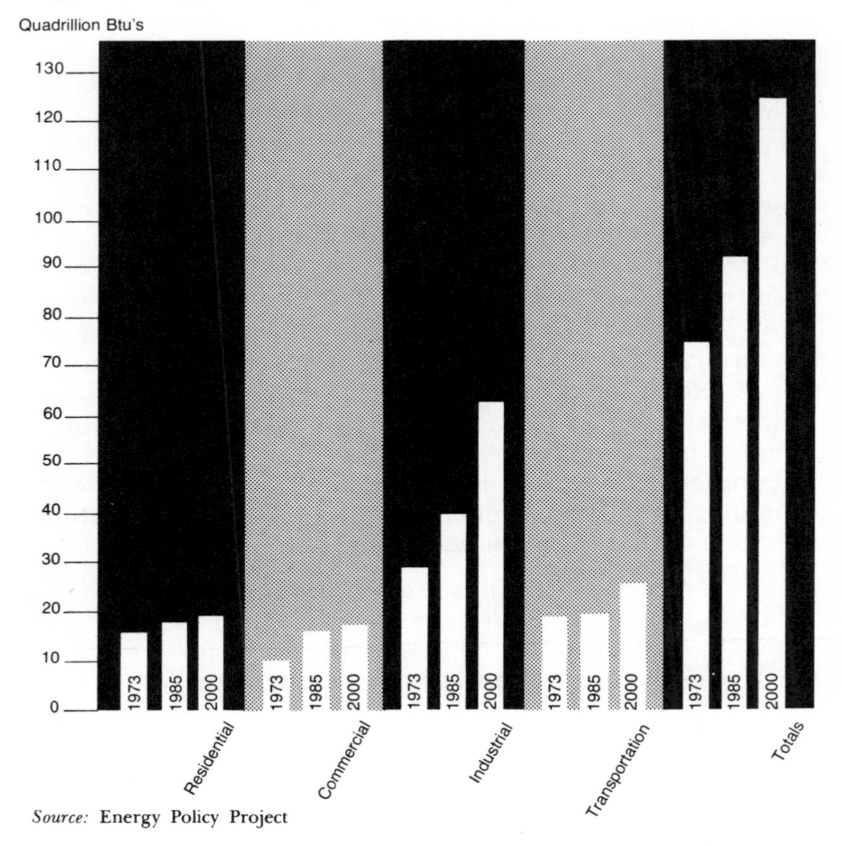

Source: Energy Policy Project

economically justified at existing prices. Should energy prices be even higher in the future, these energy conservation options would be even more attractive. Energy saving technology is introduced in this scenario at the normal rate that new capacity replaces old through retirement and growth.

The econometric model described in Appendix F provides a complementary description of the *Technical Fix* scenario. The model shows how the scenario might develop in response to rising energy prices. In our discussion, we consider both the role of energy prices and specific government policies needed to make the *Technical Fix* scenario happen.

The energy savings in the *Technical Fix* scenario fall into two principal categories. The first category involves the direct energy savings resulting from the application of energy conservation technologies (or technical fixes) at the point of energy use. Increased thermal insulation, heat pumps, improved automotive fuel economy, total energy systems, for example, fall into this category. In addition to the direct energy savings at the point of use, there are indirect energy savings in the energy processing sector —power plants, petroleum refineries, uranium enrichment plants, and the like. For example, for every Btu of electricity saved by using heat pumps instead of resistance heat, about two Btu's of energy are saved that would otherwise be lost as waste heat dissipated at the power plant. Savings tabulated for specific end uses in this chapter include both the direct and indirect savings from energy processing.

The two basic criteria used to select conservation measures are: first, that a significant energy saving should be possible; and second, that this energy saving must be achieved by using methods that will save the consumer money—that is, the conservation measures should be economical. We illustrate the potential for saving energy economically at current energy prices with three examples, each of which has major energy conservation potential.

- In the residential sector, it costs about $700 (1972 dollars) extra to fully insulate a new 1,200 square foot home—excluding unheated basement—with storm windows and heavier insulation in the walls, ceiling, and floor. If the homebuyer gets a ten year loan at 10 percent to finance this investment, his annual payment on the loan would be about $110. Comparing this house to a typically insulated house in New York, say, this investment would save about 400 to 500 gallons of fuel oil per year. This represents a dollar savings of $120 to $150 (at 30¢ a gallon for home heating oil) and a net savings of $10 to $40 a year. The insulation would produce further savings by reducing the electricity requirements for air conditioning.

- Similarly, it would add $400 to $450 (1974 dollars) to the cost of a car, to increase its fuel economy from 12 to 20 miles per gallon, without reducing the size of the car (see section on transportation). Financing the extra investment at 15 percent for five years would cost the car buyer $125 a year. But with gasoline at 55¢ a gallon, the annual fuel saving would be about $185; net savings would be $60. If the car buyer were to choose a small car that gets 20 miles per gallon, he could enjoy the full saving of $185, since the car would probably cost no more than a standard large model.

- Adding $12,000 (1972 dollars) worth of heat recuperators to an industrial furnace using natural gas at 60¢

per million Btu's would save the manufacturer enough in fuel costs that, after additional maintenance and operating costs are taken into account, annual net savings would amount to about $2300.[1] This provides a rate of return on investment of about 15 percent; at the higher natural gas prices that now prevail, this investment would be correspondingly more attractive.

Our calculations assume that in the future, users in the industrial sector will be more aware of energy costs (and therefore more responsive to using energy in an economically efficient way), and that the market imperfections that inhibit investments to save energy in the residential, commercial, and transportation sectors can be removed by specific government actions.

If there are no effective energy policies for supplementing market prices, and if most industries treat energy as an overhead cost about which they do little or nothing, then the econometric model[2] commissioned by the Project suggests that much higher prices would be necessary to achieve the overall 1.9 percent growth rate of the *Technical Fix* scenario. We believe that if the nation adopts energy conservation as a goal, and adopts the set of policies that we shall analyze in detail below, we can achieve the level of savings in the *Technical Fix* scenario, and by so doing alleviate concerns about supply, environment, and foreign policy without appreciable energy price increases.

Residential energy use

In the residential sector the greatest potential for energy conservation lies in space conditioning (which at present accounts for more than 60 percent of residential energy use) and water heating.[a] The potential savings from these two end uses account for about 90 percent of the potential energy savings in the home, as Table 6 and Figure 5 show.

Three basic and complementary approaches can be taken to save energy in space conditioning: improved building design and construction so that less heating and cooling are required; more efficient systems for heating and cooling; and, after 1985, widespread use of solar energy as a renewable resource.

---

[a] Details of energy conservation calculations appear in Appendix A. Only the highlights are presented here.

Table 6–**Potential energy savings in the residential sector** (Quadrillion Btu's)

	Technical Fix vs. Historical Growth		
	1985	*2000*	
Residential energy use in HG scenario	24	32	
Potential Savings			*Conservation Measures*
Space heat	1.0	3.1	Insulation against heat loss
	0.1	0.6	More efficient fossil fuel furnaces
	1.8	3.4	Heat pumps instead of resistance heat
	—	0.3	Solar heating
	0.1	0.2	Electric igniters instead of pilot lights
Subtotal	3.0	7.6	
Air conditioning	0.6	0.9	Improved efficiency
	0.3	0.6	Insulation against heat infiltration
Subtotal	0.9	1.5	
Water heat	0.5	1.2	Fossil fuel or solar instead of electric
	0.1	0.2	Electric igniters
	—	0.2	More efficient heaters
Subtotal	0.6	1.6	
Other	0.7	1.0	
Total savings	5.2	11.7	
Residential energy use in TF scenario	19	20	

Note: The residential sector's share of all energy processing losses are included in these numbers.

In the first instance, widespread installation of ceiling insulation, storm windows and weather stripping on existing houses, and construction of new houses that are thermally "tighter" are practical measures that are at present both technically and economically feasible.

The heat pump is a good example of a more efficient heating system. Heat pumps are like air conditioners operating in reverse: they warm the house by cooling the out-of-doors. They use mechanical energy, generally from electric motors, to bring in the "free" but low temperature energy from the natural environment and pump it up to

Figure 5–**Energy savings, residential sector Technical Fix vs. Historical Growth**

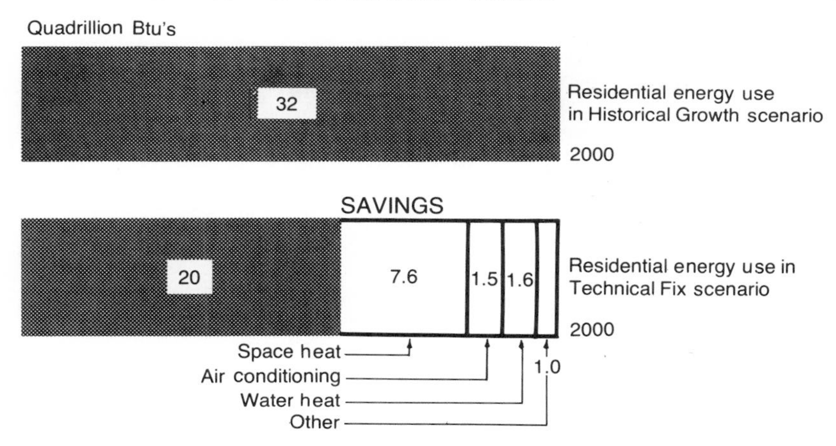

Source: Energy Policy Project

useful temperatures. For every Btu a heat pump consumes in electricity, it produces two or three in heating for a house. This compares with the one Btu of energy provided by electric resistance heaters for every Btu they consume. The use of the heat pump for space heating (and cooling when operated in reverse) is economical in most regions of the nation.

Solar heating and cooling systems can gather the sun's energy in rooftop collectors to provide much of the energy needed for buildings. These systems have already caught the public's eye through widespread publicity of the relatively few solar homes built around the country. Solar heating is technologically feasible, and additional research and development is expected to produce air conditioning systems that use solar heat, further expanding the usefulness and benefits of this approach. Because of the costs and difficulties of installing solar heating and cooling in existing homes, the Project estimates that this technology will not achieve widespread use in homes until after 1985. But by the year 2000, we expect about 10 percent of U.S. homes to be equipped with solar systems. While the savings to be realized with solar systems by the year 2000 are rather modest, a long term savings potential of ten to twenty quadrillion Btu's per year is possible perhaps a half century from now when most of the existing housing stock has been replaced.

Commercial energy use

As in the residential sector, space heating and cooling offer the largest potential for energy conservation in the commercial sector, as Table 7 and Figure 6 show.

Table 7–**Potential energy savings in the commercial sector** (Quadrillion Btu's)

Technical Fix vs. Historical Growth

	1985	*2000*	
Commercial energy use in HG scenario	16	23	
Potential Savings			*Conservation Measures*
Space heat	0.6	1.7	Heat pumps instead of resistance heat
	0.1	1.3	Total energy systems
	0.4	1.1	Insulation against heat loss
Subtotal	1.1	4.1	
Air conditioning	—	0.2	Total energy systems
	0.3	0.3	Insulation against heat loss
Subtotal	0.3	0.5	
Other	—	0.3	
Total savings	1.4	4.9	
Commercial energy use in TF scenario	15	18	

Note: The commercial sector's share of all energy processing losses are included in these numbers.

Figure 6–**Energy savings, commercial sector Technical Fix vs. Historical Growth**

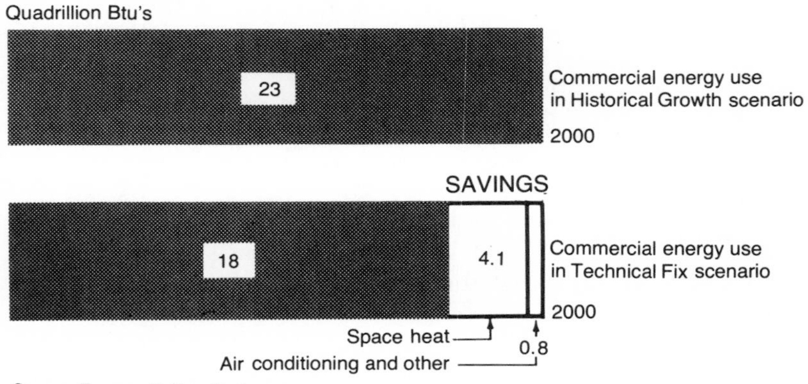

Source: Energy Policy Project

Energy conservation policies for the residential and commercial sectors

Energy conservation in the residential and commercial sectors is primarily a function of building design. This is an area where market forces do not operate effectively. Because of the structure of the industry, builders have incentives to keep first costs low, and thus forego investments in insulation and heat pumps that would be economical over the life of the building. Many owners of residential and small commercial buildings lack the money or credit to make economical investments such as adding insulation or other energy saving equipment. Furthermore, many consumers are simply unaware of energy saving opportunities—some of which do not even add to first costs.

Because the market for residential and commercial buildings does not operate to encourage energy conservation, as reports to the Project[3,4] attest, projected energy savings are likely to be achieved only by supplementing market forces with the following policies:

- Consumer education.
- Adjusting the energy price structure.
- Government actions to overcome the institutional barriers to technical innovation and economically optimal design in the building industry.

Consumer education: Making economically sound decisions requires adequate information. In an era of low energy prices and promotional rates, energy costs represented such a small fraction of building owners' and occupants' costs that saving energy did not seem worth the trouble. This is no longer the case.

A first step would be the labeling of major appliances with energy requirements and operating costs in addition to the purchase price, so that consumers can make their own judgments concerning trade-offs between higher first costs and lower operating costs. A legal requirement for labeling and the necessary testing should be enacted.

In addition, persons purchasing or renting real estate need accurate information concerning utility costs as an item of information relevant to the overall rental or purchase decision. Because of the nature of building markets, disclosure of utility costs alone will not automatically provide adequate incentives to design and build structures on a life cycle cost basis, but it will further that objective. State and local real estate laws should require disclosure of energy costs as well as other economic information to buyers and renters.

Replacing promotional rates with conservation rates: Electricity, and to a lesser extent natural gas, are now priced on a promotional basis that provides discounts for greater use and fails to reflect the higher costs of loads such as air conditioning that add to the utility system's peak load. A reform in this pricing system to reward conservation rather than wasteful use is set forth in Chapter 10 dealing with electric utilities. Such a reform is necessary to strengthen market forces by making it more economical to incorporate energy saving features into buildings and to operate them with energy conservation as a high priority.

Market imperfections in the construction industry: Many institutional characteristics of the construction and rental markets inhibit the introduction of more efficient, economically attractive building designs and technologies for heating, ventilation, and air conditioning.

Unlike most industries that manufacture energy consuming equipment such as aircraft and appliances, the construction industry is fragmented into a great many small companies. As a result, technical talent is not concentrated anywhere in the industry to develop and commercialize significant improvements in construction materials and methods. Differing local building codes and local agreements with labor unions also tend to inhibit innovation.

Because of the building industry's difficulty in gaining access to capital, it emphasizes keeping first costs as low as possible at the expense of higher operating and life cycle costs. The building owner or developer is invariably a small economic entity and is consequently dependent on outside sources of capital such as banks and savings and loan associations. He typically does not have the cash flow and retained earnings to finance capital investment. Furthermore, when the builder/developer does borrow, he typically pays very high interest rates, and when money becomes tight, construction activity invariably slows to a greater extent than other economic activities. For these and other reasons, many investments in buildings that make economic sense are not made.

Energy saving investments are also discouraged by the fact that the person who makes the final decision concerning the design of buildings is usually not the same person who operates them and pays the utility bills. Most houses, apartments, and office buildings are built by speculative builders who are under intense competitive pressure to sell or lease their buildings at the lowest apparent cost to persons who (at least until recently) had no reason to be particularly concerned about long-run costs, including utility bills. Strapped for capital and paying high interest rates, the builder/developer has had little incentive to invest a little extra to reduce operating costs when potential buyers or lessors tended to choose the properties with the lowest first cost or rental.

What can be done through government policies to improve this inefficient situation? The decentralized nature and regional specialization of the construction industry makes it inappropriate for direct federal regulation. The existing system of direct state and local regulation of building design through building codes should be maintained in view of climatic and other regional variations in building design requirements.

But the nation's building codes need to be revised and periodically updated to make energy conservation a priority objective consistent with life cycle economics based on current prices for energy. The federal government can play an important role by making the Federal Housing Authority building code a model of excellence. It can also provide technical assistance for innovation and help the industry raise capital on a more equitable competitive basis with the rest of the economy. Finally, it can create a market for technically advanced, energy saving buildings purchased for government use.

Upgrading local building codes to specify thermal performance of structures to reduce heating and cooling loads would protect the consumer who lacks the time and expertise to analyze the life cycle costs of buildings. Rather than specifying construction methods, such codes should allow maximum freedom for innovation and technical advances by building designers. The technical basis for the new codes could be strengthened through federal sponsorship of building research, as the National Bureau of Standards has already begun, and by providing technical assistance to state and local authorities through the National Conference of States on Building Codes and Standards.

Beyond this, the standards for issuance of federally backed loans by agencies such as the Federal Housing Authority, Veterans' Administration, and Small Business Administration should require life cycle costing of building designs, including analysis of how to minimize energy related costs over the life of the building. While such standards do not apply to all buildings, they influence the design of more buildings than are actually financed by government, because builders often want to ensure that their projects will be eligible for purchase by persons using these financing methods. In addition, the construction of government buildings can also be used to set examples and influence design.

Apart from direct and indirect performance standards, government at various levels has an important role to play in assisting the construction industry in gaining access to capital on terms roughly comparable to those in other sectors of the economy. Whatever measures are taken to provide capital to the housing industry generally, two specific energy saving measures are worth consideration. Homeowners and small businessmen are now unable to obtain long term loans to finance energy conserving modifications to existing homes and buildings. At a minimum, the Federal Home Loan Bank Board and the Small Business Administration should encourage lending institutions to make such long term loans. In addition, steps should be taken to provide lower income homeowners with subsidized, guaranteed loans to allow them to improve the energy efficiency of their homes.

A second step that would improve builders' and owners' access to capital for constructing more efficient buildings or improving existing ones would be for the Federal Power Commission and state utility regulatory bodies to encourage the utilities to advance the money. Under a utility-sponsored "energy conservation service," customers would pay monthly bills that include both fuel charges and the repayment of investment made to reduce fuel requirements. A California utility is experimenting with the lease of gas/solar hot water heating units,[4] and a Michigan utility will install insulation in a customers' residence and recover the cost in the regular monthly bill. The fuel savings should quickly offset the repayment charge and the consumer saves money as well as fuel.

Finally, strong measures are needed to encourage the introduction of improved technologies in the building industry. A Project study[4] recommended the establishment of regional technical assistance centers to provide engineering analyses of new technologies to contractors. In addition, government at all levels can facilitate the introduction of more efficient buildings by providing an expanded market for advanced building technologies. The federal government can also give priority to energy conservation in existing programs of research and education for the architects and engineers who design buildings.

Transportation energy use

Transportation accounts for about one-fourth of total energy use and more than half of our petroleum use. Automobiles and trucks consume 75 percent of the transportation energy. A great deal of energy waste occurs in their use. The airlines, which are the least efficient energy users, are growing most rapidly, while the railroads, which carry people most efficiently, continue to lose traffic.

The energy savings considered here and summarized in Table 8 and Figure 7 are based on the use of engineering improvements to deliver essentially the same energy services as in the *Historical Growth* scenario. They would not appreciably affect the quality of transportation services.

The automobile: About 80 percent of the potential transportation savings in 1985 and 60 percent in 2000 can come from improving the fuel economy of the automobile. At present most autos are large, high powered, and grossly inefficient. Through rather simple engineering innovations, it is technically and economically feasible to improve overall average fuel economy from the 1973 average of about 12 miles per gallon (mpg) to 20 mpg by 1985 and 25 mpg by 2000 without shifting entirely or even predominantly to small cars. Larger savings are of course possible, but achieving this entirely feasible goal would mean that gasoline

Figure 7–**Energy savings, transportation sector Technical Fix vs. Historical Growth**

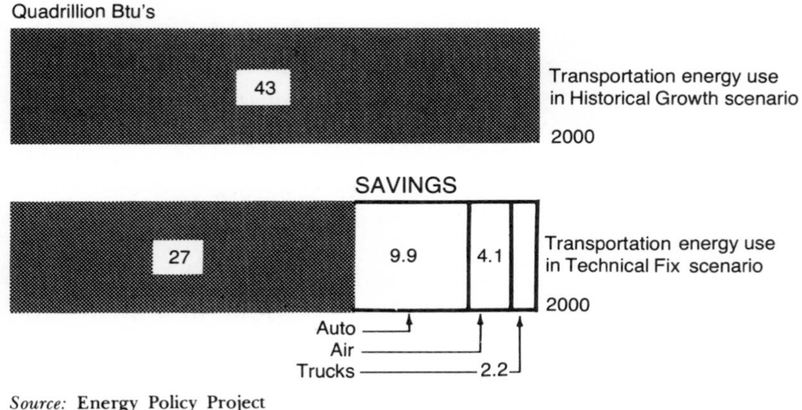

Source: Energy Policy Project

Table 8–**Potential energy savings in transportation**
(Quadrillion Btu's)

	Technical Fix vs. Historical Growth		
	1985	*2000*	
Transportation energy use in HG scenario	29	43	
Potential Savings			*Conservation Measures*
Auto	5.9	9.9	Improve fuel economy to 20 mpg by 1985 and to 25 mpg by 2000
Air	1.1	2.9	Increase passenger load factor to 67% and ton load factor to 58%
	0.2	0.5	Reduce flight speeds 6%
	—	0.4	Shift short run (less than 400 miles) passenger trips to highspeed rail
	—	0.3	Shift short run (less than 400 miles) freight to truck and rail
Subtotal	1.3	4.1	
Trucks	—	1.6	Shift gasoline fueled trucks to diesel
	0.2	0.6	Shift intercity traffic to rail: 20% in 1985, 40% in 2000
Subtotal	0.2	2.2	
Total savings	7.4	16.2	
Transportation energy use in TF scenario	22	27	

Note: The transportation sector's share of all energy processing losses are included in these numbers.

consumption in 1985 would be much less than today, with a 10 percent increase in passenger miles. And with automobiles averaging 20 mpg, travel in a personal car would compare favorably with present-day mass transportation from an energy point of view. However, mass transportation can become more energy efficient too, and electrically powered mass transportation would not require oil and thus would reduce the need for oil imports.

Many small and some medium-sized cars already can achieve 20 mpg or more. Much can be done to improve fuel economy without significantly reducing the size of the auto. For example, in the short term, fuel economy could be improved from 12 to 20 mpg by the following measures:

	Fuel Economy Improvement (percent)
• Aerodynamic drag reduction through body redesign	5
• Rolling resistance reduction through use of radial tires	10
• Better load to engine match	10 to 15
• Substitution of 300 lbs. of aluminum for 750 lbs. of steel[b]	18

As we noted earlier such improvements may increase the price a new car buyer pays by as much as $450, but the fuel savings would more than compensate for the extra investment. With such innovations, moreover, he could enjoy the benefits of increased fuel economy while keeping a fairly large car. In the *Technical Fix* scenario the consumer would have the choice of an inexpensive, efficient small car or a more expensive, but still efficient, larger one.

It should be stressed that achieving an average fuel economy of 20 mpg by 1985 means that new cars must meet improved fuel economy standards well in advance of that target date. A schedule for increasing fuel economy of new cars by model years, which would meet the overall 20 mpg average in 1985, is shown in Table 9. This schedule is based on the assumption that the cars on the road in 1985 will have the same age distribution as in 1970 and that cars of a given age are driven the same amount in 1985 as in 1970.

Further improvements in fuel economy can be realized in the longer term through changes in engine design. For example, the lightweight diesel engine is especially promising.[6] It is widely used today in Europe by taxicabs because of its remarkably high fuel economy at the low speeds characteristic of urban driving.

Truck/rail freight: In 1970 trucks accounted for about 20 percent of transportation energy (5 percent of total energy), with an average energy use efficiency (for large trucks) about one-fourth that for railroads. Energy savings can be realized by switching gasoline powered

[b]While the manufacture of aluminum requires much more energy than the manufacture of steel, a substantial net energy savings is possible by substituting aluminum for some steel. Operating the lighter weight car saves about seventeen times the extra energy required to make the aluminum.[5]

Table 9–**New car fuel economy schedule**

Model Year	*Miles Driven Annually in 1985*	*Fraction of Total Cars in 1985*	*Required Fuel Economy for New Cars (Average)*
1985	17,500	.083	22
1984	16,100	.122	22
1983	13,200	.109	22
1982	11,400	.115	22
1981	11,700	.121	22
1980	10,000	.096	20
1979	10,300	.087	20
1978	8,600	.073	18
1977	10,900	.045	15
1976	8,000	.041	15
up to 1975	6,500	.108	12

trucks to diesel or equally efficient engines, and by shifting long distance freight cargoes to rail.

Fuel economy can be improved roughly 30 percent by switching from a gasoline to a diesel engine. Significant energy savings also could be achieved if freight traffic were shifted from truck to rail.

The rails do not command a larger share of freight traffic today largely because of institutional factors that restrict productivity to levels far below those readily achievable with existing equipment.[7] A basic obstacle is the antiquated regulatory machinery of the Interstate Commerce Commission (ICC). The ICC's regulatory practices severely limit the railroads' flexibility to develop and offer economically attractive alternative services to shippers. In addition, railroad work rules, designed for another era, urgently need to be updated. This could be done without displacing present workers if the railroads' potential for growth in traffic were realized. Finally, consolidation and rationalization of some routes and greater efforts to improve competitive service could enable railroads to capture a greater share of that freight for which they offer greatest efficiency. In 1967 about 40 percent of all freight tonnage could have been hauled by either truck or rail. Trucks carried more than 80 percent of this "competitive" cargo.

The feasibility of switching freight traffic from truck to rail is further demonstrated by the fact that a substantial fraction of truck ton mileage is long haul, and there the switch is especially attractive. About 35 percent of truck ton mileage is in hauls greater than 200 miles, for which rail service could be competitive. Rising oil prices will help to make the railroads a more attractive freight option.

Air transportation: Although air transportation is extremely efficient in saving time, it is inefficient from an energy viewpoint:

- Passenger transport is twice as efficient by rail and car.

- Freight transport is more than ninety times more efficient by rail and twenty times more efficient by truck.

Air transport has not been a major energy consumer in the past. In 1970 it accounted for 8 percent of transportation energy (2 percent of total energy use), but its projected growth is spectacular. Between 1965 and 1970, air travel and air freight experienced average growth rates of 14 and 13 percent per annum respectively, and rapid growth is expected to continue (see Appendix A).

Air travel energy use can be reduced either by improving operating efficiencies or by shifting to alternative transport modes for shorter trips. The most important single measure for reducing energy requirements for air travel is to increase the load factor (seating capacity filled) for passengers. Current Civil Aeronautics Board (CAB) scheduling regulations result in an average load factor of 54 percent of capacity. The load factor can be increased up to 67 percent without appreciably reducing a passenger's chances of losing his reservation (his chances would be only one in a thousand).[8] This improvement would result in a 28 percent direct fuel savings for domestic flights and an 8 percent savings for international flights, which are already carrying fuller loads.

In the past, airplanes have travelled faster than the speed at which fuel consumption is most efficient. Reducing speeds to this level would result in a 4.5 percent fuel savings and would lengthen flight times only 6 percent—about twenty minutes for a transcontinental flight.

By the turn of the century it should be feasible to shift short haul air travel (less than 400 miles) to much more efficient rapid rail ground transport. With such a shift, door-to-door travel time would not be significantly increased. A complementary measure would be to shift short haul air freight to truck and rail.

Energy conservation policies for transportation

Government actions are critically important in the transportation sector. The national network of the transportation system gives it a quasi-utility, "common carrier" status, which has brought about government regulation in the past.

The shift in modal emphasis projected for this scenario might in part be accomplished with higher prices for energy. Higher prices could greatly encourage the switch to smaller cars and higher load factors on aircraft, and give a new competitive advantage to railroads. Additional policy actions are needed, however, to ensure achievement of the savings tabulated above. These actions include:

- Despite the fact that investment in more efficient automobiles is economically justified even if gasoline prices do not rise higher, most automobile purchases are not made on the basis of comparing life-cycle costs of different choices. Therefore, if the nation should choose for the sake of environmental quality and national security to realize the enormous savings potentially available through better automobile design, fuel economy regulations would most likely be required. The automobile industry could voluntarily make binding commitments to the fuel economy goals set forth in this report. If it did not, such regulation should take the form of a minimum standard that could be readily met by most cars (and which could be gradually raised). This could be combined with a purchase tax that increases as the average mpg decreases, and a tax credit for cars with very high fuel economy. This would permit considerable variation in vehicle design and consumer choice. The standard could be raised or lowered from time to time depending on whether average mileage was increasing at the desired rate.

- To increase the efficiency with which freight traffic is handled, ICC regulation of the railroads must be revamped to give them the necessary freedom to compete with other modes of transport and encourage productivity improvements. Such reforms should include flexible ratemaking, freedom to develop new forms of service, including offering shippers a rail/truck door-to-door service, consolidation of some roads, and improved intermodal capability. If work rules cannot be changed through labor-management negotiations, the federal government should arbitrate the necessary changes.

- Rail passenger service for short hauls (up to 400 miles) must be improved to a standard comparable to that found in other advanced industrial countries. This would require capital funding for Amtrak to upgrade equipment and improve roadbeds.

- The CAB should consider energy and economic impacts on the airlines of higher fuel costs as a factor in scheduling flight speeds and frequency, with the objective of improving load factors wherever possible without severely compromising convenience or competition.

In this scenario we have not considered the energy savings of advanced mass transit systems. While a 25 mpg auto is as efficient as today's mass transit, substantial savings could come from a shift to advanced mass transit systems that incorporate such innovations as flywheels or electric storage systems for capturing energy which is otherwise dissipated as heat in braking operations. In addition, such systems would provide a bonus for subways because capturing the braking energy greatly reduces the air conditioning load in underground stations. Besides offering such potential savings opportunities mass transit is of course fully justified in many locations as a means of reducing congestion and pollution (a major side effect of continued automobile use) and for speeding up the commuter's trip. During the next few years, while the fuel economy of cars is still only half as good as mass transit, improved bus service offers a means for large energy savings.

Industrial energy use

The industrial sector accounts for about 40 percent of energy use today. Major energy conservation opportunities are available in industry, where in the past, with a few exceptions, the interest in saving fuel has been relatively low. Energy has accounted for only about 5 percent of value added on the average, and industrial managers have been as oblivious to the opportunities for savings as have homeowners. Potential energy savings are summarized in Table 10 and Figure 8. Details regarding the calculation of these savings are given in Appendix A.

For every Btu of energy saved at the end use point in

Table 10–**Potential energy savings in the industrial sector** (Quadrillion Btu's)

	Technical Fix vs. Historical Growth		
	1985	*2000*	
Industrial energy use in HG scenario	46	87	
Potential Savings			*Conservation Measures*
Five energy intensive industries	4.3	13.1	More efficient production processes in paper, steel, aluminum, plastics and cement manufacture.
Miscellaneous process steam	0.5	3.5	Onsite industrial cogeneration of steam and electricity.
Miscellaneous direct heat	2.9	5.4	Use of heat recuperators and regenerators with direct use of fuels instead of electric resistive heat.
Other	2.5	7.4	
Total savings	10.2	29.4	
Industrial energy use in TF scenario	36	58	

Note: Only the manufacturing sector's share of energy processing losses is included above.

Figure 8–**Energy savings, industrial sector Technical Fix vs. Historical Growth**

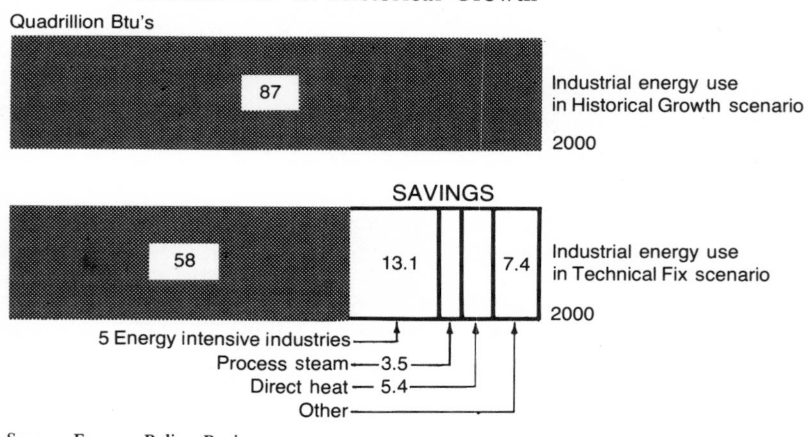

Source: Energy Policy Project

any sector of the *Technical Fix* scenario, another Btu is saved, on the average, in the industries that process the fuels. In Table 11 and Figure 9 we show the energy re-

Table 11–**Energy consumption in the energy processing sector** (Quadrillion Btu's)

	HG		*TF*		*Savings*	
Processing Sector	*1985*	*2000*	*1985*	*2000*	*1985*	*2000*
Electric power						
Generation	23.2	43.4	14.4	18.2	8.8	25.2
Transmission	1.3	2.7	0.8	1.1	0.5	1.6
Uranium enrichment	0.6	2.2	0.3	0.3	0.3	1.9
Petroleum refining	5.0	7.5	3.6	4.3	1.4	3.2
Gas processing and transport	2.9	3.4	2.6	3.2	0.3	0.2
Synthetic fuels processing	0.8	3.2	0.0	1.2	0.8	2.0
Totals	33.8	62.4	21.7	28.3	12.1	34.1

Note: The savings indicated here have been allocated to various end uses in Tables 6, 7, 8 and 10.

Figure 9–**Energy consumption for energy processing**

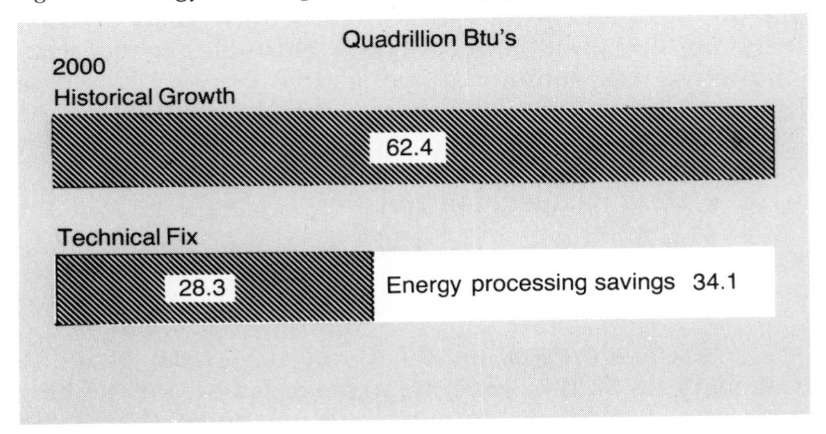

Source: Energy Policy Project

quirements of the energy-processing sector for both the *Historical Growth* and *Technical Fix* scenarios. The savings realized in the energy processing sector of the *Technical Fix* scenario are separated here to show the tremendous losses that occur in processing—an energy "use" that provides no benefits to consumers.

Especially noteworthy is the fact that the energy wasted at electric power plants in the *Historical Growth* scenario represents about one-quarter of total energy consumption in the *Historical Growth* scenario in the year 2000. About 40 percent of the total energy savings in the *Technical*

Fix scenario in the year 2000 arises from reducing these losses. This reduction occurs both through making more efficient use of electricity and through the use of processes that are more efficient in generating electricity: the total energy systems described for the commercial sector and the combined steam/electric power generating systems for industry, as described below—both of which put the waste heat to good use.

When this Project began, relatively little work had been done on energy conservation opportunities in industry. Accordingly, we have supported two studies,[1,9] to survey energy conservation opportunities in a number of key areas. Most of the savings indicated in Table 10 and described more fully in Appendix A are based on this work.

Manufacturing operations are so diverse that it is difficult to list all the energy uses where savings can be captured. In the next few years most of the savings that can be made will come from "leak plugging"—that is, measures that do not involve installing much new capital equipment. But the next few years can also be a time when industry gears up for even larger savings through technological innovation. The substantial savings that can be realized by 1985 and beyond fall into four general categories:

- More efficient steam generation.
- Heat recuperation (recovery).
- More efficient industrial processes.
- Materials recycling.

Process steam generation: About 40 percent of industrial energy is used to generate process steam—for the most part using relatively primitive technologies that make inefficient use of the potential energy in fuels. Two alternative technological options for generating steam are especially attractive: the use of the heat pump principle and combined electric power/steam generation.

In the first approach, relatively low temperature solar energy is extracted from the outdoor air, a lake or other body of water, or even a solar collector, and "pumped up" to useful temperatures for process steam generation using either an electric or engine driven heat pump. The Thermo Electron study[1] explicitly considered the use of a solar collector with an engine-driven heat pump, where both solar energy and waste heat from the diesel engine are captured for steam generation. When the sun is not shining, the system would use commercial fuel. Their study found that when the sun is shining, energy requirements for steam production could be reduced two- to threefold. The system would be economic at present fuel prices (with fuel oil at about $8.50 per barrel), if the cost of the solar collectors could be brought down to about $2.50 per square foot, which seems reasonably attainable within the next decade.

The second approach seeks to improve efficiency by producing both steam and electricity together. When electricity is produced alone, only about 30 to 40 percent of the fuel is converted to electricity; but in combined systems, about 80 percent of the fuel energy can be used to produce both steam and electricity. Large savings result when the electricity is generated near the industrial plant so that the waste heat that would pose a thermal pollution at a central power station can be put to use for industrial process steam. The net savings in total energy requirements for steam and electricity can be about 30 percent. The institutional impediments to combined power/steam generating systems are discussed in the policy section below.

Heat recuperation: Process heat requirements (other than steam) are also large—accounting for nearly 30 percent of industrial energy use. A significant fraction of this heat is lost today to exhaust gases or to materials in process. Through use of heat recuperators or regenerators that return some of this otherwise wasted energy to process, fuel consumption can be reduced 20 to 25 percent. According to the Thermo Electron study, such devices are a good investment and can generally be installed in both new and existing plants.

Historical growth in industrial energy use involves substantial use of electric resistance devices to furnish process heat. Replacing these devices with heat from direct burning of fuels, plus the use of heat recuperating equipment, yields important fuel savings (See Appendix A).

More efficient industrial processes: Improving the efficiency of industrial processes can produce substantial energy savings. Consider these two examples. Primary aluminum production today requires 190 million Btu's to make one ton of aluminum, which places aluminum among the most energy intensive industries. A new Alcoa "chloride" process would reduce fuel needs for primary aluminum production to 131 million Btu's per ton—a saving of 30 percent. In the paper industry a process developed to reduce water requirements fourfold—developed primarily to reduce water pollution—also cuts energy requirements in half. Similar savings are available in cement making. Given an emphasis on energy conservation, it is likely that many such process changes will emerge in the future.

Metals recycling: Primary materials recycling leads to substantial energy savings[c]—especially for primary metals, where, for example, recycled aluminum requires only 5 percent as much energy to process as primary aluminum. According to a study done for the Project,[10] it is economical today to recover scrap metals from municipal wastes wherever incineration is the only practical or acceptable disposal method for solid wastes. With rising fuel prices, municipal waste incineration will continue to become a supplemental energy source, thereby making metals recycling more attractive. Given favorable policies for recycling, the prospects are good that over the next decade we will see the development of a large scale metals recycling industry in the United States.

In addition to the savings opportunities discussed here, which for the most part require significant capital investment, there are many short term measures requiring little or no capital investment. Improving present practices through "leak plugging" can achieve energy savings, generally in the range of 10 to 15 percent in the short term.

Policies for industrial energy conservation

The price of energy (and the fear of shortages signalled by higher prices) will be a major force encouraging conservation measures in industry. Indeed, the industrial sector is probably the most responsive to the price of energy. Moreover, the capability for the analysis and engineering that go into industrial conservation technologies is to be found within industry itself. Hence, a general government policy of regulation or compulsory energy performance standards would appear to be unnecessary, difficult to administer, and perhaps counterproductive.

Yet it would be a mistake to assume that these savings will automatically take place. Government policy does affect

---

[c]An important exception is the recycling of glass. Recycled glass requires about as much energy as glass from virgin raw materials (i.e., sand). Energy savings are captured from glass container reuse rather than the recycling of glass.[14]

the price industry pays for energy. In the past, through promotional rates by utilities, large industrial users got bargain prices. In 1971, for example, the average price of electricity to industrial users in nine major geographic regions was 1.1 cents per kilowatt hour (kwh), compared to 2.7 cents per kwh for residential users.[11] Some difference is, no doubt, justified because industrial customers usually are served directly at higher voltage and do not require distribution service. But changing the rate structure to eliminate promotional rates based on historical cost trends that are now reversed (as discussed in Chapter 10) would make the average price paid by different consumer groups more nearly equal. In addition to rate structure changes, taxes on industrial energy use aimed at internalizing environmental or other external costs could be effective in encouraging investments that would reduce energy requirements.

There are two other areas where institutional impediments to industrial energy conservation must be dealt with by government. One is combined steam/electricity production; the other is metals recycling.

In order to achieve the large savings inherent in combining electric power generation with using the "waste" heat as process steam, electric generation must take place on the site or close to the industry using the waste heat, because it is uneconomical to transport steam over long distances. In the past, many industries generated their own electricity and steam, but the economies of scale combined with promotional rates have enabled the utilities to capture most of the industrial electricity market. The time has now come largely to reverse that trend, in order to save both energy and money for industrial consumers at current fuel prices.

While industries could locate adjacent to the larger utility power stations in order to use byproduct steam, these so-called energy-industrial parks require intensive planning and coordination. For many industries, other factors are overriding. Perhaps most important, electric utilities today require eight to ten years to bring a new power station on line, while user industries typically have much shorter lead times. This difference in the planning period discourages dual energy use because potential user industries are hesitant to commit themselves as far in advance as a utility requires.

Therefore, it appears that unless we develop a more centralized industrial planning system than the United States has today, the energy conservation potential of the energy-industrial park will be difficult to realize on a large scale. But, as we point out in Chapter 8, the large "nuclear park" idea appears to be a desirable approach to redirect nuclear power growth in the future so as to reduce many nuclear environmental risks. As such installations could produce enormous quantities of low cost byproduct steam, economics may provide sufficient incentive to overcome the institutional obstacles.

Nevertheless, there are good reasons why most industries should consider self-generation of electricity, either at new locations or at existing sites, wherever feasible. Such on-site power generation as a byproduct of steam production at industrial installations appears to be practical and economical on a wide scale. A major obstacle to wide application of this concept appears to be the uncertainty about interconnections with the utility distribution system. As discussed in Chapter 10, utility regulations should require interconnections at reasonable rates so that on-site power generation can have a significant impact on industrial energy use. An incentive for moving in this direction is that with on-site power generation, the industrial sector's requirements for central station power by the year 2000 can be reduced 30 percent or more below that projected for the *Historical Growth* scenario—corresponding to about 265,000 Mw(e) in central station generating capacity.[d] Because of long lead times, most of the savings will come only after 1985, but the policy must be implemented at once.

Metals recycling is controlled primarily by the marketplace for scrap as compared to virgin metals. But there are at least two major government disincentives that tip market forces against recycling: federal income tax advantages for virgin metal ores and railroad transportation rates. Depletion allowances for virgin metal ores are about 15 percent. As a result of these tax provisions, the mineral ore–based industries enjoy a much lower effective tax rate than other manufacturing industries, including those industries engaged in recycling of metals.

Railroads are the chief means of transporting scrap metal from processors and dealers to the user mills. Railroad rates, which are controlled by the federal government, needlessly inflate the transportation costs of recycled metals. Existing rail transport rates for scrap iron amount to about

---

[d]This assumes that about 70 percent of industrial process steam production will involve electric power generation as a byproduct.

$4.50 per ton (1969 data), compared to about $3.50 for the equivalent amount of iron ore and coal.

The elimination of the policies favoring the use of virgin materials over scrap—or at least giving comparable treatment for scrap materials—would provide a much needed incentive to resource recovery from urban wastes. In addition, federal R & D programs could spur recycling through demonstration projects. Federal guidelines requiring manufacturers to make it easier for the metals in their products to be separated from each other at disposal would also help.

The federal government can encourage conservation in the industrial sector by greatly expanding research and development on processes that save energy and improve economic efficiency. Many such processes would be difficult to patent, hence there may be inadequate incentive for individual companies to develop them. At the same time, because such processes would be improvements on the verge of commercial viability, it is essential that the user industries be involved in their development. A possible arrangement to encourage technical developments would be the establishment of collective R & D organizations within industries, such as the existing Electric Power Research Institute. Government action would be needed to lead and monitor such activities, to provide initial funding, and perhaps to ensure that antitrust laws were upheld. This approach to R & D is discussed in detail in a report to the Project.[12]

Energy conservation and the economy

There has been much speculation concerning the relationship between energy consumption and gross national product (GNP) as well as employment. The DRI econometric model mentioned earlier was used by the Project to simulate these relationships.

To estimate the economic impact, the model used increasing energy price trends. (See Appendix F.) The energy requirements projected by the model were in line with the technical estimates described in foregoing sections.

The impact on the economy's total production is predicted to be surprisingly small. The economic impact of gradually reducing the long term growth rate of energy consumption is fundamentally different from the impact of a sudden and unexpected interruption in energy supplies, such as resulted from the Arab oil embargo. The model shows that over the long run, production processes would be able to substitute other inputs such as labor, capital, and materials for the reduced energy consumption to achieve the desired savings.

The model predicts that, with the exception of the energy industry, output in major sectors of the economy in 1985 and 2000 would be within a percentage point or two of the *Historical Growth* scenario projections—insignificant differences in a long term projection.

Similarly, the cumulative reduction in GNP is small, about one and a half percent less in 1985 and four percent less in 2000. About three-quarters of this reduction in GNP is a direct result of slower growth in the energy industry, which is compensated for by more efficient consumption technology.

Employment is an area of particular concern in relation to energy supplies, as was demonstrated during the Arab oil embargo. The model estimates strongly suggest that a long term slowdown in the energy growth rate actually has the effect of increasing employment by a small amount. This occurs because the higher prices for energy in this scenario lead to small modifications in productive processes over time, which tend to substitute labor for energy. By 2000, employment in this scenario is about one and a half percent higher than in the *Historical Growth* projection. (See Figures 10, 11, and 12.)

Capital requirements

One aspect of economic planning for the energy future that has received considerable attention recently is the problem of raising capital. Dire predictions have been made by energy companies and financial institutions concerning the difficulty of raising the capital required to provide energy supplies. However, the projections underlying the estimated capital requirements assumed that energy consumption would grow at approximately the historical rate. Our calculations show that growth at the *Technical Fix* rate substantially reduces the total investment required in energy supply and consumption from the *Historical Growth* scenario. Investment required through 2000 for the lower rate of supply growth in this scenario plus the incremental investment in more efficient consumption technology amounts to about $300 billion *less* than the investment

Figure 10–**Gross national product: Technical Fix vs. Historical Growth**

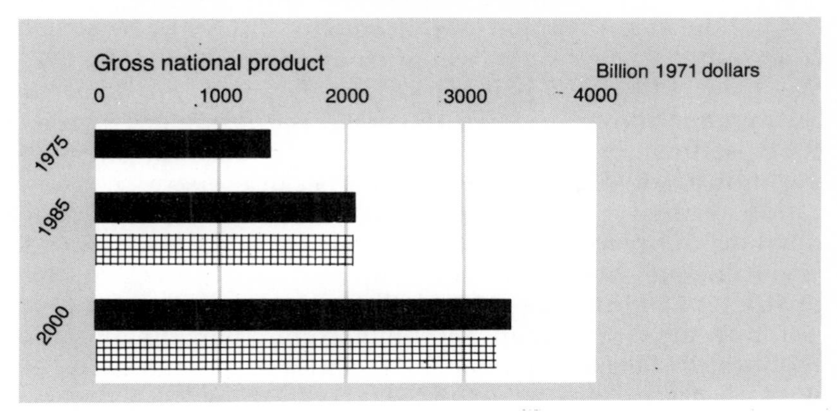

Sources: Energy Policy Project and Data Resources, Inc.

Figure 11–**Employment: Technical Fix vs. Historical Growth**

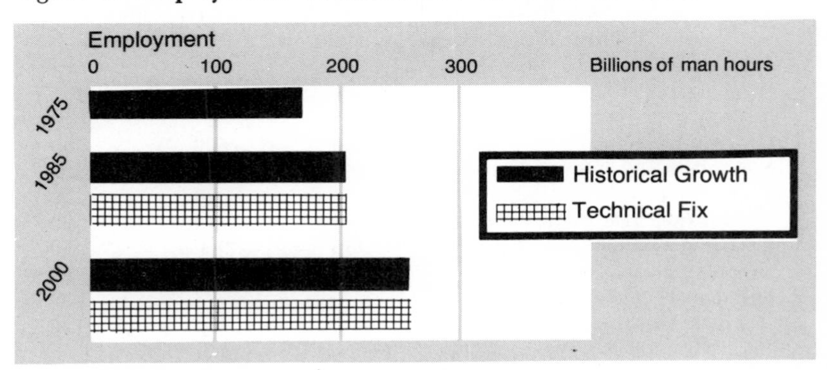

Sources: Energy Policy Project and Data Resources, Inc.

Figure 12–**Energy consumption: Technical Fix vs. Historical Growth**

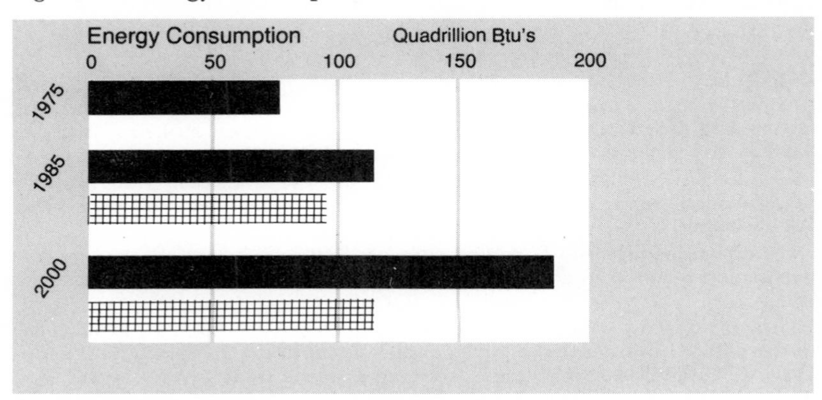

Sources: Energy Policy Project and Data Resources, Inc.

required for supply alone in the *Historical Growth* scenario. The details of the calculations are set forth in Appendix B.

The total capital requirements for the *Historical Growth* and *Technical Fix* scenarios are shown in Table 12. We have added 20 percent to the *Technical Fix* total to account for the infrastructure capital requirements such as R&D, setting up new industries, and modification of existing industries. This has been done because energy conservation technology is not as well developed as energy production technology, and some general additional capital requirements should, therefore, be anticipated. Even so, the total capital required—$400 billion—for energy conservation appears to be about $300 billion less than the amount required to develop the energy facilities if the energy were not conserved.

Table 12–**Cumulative capital requirements (1975–2000) for the energy industry, in the Historical Growth and Technical Fix scenarios** (billions of 1970 dollars)

Item	*Historical Growth*[a]	*Technical Fix*	*Difference*
1. Domestic oil and gas[b]	750	600	150
2. Natural gas pipeline	150	125	25
3. Coal production and transport	70	20	50
4. Nuclear fuel cycle	30	10	20
5. Electric generation	750	300	450
6. Supply subtotal	1,750	1,055	695
7. Residential/commercial energy conservation[c]	—	170	−170
8. Transportation energy conservation[c,d]	—	—	—
9. Industrial energy conservation[c]	—	170	−170
10. 20% for industry infrastructure	—	70	−70
11. Conservation subtotal	—	410	−410
Total (6) + (11)	1,750	1,465	285

[a] The Hass study[13] estimates capital needs to 1985. These have been extrapolated to 2000. A linear extrapolation is used for the petroleum, natural gas, and coal sectors, and the extrapolation of electric generation is based on new capacity at $500/kw for generation, transmission, and distribution.

[b] Includes oil and natural gas, synthetic oil and gas from coal, and shale oil. Petrochemical plants, natural gas transmission, marketing, and exploration costs are excluded.

[c] The capital requirements listed give estimates of the additional energy conservation capital required to achieve the *Technical Fix* scenario. See Appendix B for details.

[d] The additional investments for high speed intercity rail transport are about equal to the reduced investments in airplanes and trucks in the *Technical Fix* scenario. Little or no additional investment is required in the transportation sector. See Appendix B.

Our study of capital requirements for the energy industry[13] estimates that the share of available capital that will be used by the energy industry would increase from the present 21 percent to 27 percent in 1985 in the *Historical Growth* scenario. A continuation of such a historical trend would lead to a 30 percent share of all capital for energy supply by the year 2000. Under such conditions, capital could become scarce for other industries—which "wouldn't give up without a fight"—and lead to rising interest rates. With the lower capital requirements of the *Technical Fix* scenario, the share of investment in the energy sector in the *Technical Fix* scenario would be about 20 percent by 2000, or about what it has been in recent years. The result thus would be to alleviate the problem of "tight money" that might otherwise exist.

If the advantages of investing in energy conservation technology rather than supply technology are so apparent, one might ask, won't normal economic forces bring about this choice more or less automatically? The answer is that the market will certainly tend to encourage such investment. However, the sectors in which this investment will take place, such as railroads, housing, and materials industries, have been unable to gain access to capital on a basis comparable to the energy supply industries in recent years because of government regulation, fragmentation of the industry, and low rates of return on investment. Certainly, funds have not been available for developing new energy saving technology. A major thrust of new government energy policies should be to enable these sectors of the economy to compete for capital for energy conservation on a more equal basis with the energy supply sector. Another objective would be to earmark a larger share of federal research and development funds to energy conservation so that the commercial feasibility of technologies can be demonstrated.

Energy supply strategies

A basic advantage of the *Technical Fix* scenario is that through energy conservation, this country gains considerable flexibility in putting together an energy supply mix.[e] It is important to emphasize, however, that even the lower rate of growth in this scenario requires substantial addi-

---

[e]See Energy Supply Notes, Appendix C.

tional energy supplies, and expansion of a number of sources will be required. With the lower growth rate, however, it is possible to forego development of some major energy sources, or alternately, to meet demand by expanding various sources at about half the rate required in the *Historical Growth* scenario.

Although other variations are possible, we will examine Self-sufficiency and Environmental Protection as the two basic supply strategies for the *Technical Fix* scenario. (See Tables 13 and 14, and Figure 13.)

Table 13–**Technical Fix energy supplies**
(Quadrillion Btu's)

	Actual	*Self-sufficiency*		*Environmental Protection*	
	1973	*1985*	*2000*	*1985*	*2000*
Domestic oil	22	30	36	29	35
Shale oil	0	1	3	0	1
Synthetic liquids from coal	0	0	3	0	3
Imported oil	12	6	6	12	12
Nuclear	1	8	11	5	3
Coal (except synthetics)	13	16	22	14	22
Domestic gas	23	27	32	26	32
Synthetic gas from coal	0	0	1	0	0
Imported gas	1	1	0	2	4
Hydro	3	3	4	3	4
Geothermal	0	0	2	0	2
Other	0	0	2	1	4
Conversion losses from coal synthetics	0	0	2	0	1
Totals	75	92	124	92	123

Table 14–**Fuels for central station electric power, TF scenario** (Quadrillion Btu's)

	Actual	*Self-sufficiency*		*Environmental Protection*	
	1973	*1985*	*2000*	*1985*	*2000*
Coal	8.7	10	11	9	12
Nuclear	0.9	8	11	5	3
Oil and gas	7.3	3	1	7	8
Hydro	2.9	3	4	3	4
Geothermal	—	0	2	0	2
Other	—	0	2	0	2
Totals	19.8	24	31	24	31

An important feature of both supply options in this scenario is that a large fraction of the electricity used in 2000 comes from decentralized sources—commercial total energy systems and industrial on-site generation.

Figure 13–**Energy supply for Technical Fix**

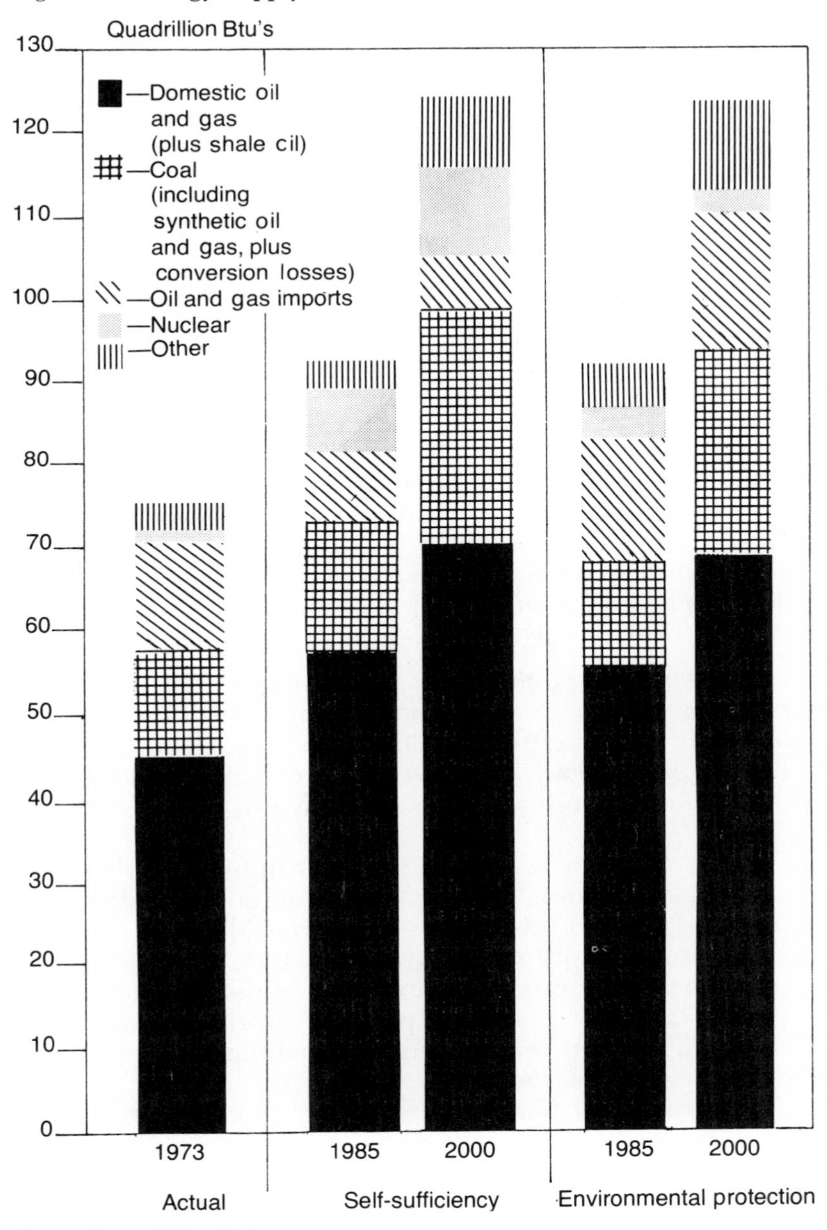

Source: Energy Policy Project

Self-sufficiency: In this option, the objective is to cut imports in half, from the present level of about six million barrels per day to three million barrels per day for the period 1985–2000. Half the growth in this option would come from nuclear power and coal. But by 1985, we would need about 80 percent of the nuclear power that *Historical Growth* would require; by the year 2000, less than 30 percent. Coal use would grow about as fast as total energy use until 1985, with more rapid growth in the period beyond. By 2000, coal production would be about one billion tons per year, compared to 600 million tons in 1973.

Most of the rest of energy growth would come from domestic gas and oil (including shale oil), which would expand at a steady rate, comparable to the overall energy growth rate. Over the next decade, secondary and tertiary recovery from existing wells could be a major source of oil. Petroleum company responses to a Project questionnaire estimated that by 1985, these advanced oil recovery methods could add some 20 to 30 billion barrels[f] to proved reserves if the price of crude were $6 to $8 a barrel in 1972 dollars. For that reason, sustaining this rate of expansion will not require major inroads in presently undeveloped offshore provinces before 1985, with most of the expansion coming from secondary and tertiary recovery from existing fields, from Alaska, and from additional offshore development in the Gulf of Mexico.

Environmental protection: The thrust of this supply mix is to minimize demands on environmentally controversial sources of energy: developments in presently undeveloped offshore areas; in western coal and shale where water is scarce and reclamation difficult; and in nuclear power.

For the near term, this means using oil supplies that are available from comparatively acceptable domestic sources: additional offshore discoveries in the Gulf of Mexico; existing onshore discoveries in Alaska; and secondary and tertiary recovery from existing fields, which can be expected to offset declining primary production in these areas. In addition, oil imports are maintained at current levels, but are not increased.

In the latter part of the century, maintaining growth in oil and gas production as indicated by Table 13 will require some judicious development of new provinces such as the Gulf of Alaska and the Atlantic and Pacific offshore regions. However, the delay that the reduced rate of growth permits in developing these areas will buy time to permit

---

[f] The estimates ranged from 10 billion to 55 billion barrels at the $6 to $8 a barrel price (in 1972 dollars).

development in better harmony with land use and other environmental objectives.

In this option, growth in coal production is curtailed until after 1985. Production could be limited to underground mines and strip mines in areas where reclamation is feasible. In addition, production of synthetic oil and gas does not take place before 1985, thus further reducing the potential for major regional development problems in the west. After 1985, coal production could resume growth; by this time advances in pollution control technology for small particles and sulfur oxides will, it is hoped, allow the growth to include higher sulfur coal from regions where reclamation is readily accomplished.

The environmental protection option also allows for the possibility of curtailing nuclear power growth. Alternately, coal production could level off and nuclear growth resume, if research resolves the various issues for this form of power, although the figures in Tables 13 and 14 are based on the assumption that nuclear fission will not be available as a source of new supplies in this period.

In the period after 1985, some of the growth can be met from environmentally superior unconventional sources—specifically, some solar energy, energy from various forms of organic wastes, and perhaps geothermal energy. By this time, too, it should be possible to implement a modest oil shale development program using *in situ* technology to reduce environmental impacts.

The precise mix of fuels in an environmentally oriented supply program such as this is impossible to foresee, but it is clear that the reduced rate of growth in the *Technical Fix* scenario allows much greater selectivity in choosing energy supplies and more time to perform research and planning functions to ensure that the development that does take place, is performed in a manner consistent with social objectives as well as energy supply.

CHAPTER 4

A Zero Energy Growth scenario

Growth was once a universally applicable solution to economic and social problems, but in recent years growth has itself come to be regarded as a problem. As we have become aware of the need for controlling population and protecting the natural environment, we have accepted the fairly radical idea that many forms of growth can be destructive of human values. Of course, the debate over whether growth should be limited and how is unending.[1]

A social and economic order capable of living indefinitely in harmony with its natural environment may be said to possess "sustainability." A prime characteristic of a "sustainable society" is a fairly constant level of energy consumption. In this chapter we examine both the desirability and the feasibility of levelling off the U.S. energy consumption in this century.

It is important to say at the outset that zero energy

growth does not mean zero economic growth. In Chapter 6 basic economic arguments are laid out showing why economic growth can be uncoupled from energy growth. Our own research confirms that it appears feasible to achieve zero energy growth after 1985, while economic growth continues at much the same pace as in the higher energy growth scenarios. The mix of the economy would of course be different. In our *Zero Energy Growth* future (*ZEG*) there would be a greater emphasis on services —education, health care, day care, cultural activities, urban amenities such as parks—which generally require much less energy per dollar than heavy industrial activities or primary metals processing, whose growth would be deemphasized. Although production of materials would be much greater in 2000 than it is today, production of material "things" would be somewhat lower by 2000 in *ZEG* than in the *Historical Growth* or *Technical Fix* scenarios. This does not mean that people would lack the valued material amenities of the higher energy growth scenarios. Rather, in *ZEG* there would be a premium placed on durability and quality of consumer goods, so that production each year could be lower. Also materials substitutions would be encouraged. As a prime example, throwaway cans and bottles would be discouraged in favor of reusable containers.

Why Zero Energy Growth?

ZEG is actually part of achieving a sustainable economy. If and when *ZEG* occurs, however, it will most likely be as a consequence of several energy-related social concerns. Policies adopted in response to these concerns—if sufficiently stringent—will raise the cost of energy and limit its availability sufficiently for energy consumption to level off.

We emphasize the critical point that *ZEG* is *not* considered as an end in itself, but rather as a consequence of policies aimed at other specific social objectives. Some observers might find a single one of these objectives important enough to justify limiting energy growth; others might find none or all of them sufficiently compelling. Regardless of individual views and priorities, we believe the nation should begin now to examine the social and economic consequences of moving toward *ZEG*.

Some of the reasons for limiting energy growth are examined below.

Limiting regional development: As we have seen in the previous chapter, a policy of "good housekeeping" or technical fixes could delay the need for development of troublesome energy sources for the rest of this century. But an energy saving program that simply squeezes the waste out of the ongoing pattern of growth can only buy time. By the turn of the century, unless the problems of nuclear fission are resolved, or unless solar energy or fusion power become feasible—and these are very big "ifs"—rapid growth in fossil fuel use is likely to resume. The phrase "dig we must" will carry the day.

If our society should be forced to exploit its remaining fuel sources at a furious pace, we will confront very serious land use problems. Especially significant are: offshore oil and gas development, particularly off the Atlantic and Pacific coasts and in the Gulf of Alaska; oil shale development; and western coal development.

Offshore oil and gas development increases the risk of oil spills and the potential threat to coastal areas. Perhaps more important, it is likely to bring with it an initial rapid buildup of supply and support industries, and subsequently refineries and petrochemical complexes and associated secondary industries. The industrialization of the remaining sea coasts will foreclose their use for recreation and wildlife habitat; this is a strong reason for restricting such development.

Western coal development raises the prospect of massive surface mining ventures in areas where reclamation is uncertain. Such development would bring forced rapid growth to the relatively undeveloped West. The construction of various coal processing plants, at first for "mine-mouth" electrical generation and later for gasification and liquefaction of coal, would have even greater regional impact. These plants would bring in far more population than would the coal mining alone. This, in turn, would require the diversion of large amounts of water from agriculture and other uses in a region where water is precious.

Oil shale and coal pose similar development problems, particularly because shale refineries would probably have to be built in the areas developed. In addition, waste disposal on the surface would be a major problem. Hopefully, the development of an *in situ* process for the extraction of oil from shale underground could solve the worst of the surface mining and waste disposal problems, but other environmental concerns such as disruption of underground water supplies would persist.

In the final analysis, even if the specific problems of large scale energy developments can be solved, there is growing opposition to further industrialization of areas that are still relatively undeveloped, are safe for wildlife, open to vacationers, and are attractive settings in which people can escape from the bustle of the city life and enjoy nature. A push for zero energy growth, then, can stem from a general desire to save what is left before it is too late.

Pollution reduction: Most energy-related air pollutants such as sulfur dioxide, small particles, hydrocarbons, trace metals, nitrogen oxides, and carbon monoxide, are likely to be controllable in the future through improved technology.

However, various factors have slowed the use of pollution control technologies, and the control of some pollutants such as submicron particles may be difficult and costly. Reduced energy consumption offers a way to reduce total pollutant output. If energy consumption is stabilized, it is more likely that we will achieve a desired level of air quality, and we can do so at far less expense.

Avoiding catastrophic accidents in energy supply systems: In the future, a society powered by nuclear reactors could be endangered by a reactor accident that would release large quantities of radioactive material. Other major dangers associated with the large scale application of fission power (described at length in Chapter 8) are the possibility that nuclear materials might be stolen for purposes of blackmail or destruction, and the possibility of accidental release of nuclear material either in transit or as nuclear wastes are stored. Safe operation of extensive nuclear fission facilities will require the creation of institutions with longevity and reliability unparalleled in human history. A *ZEG* future provides the option of minimizing and even avoiding these risks.

There are other less dramatic but potentially very serious problems that might be alleviated or completely avoided by *ZEG*. Large scale development of the Arctic to exploit the oil resources there could lead to oil spills with serious consequences—both for the sensitive Arctic marine ecology and possibly for global climate as well.[2] Other dangers arise from the possible explosions of liquefied natural gas tankers and natural gas pipelines, and from coal mine disasters.

A *ZEG* future could permit a slower and safer pace of energy development that would minimize and perhaps even eliminate many of these risks.

World development considerations: Energy consumption throughout the rest of the world is growing at a higher rate than in the United States. Economic development is moving much of the earth's population toward a more energy-intensive standard of living. When the United States buys its energy from world markets, it competes with developing nations for energy resources, driving prices higher and hampering the growth of the developing nations. Reasonably priced energy supplies are a necessary condition for their progress. As the world's leading energy consumer, the United States should set an example concerning "how much is enough." If the United States were to adopt a *ZEG* policy, it would be easier to reduce imports, thereby easing the pressure on world energy supplies and making it easier for developing nations to grow.

Avoiding climate alterations: As pointed out in Chapter 8 there are indications that the burning of fossil fuels may lead in the near future to global climatic change. In fact, there is some evidence that the use of fossil fuels may be implicated even today as a causative factor in equatorial drought. While present scientific knowledge is inadequate to say with a high degree of certainty that climatic disruption from fossil fuel use is imminent, the potential for triggering climatic change is to some people a compelling reason for moving away from fossil fuel use in the United States and other developed countries of the world. A *ZEG* future would facilitate such a move.

Decentralizing technology: There is also rising concern about the estrangement between citizens and the social and economic institutions that are supposed to serve them. The argument is that our institutions, and the technologies that supply energy and other goods, have grown too large to be responsive to the needs of individuals and society in general. Instead, these bureaucracies appear to pursue growth as a means to increase their own power.[3] In the field of energy, many feel that the oil industry has grown beyond the influence of both the citizen-consumer and the government.

One possible response to this condition in the area of energy would be to redirect future growth away from increasing centralization of energy supply and toward decentralization and smaller scale energy systems, or toward systems less dependent on centralized bureaucracies. For example, design of new communities to eliminate much of the need for transportation through close location of work and residential areas and the development of walking and bicycle paths would serve to reduce the communities' dependence on oil companies. Moreover, if energy were supplied in these communities with total energy systems (in part fueled with organic wastes from municipal refuse) the bureaucracy for meeting energy needs could be decentralized and brought more easily under the influence of local and individual decisions.

In addition, when energy is supplied by many small, independent systems rather than a few large ones, society gains added protection against the consequences of a major technical failure such as the Northeast blackout, or a political-economic threat such as the Arab oil embargo.

Of course, central station power plants and oil refineries would not be eliminated. But *ZEG* would enable small scale energy sources to supply a greater part of the total demand.

Changing attitudes and social values: In recent years the traditional American patterns of growth and materialism have come under some questioning.[4] This viewpoint is that American society has concentrated too much attention on making and acquiring ever greater quantities of goods and has neglected the community and nonmaterial needs of people. There is considerable evidence suggesting that, as persons achieve higher levels of physical well-being, they are motivated increasingly by recreational, cultural, and emotional needs.[5]

There is also evidence that a substantial number of Americans are or will soon be reaching the "saturation" level in possession of material goods. On the basis of present trends, our scenarios assume that most major appliances will be in *all* U.S. homes within ten years. The percentage of "satiated" Americans will increase as incomes generally increase. A trend toward consumer saturation has enormous implications for the future of our economy and society as well as for energy policy. If more and more people actually become interested in activities and services that require relatively less energy, the growth in energy consumption could slow drastically, and perhaps even stop.

Economic impacts of Zero Energy Growth

Merely to discuss "zero energy growth" is to unleash a torrent of indignant advertising paid for by major industrial interests which benefit from growth in energy consumption. A typical utility company ad shows a bell-bottomed, well-heeled protester carrying a sign:

"Generate Less Energy."

"Sure," the ad replies. "And generate galloping unemployment."

Another utility company ad shows an embattled housewife with her arms around her washing machine. The headline: "Try telling the lady she'll have to start washing by hand."

Is there any truth to this scare advertising intended to perpetuate the seemingly inexorable growth in U.S. energy consumption? The answer is no; the ads are grossly misleading because they fail to distinguish between energy-saving measures that really help avoid shortages and the very serious problems of unexpected, sudden shortages. As we witnessed during the recent oil embargo, and earlier in the Northeast blackout, the unexpected interruption of energy supplies certainly leads to social and economic disruption. But insulating homes and buildings and making cars that get better mileage are no threat to anyone—except perhaps to the energy company salesmen.

The *ZEG* scenario is an attempt to avoid shortages by a gradual tapering off of energy growth, which occurs over a period of decades. Such a gradual reduction in the growth rate would occur as part of a conscious policy, and energy producers and consumers would be able to plan accordingly, thus avoiding the factor of surprise which has caused so much difficulty in recent supply interruptions. Furthermore, *ZEG* involves curtailing energy growth at a level higher than at present—perhaps 10 percent higher per capita. Actually, the level of goods and services provided by energy could increase much more than this aggregate figure suggests, because energy would be used in *ZEG* much more efficiently than it is today.

If we were to adopt this goal, it would not come as a surprise. There would be years to adjust attitudes, life styles, and policies. For example, if homeowners know in advance that energy will become gradually more scarce and expensive, they will have time to insulate their homes. They can consider gas mileage more carefully when buying new cars. Businessmen planning to buy new equipment will be able to consider energy in their purchases more carefully and to order equipment offering both energy and dollar savings.

Under such a policy there is reason to believe that in addition to the shift toward energy-conserving devices by consumers and businesses, there will probably be a modest redirection in the mix of economic growth. Different products and services in the economy require very different amounts of energy to produce the same dollar value of output. In Table 15 a variety of goods and services commonly purchased by consumers are compared to show the wide differences in consumption of energy.

Table 15–**Energy input for consumer expenditures on selected goods and services**

		Thousand Btu's per Dollar[a]	
1.	Airline transportation	130.9	Partial substitutes (1–2)
2.	Railway and sleeping car travel	59.6	
3.	Kitchen and household appliances	58.9	
4.	New and used cars	55.7	Partial substitutes (4–5)
5.	Auto repair and maintenance	23.5	
6.	Drug preparations and sundries	52.5	
7.	Radio and TV receivers	43.4	Partial substitutes (7–8)
8.	Radio and TV repair	28.1	
9.	Magazines and newspapers	42.2	
10.	Food purchases	41.1	
11.	Private higher education	34.8	
12.	Women's and children's clothing	33.1	
13.	Health insurance	22.0	
14.	Theaters and opera	15.4	
15.	Physicians	10.3	

[a] For 1971.

Source: Robert Herendeen and Anthony Sebald, "The Dollar, Energy, and Employment Impacts of Certain Consumer Options," draft report to the Energy Policy Project, April 1974.

If energy were to become more scarce relative to other inputs in the economy the price of energy-intensive goods and services would rise more rapidly than the price of nonenergy intensive ones. There would then be demand substitution of one for another. For example, the nation's economy might be expected to shift its growth somewhat from manufacturing recreational vehicles and large cars to patronizing the performing arts and manufacturing smaller cars.

The Zero Energy Growth economy

Having mentioned the type of substitutions that consumers make on a modest scale in a *ZEG* economy, let us now turn to overall effects. For instance, how much substitution of services for manufactured goods is required by 2000 under *ZEG*? What are the overall effects on GNP and employment impacts of a *ZEG* policy?

As we have noted in Chapters 2 and 3, the Project's econometric model simulates the growth of the economy and the interactions between the energy sector and other sectors.[6] This model shows that energy use could level off at about 100 quadrillion Btu's per year after 1985 with prices similar to those in the *Technical Fix* scenario, plus an energy tax that increases from about 3 percent of the price of energy in 1985 to approximately 15 percent by 2000. The application of a sales tax on energy as used in this simulation is strictly illustrative of how taxes could be used to control energy growth. The tax itself could be applied in any number of ways: as a resource depletion or severance tax; as a Btu tax; as very high pollutant emissions taxes; or in other ways. The energy tax could of course be a substitute for other federal taxes and not result in any greater burden on taxpayers.

The model indicates that the lower level of energy consumption in *ZEG* gives rise to a slight reduction in real GNP. However, this reduction is mostly due to reduced output in the energy sector. In any case the effect is small; by 2000 GNP is reduced less than 4 percent from the *Historical Growth* level, despite the fact that energy consumption is only slightly more than half that of *Historical Growth.* The composition of this slightly smaller GNP is also modestly different from the other scenarios. The output of the service sector is up by about 1 percent compared to the *Historical Growth* economic projections. The other sectors, being more energy-intensive, are off between 1 and 3 percent, except for the energy sector: its contribution to national output is off about 60 percent compared with the *Historical Growth* scenario. (See Appendix F for details.)

Thus we see that the productive capacity of the country should be able to respond readily to reduced energy availability by growing in a pattern slightly different from that anticipated under *Historical Growth.* It should be emphasized that the figures do not show a contraction of any major economic sector (including the energy sector) because all sectors grow considerably from their present levels. Instead we see slightly different growth *rates* for various economic sectors than is projected under *Historical Growth.*

On the demand side, the *ZEG* program outlined here involves a shift on about 1 percent of GNP from personal consumption to the government sector. Aside from this shift, the other categories of demand are off slightly, about in line with the 4 percent overall loss of GNP.

For perspective it should be noted that the losses in consumption under *ZEG* are only relative to the *Historical Growth* projections. The absolute levels of consumption under all scenarios are much higher than at present. For example, the dollar value of personal consumption under *Historical Growth* in 2000 is about 140 percent greater than the 1975 projection used in all scenarios; personal consumption in *ZEG* is about 125 percent higher than the projected 1975 level.

Employment: There is much talk—and considerable anxiety—about the supposedly close and unbreakable relationship between energy consumption and employment. Both the econometric model and the analytical work of the Project Staff (Chapter 6) reveal that such commonly held fears are unfounded. While it is true that a sudden and unexpected energy shortage can cause, and has caused, major unemployment, our conclusion is that a long-term slowing of energy growth signalled by clear policy commitments, slowly rising prices, and appropriate compensatory policies could actually increase employment.

Just as different sectors of the economy require different energy inputs for a given dollar value of output (see Table 15), various sectors of the economy require different amounts of labor input for a dollar of output. Not too surprisingly, economic activities with low energy requirements—largely those in the service sector—are also those with large labor inputs. In addition, marginal substitution of labor for capital and energy in manufacturing also contributes to higher employment in this scenario. Thus, while some jobs in this scenario are different from those in the other scenarios, it is likely that there would be more overall employment opportunities in *ZEG* compared to other scenarios.

The higher prices of energy in this scenario lead to the substitution of a slightly greater fraction of economic activities that use less energy and more labor for more energy-intensive products, and also the substitution of slightly more labor-intensive processes in manufacturing. The overall result is an increase in the demand for labor, with the man hours worked increasing by more than 3 percent in 2000 over the *Historical Growth* scenario. The greatest increase in employment is in the services and government category, where employment is up almost 7 percent over the *Historical Growth* scenario. Other sectors show lesser increases in employment, except for the transportation sector, in which there is a decline. Figures 14, 15,

Figure 14–**Gross national product: Historical Growth, Technical Fix, and ZEG**

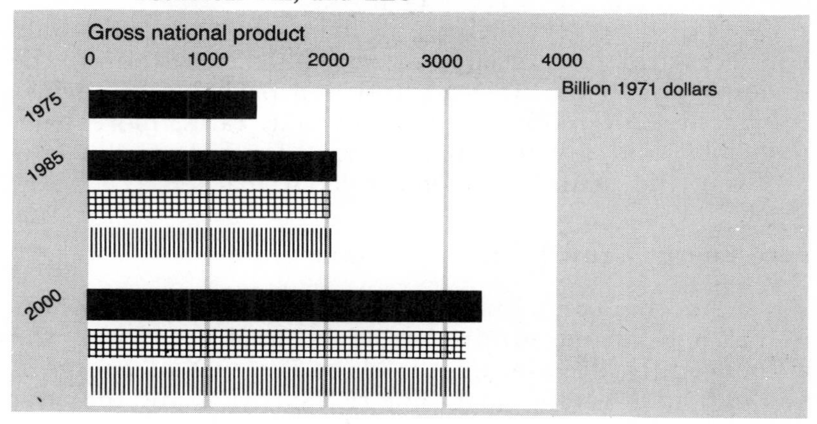

Figure 15–**Employment: HG, TF, and ZEG**

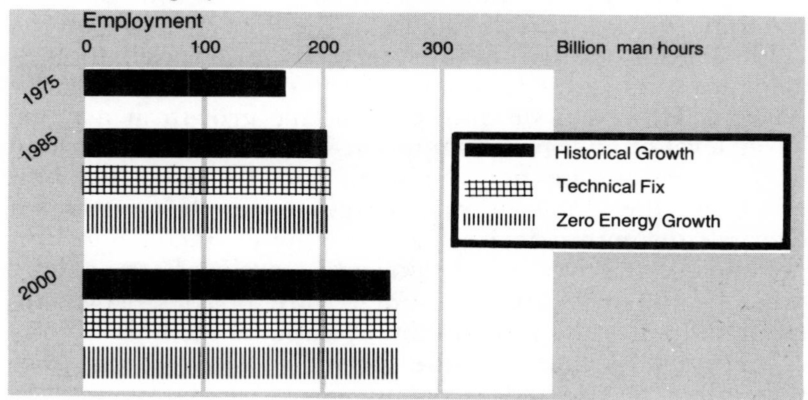

Figure 16–**Energy consumption: HG, TF, and ZEG**

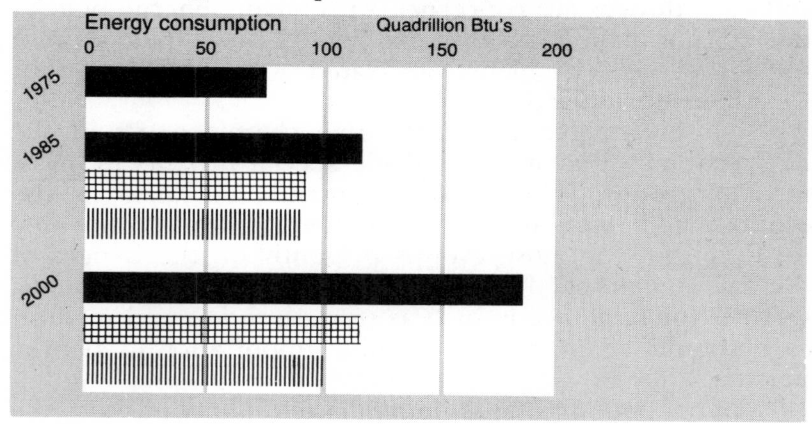

Sources: Energy Policy Project and Data Resources, Inc.

and 16 show the projected GNP, employment, and energy consumption in all three scenarios.

The 3 percent increase in man hours worked would not necessarily be translated directly into a reduction in the unemployment rate. Instead, it would probably result in a combination of reduced unemployment rates, more part-time jobs, and greater opportunities for groups not currently in the mainstream of the labor market.

Zero Energy Growth and low income groups

Another commonly held assumption concerning *ZEG* is that low income groups would be unable to enjoy the affluence afforded the middle class. Less use of energy is said to hold down the poor. The rich stay rich, and the poor stay poor. But it is not necessarily so.

There is of course a valid connection between economic growth and the ability of the poor to move up the ladder, even though today's poverty level population is evidence that growth alone offers no cure for relative poverty. However, sustaining economic growth at a much lower level of energy consumption—as would be the case in *ZEG*—threatens neither the poor nor the rich. Just how much of the *ZEG* energy budget (and the output of energy-consuming devices) goes to lower income groups depends on government policies unrelated to energy policy—principally the tax and welfare policies, which directly influence the distribution of national income. Energy related policies could include specific subsidies to the poor that might be enacted in conjunction with the energy tax —say, "energy stamps," similar in purpose to food stamps. But such devices are concerned with having energy policies that do not make matters worse for the poor. They cannot solve the basic problems associated with distribution of wealth in our society.

Interestingly, there are some characteristics of the *ZEG* economy that might provide special benefits to lower income groups. The greater demand for labor in the economy to replace higher priced energy should contribute to more jobs for those groups presently on the fringes of the labor market. In addition, the emphasis on public services for basic needs such as transportation and medical care should be of greater relative importance to poorer persons.

Economic activity is not restricted in any of the three

scenarios. The different levels of energy growth would not materially change the share of national income received by any economic group one way or another. Relative poverty would likely still exist in any energy scenario unless more direct measures are taken to remedy it.

Transition problems: ZEG in comparison with other scenarios

All three scenarios involve transitions and socioeconomic changes. *ZEG* would not upset growth in economic output and employment, provided it were carried out gradually, over a period of decades. Indeed, the changes contemplated here are probably less severe than many transitions the economy has undergone in the past. The economic adjustments involved here are minor compared with the dislocations that occurred when 300,000 coal miners lost their jobs after 1947; or when a million aerospace workers were discharged between 1967 and 1971; or when thirteen million persons left farms for cities between 1950 and 1970.

This is not to ignore the possibility of local social and economic dislocations. For example, if shortsighted planning or reluctant response by automobile companies does not produce small cars as rapidly as government policies and/or the market demands, workers on large car assembly lines would in the future, as in the recent past, be laid off until the assembly lines are converted. Some suppliers also would be temporarily affected.

Obviously both industry and government policies should be sensitive to such transition problems. With foresight and planning, workers can be given on-the-job training, and instead of closing a factory it could be converted to a less energy intensive line of goods (smaller cars, for example). As we have seen, the employment effects of a *ZEG* scenario would be positive on the whole. There would be new jobs in the fields of mass transit construction, health care, and other public services.

The other scenarios involve dislocations and transition problems too. Historical growth has been interrupted through the years by periods of recession and high unemployment. In addition, historical growth emphasizes expansion of energy intensive areas which have low numbers of jobs per dollar of output. These industries substitute energy for labor.

A note about the very long term

Technical progress and changes in a society's kind of output can dramatically lower the level of energy required. Ultimately, however, there are minimum amounts of energy required by the laws of physics to perform certain functions, such as materials processing. Without them, the materials simply are not produced. As this minimum is approached in any given process, further efficiency gains would become harder to achieve, and economic growth would have to slow down and ultimately cease if energy use is stable.

But our calculations show that there is a great deal of room for lowering the number of Btu's required per dollar of GNP. Our survey of technology suggests that even through 2000 we cannot expect to achieve sufficient gains in efficiency to reach the limits of what is possible for industrial processes. Long-term research and development can bring many more economical opportunities for conserving energy.

Aluminum is a good example. The average energy required to produce a ton of aluminum (1968 data) is 190 million Btu's. By using technology available in 1973, a ton of aluminum could be produced with 152 million Btu's. But the theoretical minimum is very much lower, about 25 million Btu's per ton. Furthermore, each year we could learn to use aluminum and other materials more efficiently in consumer products, thereby stretching further the energy used to make it.

It is thus useful to think of a *ZEG* society as one in which continuing gains are made in the efficiency of energy consumed in industrial activities and in transportation that offset the growth in energy required for services, agriculture, and other sectors of the economy. The increased efficiencies would provide the energy to sustain economic growth. We can thus continue to increase per capita GNP for many decades after zero energy growth is achieved —although, of course, in time the limits of increased productivity must be reached. By then, some time in the 21st century, zero population growth can be achieved, and it may be that society will be ready to return to the normal state over the centuries—one of stability rather than growth. Of course by then technological break-throughs may resolve resource and environmental concerns sufficiently that growth can be resumed for a while. At least there would be that choice. Or breakthroughs in energy conservation technologies may make reducing energy consumption attractive. A *ZEG* society need not be locked in to a particular level of energy use forever.

Policies for ZEG

The *ZEG* scenario closely follows the *Technical Fix* scenario out to 1985, after which time energy growth levels off gradually at about 100 quadrillion Btu's. The policies needed to bring about *ZEG* include all the major policies of the *Technical Fix* scenario plus specific economic policies needed to bring about a shift in the mix of GNP. Specifically, the major policy actions needed are as follows.

- An energy sales tax.[a] This tax would be imposed gradually, on a predetermined schedule, so that purchasers of energy-consuming equipment could plan accordingly. The tax would begin in 1985 at 3 percent of the retail price of energy, and increase a fraction of 1 percent each year to about 15 percent in the year 2000. The tax would raise the price of energy intensive goods and services relative to nonenergy intensive activities and thus would use traditional market mechanisms to reduce energy consumption.

- Imposed independently of other policies, such an energy sales tax would be both regressive in its impact on consumers (i.e., its impact felt more by lower income groups than by the rich) and a restraining influence on economic growth. Other policies would have to be adopted to offset these effects. A reduction of other federal taxes, or increases in federal payments, especially for lower income citizens, would be an obvious part of such an energy tax proposal. Some of the funds brought into the treasury could be directed toward public services that would facilitate and enhance zero energy growth such as:

    Public transportation
    Health care

[a] In practice, the taxes which bring about *ZEG* would most likely take the form of specific levies aimed at restricting growth in the specific activities mentioned earlier—e.g., state severance taxes to restrict development in certain areas, or taxes on CO_2 emissions. However, in the absence of clear indications of the most likely direct motivation for *ZEG*, and for purposes of studying the economic impacts of *ZEG*, a sales tax on energy generally will be the policy analyzed here.

Housing
Urban amenities, including clean streets and parks
Law enforcement
Education
Cultural activities
Day care
Nursing homes and other old age benefits.

• Increased automobile gas mileage is essential to the achievement of energy savings in the transportation sector. The energy tax would doubtless move consumers in the right direction. Since automobiles are not bought on the basis of life cycle cost analysis, and since the automobile industry resists significant changes in vehicle design, additional action is required to ensure that savings are achieved. A legal performance standard or a heavy tax imposed on inefficient automobiles (perhaps with a credit on very efficient ones) as part of the purchase price would be required to achieve an *average* vehicle efficiency of 20 mpg by 1985 and 33 mpg by 2000. The 33 mpg target for the year 2000 would achieve considerable savings over the 25 mpg goal in the *Technical Fix* scenario, but of course a much greater effort is required to realize it.

• Tightening of building codes, lending requirements, and improved capital availability to ensure optimum building design. These policies are the same as in the *Technical Fix* scenario. We feel that they are needed in *ZEG,* even with the added incentive of higher energy prices, because institutional constraints in the building construction industry preclude the person who pays the utility bills from specifying energy conserving features in the initial design. As discussed in the previous chapter, buildings in the United States simply are not built or purchased with total life cycle costs as the basis of the selling price.

In addition to these four main policies for reaching zero energy growth, a number of less important assumptions were made, including

• Expansion of urban mass transit systems and development of a system of bikeways.

• Implementation of airline energy conservation regulations to raise load factors and slightly reduce cruising speed, which are to the airlines' financial advantage in the light of fuel prices.

• Upgrading of rail service, including fast passenger

service for short haul intercity runs. Granting of increased flexibility to the railroads in their rates, and elimination of the numerous institutional barriers to improved railroad productivity (described in the previous chapter).

• Elimination of all depletion allowances on virgin ore and of discriminatory freight rates, which are lower for ore than for scrap. These rates currently discourage recycling of energy intensive materials. Recycling would also be encouraged by requirements to "mine" urban wastes for scrap materials (as well as for their energy value) and by federally funded demonstrations of recycling technology.

• Implementation of an aggressive government program to ensure research and development of technological improvements in energy consumption, which would become economically attractive through the energy tax. Industry should be involved from the beginning, using flexible policies depending upon the circumstances.[7]

• An encouraging U.S. government attitude toward American investments in foreign countries, which would have the effect of shifting some of the growth in energy intensive industries to economically favorable areas.

Where would the energy go in a zero growth scenario?

We have developed an "energy budget" to show how the nation might use the energy available in a *ZEG* scenario (see Table 16 and Figure 17). Only an overview of this budget is presented in this chapter. Details appear in Appendix A. This energy budget illustrates what is possible under the constraint of zero energy growth. It is not the only way to keep America going, or necessarily the best way. In practice, the distribution of the available supply of energy under a zero growth energy policy would be determined by government actions and the marketplace.

Transportation: The transportation pattern in *ZEG* differs from *Technical Fix* in that it assumes slower growth in air travel, greater use of railroads for passengers, even more efficient cars, and a de-emphasis of auto use in urban areas.

In this scenario, by the year 2000, people would travel by all modes—air, car, rail and bus—about 25 percent more than they do today, but somewhat less (by 15 to 20 percent) than in the other scenarios. (See Table 17.) They

Table 16–**Energy consumption in the ZEG scenario**
(Quadrillion Btu's)

	1973			*1985*			*2000*		
	Fuel	*Elect.*	*Total*	*Fuel*	*Elect.*	*Total*	*Fuel*	*Elect.*	*Total*
Residential	9.4	2.3	16.3	8.9	2.9	17.3	7.2	3.6	17.0
Commercial	6.2	1.4	10.4	9.0	1.9	14.5	11.5	2.7	18.8
Industrial	21.4	2.7	29.5	28.9	3.1	37.9	33.0	5.2	47.0
Transportation	18.8	—	18.8	18.4	—	18.4	17.2	—	17.2
Totals	55.8	6.4	75.0	65.2	7.9	88.1	68.9	11.5	100.0

Note: Totals for each sector include waste heat produced at the power plants in generating electricity. See footnote to Table 1.

Figure 17–**Energy consumption for ZEG scenario**

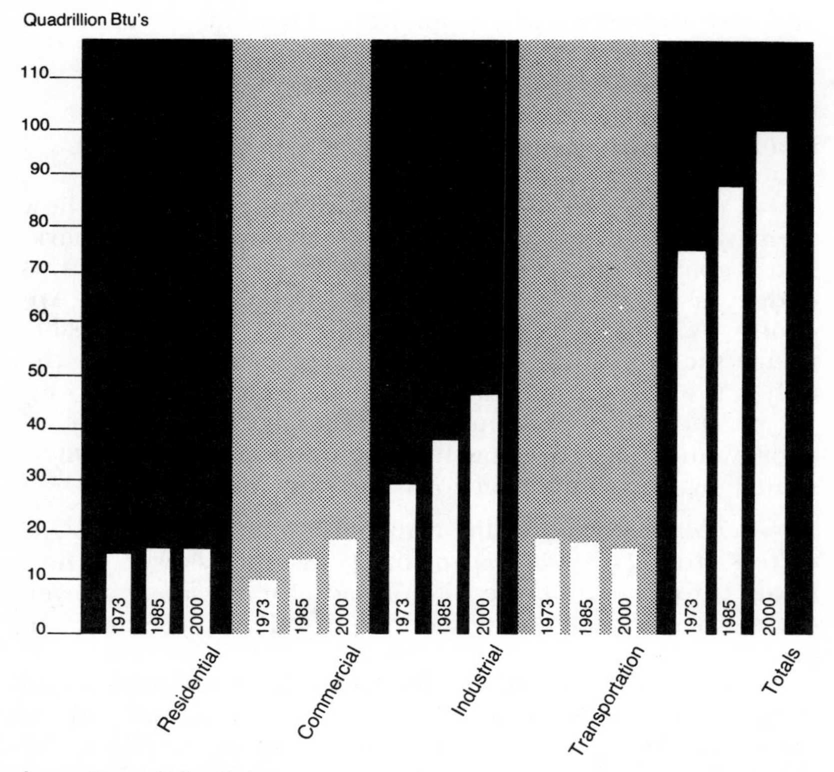

Source: Energy Policy Project

Table 17–**Travel in scenarios** (passenger miles per capita)

		HG		*TF*		*ZEG*	
	1970	*1985*	*2000*	*1985*	*2000*	*1985*	*2000*
Urban							
Auto	3140	4130	4440	4130	4440	3720	2750
Bus	120	130	150	130	150	540	1260
Total urban	3260	4260	4590	4260	4590	4260	4010
Rural auto	5070	5020	4970	5020	4970	5020	4820
Intercity bus	120	120	105	120	105	120	105
Air	780	2090	4190	2090	3905	1400	1955
Rail	60	80	125	80	410	150	590
Total Intercity	6030	7310	9390	7310	9390	6690	7470
Total Travel	9,290	11,600	14,000	11,600	14,000	11,000	11,500

would travel by air about two and a half times as much as in 1970, but about half as much as in *Historical Growth* and *Technical Fix*.

To partially offset the slower growth in travel by air, a major upgrading of rail service is a key feature of this scenario. High speed rail service will need to be provided between cities up to 400 miles apart. This high quality service is anticipated to take up the bulk of short haul passenger travel, leaving the long haul service to aircraft. Passenger rail travel is estimated to increase at a rate of 10 percent per year after 1975.

The slowdown in growth of air transport (for both people and packages) produces more than half the energy savings in transportation for *ZEG* compared with *Technical Fix*. From a policy point of view, this is expected to be accomplished through the effect of the energy tax on air transport prices as well as through the regulatory and financial measures needed to make the railroads an attractive alternative to air travel.

The other major change in *ZEG* relative to *Technical Fix* in the transportation sector concerns the automobile. Savings can be achieved by use of more efficient autos; by a shift to more efficient transportation modes; or by reducing the need to travel by making possible environments in which the commuting distance is shortened.

To achieve the reductions necessary for zero energy growth, the average fuel economy would have to be increased from a 1970 average of 13.6 mpg to 20 mpg in 1985 and 33 mpg in 2000. This would require smaller cars

Table 18–**Potential energy savings in transportation**
(Quadrillion Btu's)

Zero Energy Growth vs. Technical Fix

	1985	*2000*	
Transportation energy use in TF scenario	21	27	
Potential Savings			*Conservation Measures*
Auto	—	1.9	Improve fuel economy to 33 mpg by 2000
	0.2	0.1	Shift urban traffic to buses: 10% by 1985; 25% by 2000
	—	0.2	Expand new communities
	—	0.3	Shift 10% of urban traffic to walkways and bikeways by 2000
Subtotal	0.2	2.5	
Air	1.2	4.5	Assume slower growth in air transport (3% per year for passenger travel and 6% per year for freight)
Trucks	—	0.8	15% reduction in freight hauling requirements by 2000
Rail	—	−0.1	Passenger transport increases at 10% per year after 1975
		0.3	15% reduction in rail freight requirements by 2000
Total Savings	1.4	8.0	
Transportation energy use in ZEG scenario	20	19	

Note: The transportation sector's share of energy processing losses are included in these numbers.

Figure 18–**Energy savings, transportation sector: ZEG vs. Technical Fix**

Quadrillion Btu's

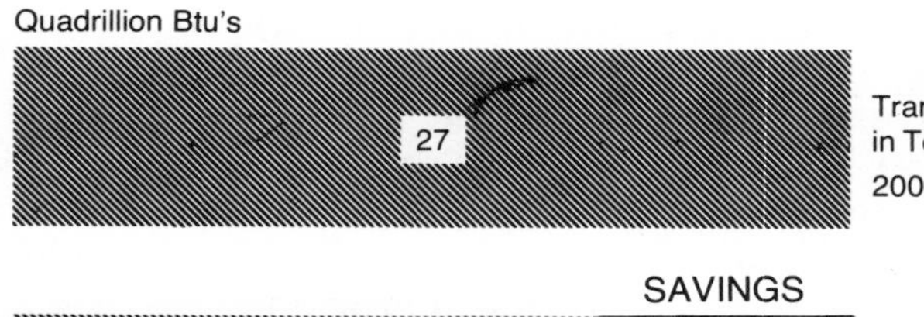

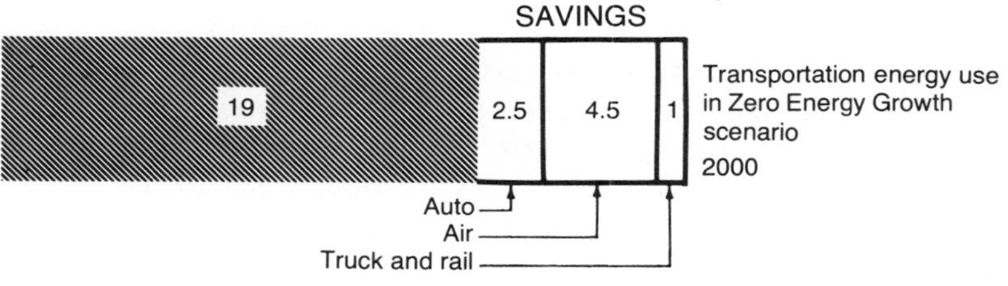

Source: Energy Policy Project

on average than at present.[b] Toward the latter part of the century, engine and body design changes could make it possible to meet these fuel economy goals with cars larger than subcompacts.

The other conservation measures affecting auto travel, which are included in the *ZEG* energy budget to illustrate their relative importance (or, more accurately, their relative unimportance), include:

- Switching 10 percent of the urban auto travel projected in the *Historical Growth* case to buses by 1985, 25 percent by 2000. This would happen through upgrading service, using new buses, and having special bus lanes in congested areas.

- Accelerated building of new communities until about 6 percent of the population lived in such communities by 2000. We assume that the proximity of work to homes would allow residents of these communities to cut their auto travel in half.

- Elimination of 10 percent of urban auto traffic by 2000 through shifting commuters to bikeways and walkways.

The use of buses increases in *ZEG* over both the other scenarios, but the energy savings achieved by shifting people from efficient autos to buses is small (see Table 18 and Figure 18). Energy used by trucks and railroads for freight hauling is down 15 percent from the *Historical Growth* scenario levels by 2000 because of the reduced output of the manufacturing sector. All other transportation energy demands would be the same as in the *Technical Fix* scenario. The cumulative effects of these measures taken in the transportation sector are indicated in Table 18.

Residential: In the residential sector, the *ZEG* scenario looks very much like the *Technical Fix* scenario. As shown in Table 19, energy supply is sufficient for major appliances to reach 100 percent saturation levels by 2000 for all households, including low income households. Heating and cooling requirements are about the same as in the *Technical Fix* scenario, but there are additional savings from fewer new appliances as shown in Table 20 and Figure 19. The energy needed in homes is only half that projected in the *Historical Growth* scenario.

---

[b] It should be emphasized that there is no *technological* problem to achieving these economies. The Honda Civic with CVCC engine, scheduled to be marketed in this country in 1975, delivers over 30 mpg while meeting the original 1975 Clean Air Act emission standards. It is, of course, a subcompact design.

Table 19–**Appliance saturation levels** (percent unless otherwise specified)

Appliance	1970[a]	1985 HG	1985 TF	1985 ZEG	2000 HG	2000 TF	2000 ZEG
1. Space heat	100	100	100	100	100	100	100
2. Air conditioning (total)	35	100[b]	100[b]	100[b]	100	100	100
(a) Room	25	50	50	50	0	0	0
(b) Central	10	50	50	50	100	100	100
3. Water heat	100	100	100	100	100	100	100
4. Refrigerators (total)	100	100	100	100	100	100	100
(a) Regular	—	0	0	0	0	0	0
(b) Frost free	—	100	100	100	100	100	100
5. Lighting	100	100	100	100	100	100	100
6. Cooking ranges (total)	100	100	100	100	100	100	100
(a) Fossil fuel	50	40	40	40	30	30	30
(b) Electric	50	60	60	60	70	70	70
7. Dishwashers	25	100	100	100	100	100	100
8. Clothes dryer[c]	40	50	50	50	60	60	60
9. Clothes washers	60	80	80	80	100	100	100
10. Freezers	30	60	60	60	100	100	100
11. Portable appliances[d] (relative units)[e]	1	2	2	2	3	3	2
12. Unknown appliances (relative units)[e]	0	1	1	0.5	2	2	1

[a] Saturation numbers for 1970 rounded to the nearest 5 percent.

[b] Half the households in 1985 have room air conditioners, and the other half central air conditioning. Of the 40 million households with room air conditioners, 20 million are assumed to have one and 20 million two.

[c] Saturation is not 100 percent due to increased trend to multiple unit housing structures with common drying facilities.

[d] Portable appliances consist of things like TV, vacuum cleaners, electric irons, toasters, electric shavers. The saturation of these appliances varies a great deal from very low to near 100 percent.

[e] The relative units in which saturation is measured reflect appliance electricity consumption.

Commercial: This is the only sector where *ZEG* would mean more energy consumption instead of less, because it includes all nonresidential and nonfactory buildings such as business and government offices, educational and medical facilities, stores, and repair shops. Since these services would be larger in this scenario, more energy would be required, as shown in Table 21 and Figure 20. Provision is made for "miscellaneous" energy needs to cover the additional equipment in the commercial sector (compared to other scenarios) such as equipment in shops that

Table 20–**Potential energy savings in the residential sector**
(Quadrillion Btu's)

	Zero Energy Growth vs. Technical Fix		
	1985	*2000*	
Residential energy use in TF scenario	19	20	
Potential savings			*Conservation Measures*
Miscellaneous Household appliances	—	0.8	Maintain saturation levels for miscellaneous portable appliances at 1985 levels
Presently unknown household appliances	0.7	1.6	Reduce rate of introducing presently unknown appliances
Total savings	0.7	2.4	
Residential energy use in ZEG scenario	18	18	

Note: The residential sector's share of all energy processing losses is included in these numbers.

repair household devices, medical equipment, teaching machines, office machines, and computers.

However, *ZEG*'s energy sales tax, coupled with tough enforcement of building codes, would affect the design and operation of commercial buildings. They are assumed to be as efficient in using energy as those in the *Technical Fix* scenario.

Figure 19–**Energy savings, residential sector: ZEG vs. Technical Fix**

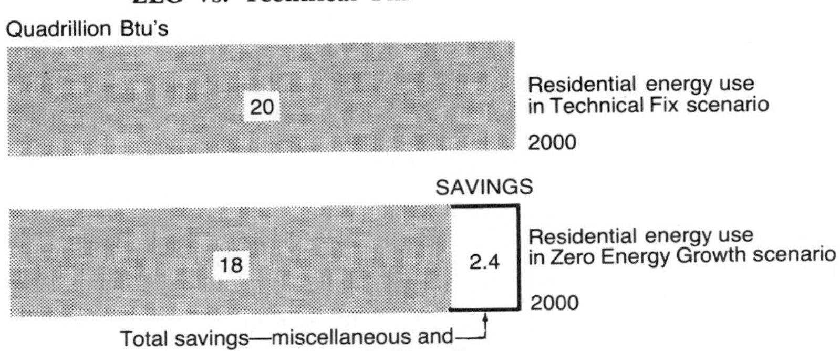

Source: Energy Policy Project

Table 21–**Additional energy requirements in the commercial sector** (Quadrillion Btu's)

	Zero Energy Growth vs. Technical Fix		
	1985	*2000*	
Commercial energy use in TF scenario	15	18	
Additional requirements			*Reasons for Changes*
Space conditioning	0.2	1.0	Greater employment in the service sector
Miscellaneous and presently unknown uses	0.6	1.4	
Road oil and asphalt	—	−0.3	Fewer (and smaller) cars
Total additional requirements	0.8	2.1	
Commercial energy use in ZEG scenario	16	20	

Note: The commercial sector's share of all energy processing losses is included in these numbers.

Industrial: Even in this scenario, manufacturing would continue to grow and reach a higher level than today, both on an absolute and a per capita basis. However, the manufacturing sector would not grow as much as in the other scenarios.

Our econometric model is not designed to tell precisely where the growth in manufacturing would be slowed. However, we prepared an industrial energy budget for the *ZEG* scenario that illustrates a plausible set of industrial energy savings.

Figure 20–**Extra energy requirements, commercial sector: ZEG vs. Technical Fix**

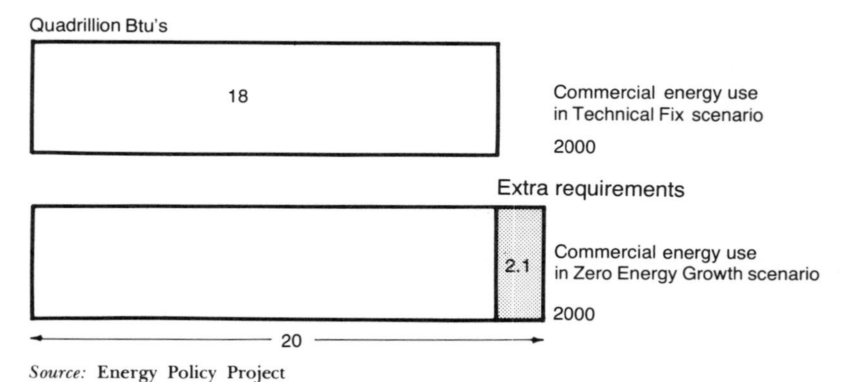

Source: Energy Policy Project

Since the energy sales tax would raise the price of energy intensive materials relative to other things, less material would be used in products and growth in output would decline. Industries would grow more or less slowly depending upon whether they were energy intensive and upon the buyers response to the prices of their products. Slower growth in the production of three especially energy intensive materials is shown in Table 22. Reduced production of these materials accounts for about half of industrial energy savings vis-a-vis *Technical Fix*.

Table 22–**Production of selected energy intensive materials in ZEG scenario**

	Production Increase Over 1975 Levels (percent)		*Production Decrease Below HG Levels* (percent)	
	1985	*2000*	*1985*	*2000*
Plastics	110	160	30	60
Aluminum	30	130	20	30
Steel	10	20	10	35

In addition to the savings in these specific energy intensive industries, achieving *ZEG* will also involve slower growth relative to *Historical Growth* and *Technical Fix* in the rest of the industrial sector. (See Table 23 and Figure 21.) These savings will come from increased efficiency in the use of energy in response to higher prices, and a slight shift in the economy to require somewhat less industrial production than in other scenarios.

Energy supplies for a Zero Energy Growth scenario

The energy supplies required for *ZEG* are not simply scaled down versions of the supply schedules for higher growth scenarios. Some of the motivations that curtail growth in demand are reflected in the supply mix for *ZEG*.

A decision to level off energy consumption a decade hence might stem in part from a desire to avoid development that causes serious environmental problems. This means avoiding the Atlantic and Pacific coasts, oil shale, and much western coal. It also means avoiding the expansion of nuclear power. Similarly, concern over climatic alterations from burning fossil fuels would motivate a limit on the

Table 23–**Potential energy savings in the industrial sector** (Quadrillion Btu's)

	Zero Energy Growth vs. Technical Fix *1985*	*2000*	
Industrial energy use in TF scenario	36	58	
Potential savings			*Conservation Measures*
Aluminum	0.2	0.3	Ban aluminum cans
	—	0.3	Recycle 75% of available old scrap (compared to 50% in *TF*)
Steel	0.4	1.0	Reduce growth in steel output from 2.5 to 1.5% per year
Plastics	1.5	4.7	Reduce growth in plastics output (to 2.7% per year for 1985–2000)
Other	—	8.4	General shift in industrial mix to less energy intensive activities
Total savings	2.1	14.7	
Industrial energy use in ZEG scenario	34	43	

Note: Only the manufacturing sector's share of energy processing losses are included here.

growth in fossil fuels. Further, a concern over the "big brother" syndrome would lead to the de-emphasis of large energy technologies in favor of small scale total energy systems, roof top solar systems, organic waste energy systems, and wind power. And use of solar energy could help alleviate chronic air pollution.

Figure 21–**Energy savings, industrial sector: ZEG vs. Technical Fix**

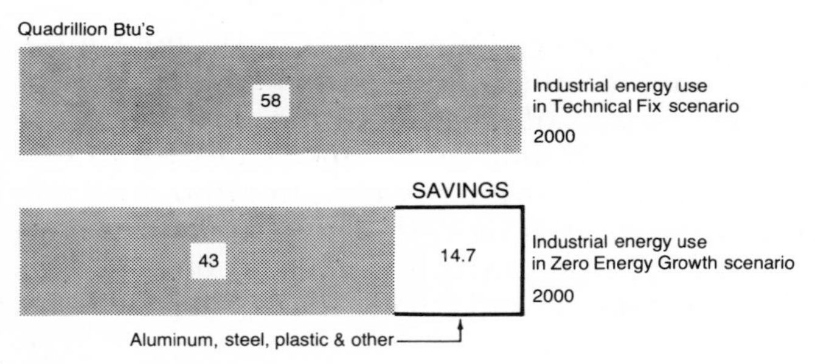

Source: Energy Policy Project

Renovating our energy supply system

Such unconventional energy sources could assume a major energy supply role in the future. But the bulk of our supply will come from the fossil fuels in this century even in the *ZEG* scenario. (See Energy Supply Notes, Appendix C.) It simply takes a good many years to move a new technology from a laboratory test to a viable technology, and then to install enough capacity to provide a substantial amount of the nation's energy needs. For example, nuclear fission was demonstrated in a laboratory in 1942; the first prototype nuclear electric plant was completed in 1956; and the first full scale commercial reactor began operation in 1967. Yet today nuclear power supplies only 5 percent of our electricity requirements.

But all new technologies need not take so long to become practical. The rate at which a new supply source becomes important is critically dependent upon its economic potential, the institutional barriers that may inhibit its use, and whether government actively intervenes on its behalf. To move toward an environmentally satisfactory, renewable energy supply system under *ZEG*, two broad approaches seem plausible.

A sensible approach would involve major government support of R&D to develop nonpolluting and renewable energy sources; elimination of institutional barriers that might prevent use of those which appear economically competitive; and assistance to small entrepreneurs who seek to develop renewable, low polluting energy sources so that they can compete with large energy companies. In addition, the energy sales tax proposed in this scenario could be applied only to nonrenewable energy sources. We believe such an effort, carried out with a strong commitment, could achieve the following results.

Solar energy: Solar energy could affect supply in two major ways. First, rooftop solar collectors can provide space heating and water heating, and, in the 1980s, cooling as well. Second, central station solar power could begin to provide electricity to substitute for coal, nuclear, or other conventional sources by 1985 and beyond.

Rooftop units are probably economically competitive today in some regions and require relatively little technological development. However, there are the numerous nontechnical barriers confronting such a new technology, and in order to keep costs down, such systems generally must be part of the original buildings. Short of a policy requiring retrofitting of solar rooftop units, the potential for replacing fossil fuels and electricity with solar units is limited by the turnover in the housing stock.

Still, if the government were to remove nontechnical barriers to innovation in the construction industry, and impose the energy sales tax on fuels but not on solar energy, perhaps one-quarter of all new housing units built in the 1985–2000 period could use solar energy for space heating, supplemented by conventional energy sources. This would save about 0.3 quadrillion Btu's by 2000.

If, as a result of economic incentives and public policy, solar units were used in new commercial and residential construction (except in areas where the climate does not permit), perhaps one-third of all new construction would use solar energy and about 1 quadrillion Btu's might be saved by solar rooftop units by 2000.

Central station solar power is further away from realization. Technical barriers, and engineering and economic problems make it unlikely that without a special effort such systems would be commercially significant by 2000. However, a national commitment, including an urgent, well funded effort to develop competitive solar energy systems, could result in commercial feasibility of a demonstration plant by 1985; with public policies that favored solar energy it would be possible to complete solar central station generation systems with perhaps 20,000 Mw capacity (or about 1 quadrillion Btu's) by the year 2000.

Organic sources: In this period, organic energy sources are likely first to be developed by burning garbage for fuel and by processing agricultural wastes into methane.[8] Some cities are already generating electricity from urban refuse. The combined effect of exempting organic sources from the energy tax and the increasing problem of disposing of urban waste could lead to a significant use of organic wastes for generation of electricity in cities by 2000, perhaps as much as 1 quadrillion Btu's.

A broad and intense development program in this area could also lead to use of animal waste from feedlots (presenting a water pollution problem) and crop residues, which would be transformed into methane gas and used like natural gas. Some R&D is needed, along with a modest investment in pilot plants and system designs to bring this resource to the marketplace. About 3 quadrillion additional Btu's from farm wastes could be produced by 2000.

In addition to encouraging the development of unconventional supplies, limited expansion of conventional energy sources is envisioned in this scenario:

- Additional development of oil and gas onshore and in the Gulf of Mexico (not, however, off the Atlantic, Alaskan, and Pacific Coasts);
- No substantial development of a synthetic oil and gas industry or a shale industry in currently undeveloped regions;
- Oil imports at somewhat below present levels;
- Gas imports primarily from Canada above present levels;
- No nuclear plants beyond those presently operating or under construction;
- Coal expansion strictly limited to regions where reclamation is feasible.

The energy supplies in this scenario are summarized in Tables 24 and 25. A comparison of energy supply for *ZEG* with supply patterns for *Historical Growth* (Domestic Oil and Gas case) and *Technical Fix* (Environmental Protection case) is shown in Figure 22.

Evaluation of Zero Energy Growth and comparison with other scenarios

An intensive program for developing nonrenewable sources, combined with the zero energy growth demand policies, would mean the following:

- Curtailed growth in U.S. fossil fuel use would ease global climatic problems.
- Nuclear risks would be minimized by curtailing growth of nuclear power.
- Air pollution problems would be more easily controlled with slower growth in fossil fuels; growth in fossil fuel consumption would be limited to about 20 percent over current levels for oil and gas, and about 40 percent for coal.
- No major regions of the country that are presently undeveloped would be devoted to energy production.
- Many urban waste and water pollution problems would be ameliorated.
- Cleaner, renewable sources, such as solar energy and organic wastes, could contribute to post-1985 energy

Figure 22–**Energy supply for Historical Growth, Technical Fix, and ZEG**

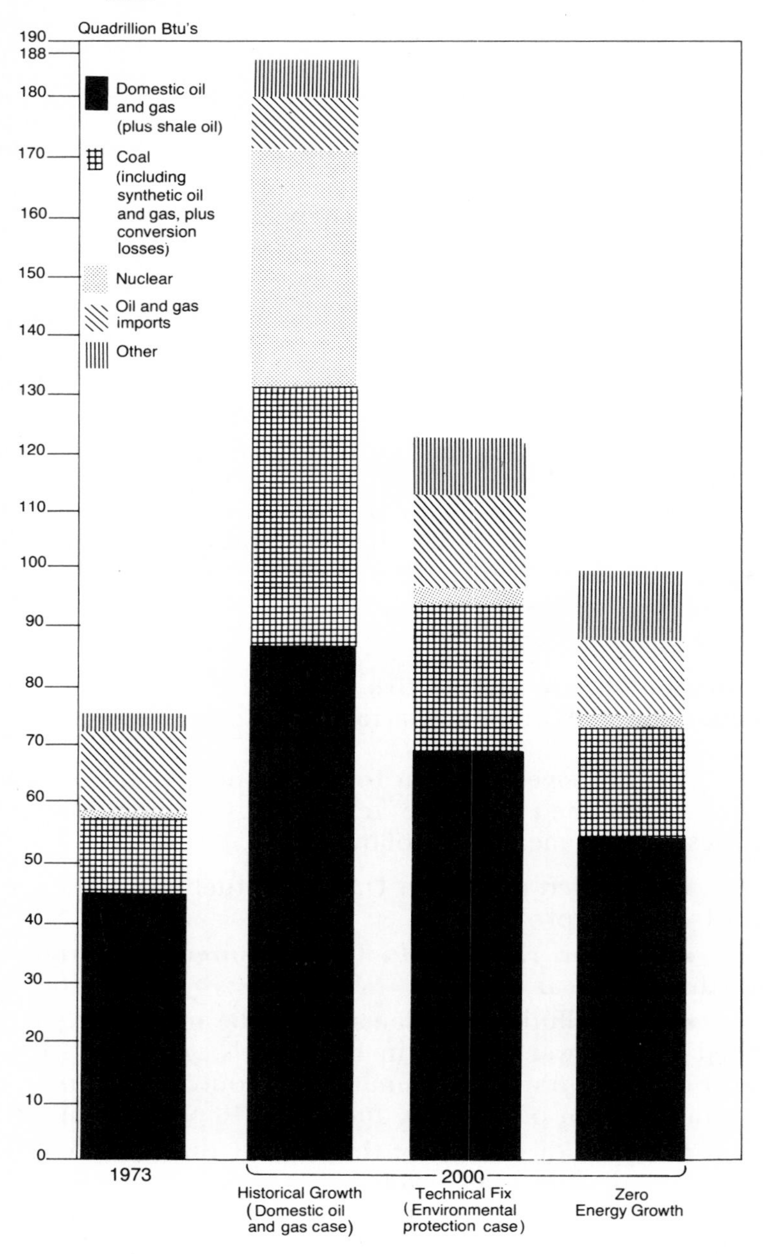

Source: Energy Policy Project

Table 24–**ZEG energy supplies** (Quadrillion Btu's)

	Actual 1973	*1985*	*2000*
Domestic oil	22	28	30
Shale oil	0	0	0
Synthetic liquids from coal	0	0	0
Imported oil	12	9	9
Nuclear	1	5	3
Coal (except synthetics)	13	14	18
Domestic gas	23	25	25
Synthetic gas from coal	0	0	0
Imported gas	1	2	4
Hydro	3	3	4
Geothermal	0	1	2
Other	0	1	5
Conversion losses from coal synthetics	0	0	0
Totals	75	88	100

Table 25–**Fuels for central station electric power, ZEG scenario** (Quadrillion Btu's)

	Actual 1973	*1985*	*2000*
Coal	8.7	9	12
Nuclear	0.9	5	3
Oil and Gas	7.3	5	6
Hydro	2.9	3	4
Geothermal	—	1	2
Other	—	0	4
Totals	19.8	23	31

growth and be ready to take over a substantial share of energy requirements in the next century.

While the precise level of energy demand resulting from the suggested policies could vary considerably from our calculations, the broad conclusions remain valid. Energy growth can level out without devastating economic effects if it is done carefully, over a long enough period. Energy prices would need to be higher—but not prohibitively higher—than in other scenarios. And the money instead of going for increased energy company income, could be used for funding growth in desired public services.

Finally, renewable energy sources could begin to make substantial contributions to energy requirements by the end of this century and could become major sources for fueling the next century. This would be no mean accomplishment.

CHAPTER 5

The American energy consumer: rich, poor, and in-between

Early in the planning of the Energy Policy Project, it became apparent that very little was actually known about how much energy Americans use in their everyday lives, and how their energy consumption relates to socio-economic conditions—especially income. In order to evaluate future energy policies, it is necessary to know how much energy people of different income groups use, what they pay for it (in absolute terms and as a percentage of total income), and the ways in which they consume it.

The Washington Center for Metropolitan Studies (WCMS) conducted a study for the Project to investigate the relationship between energy use and various socio-economic factors. The study was based on two national surveys.[a] The

[a] The first survey, done in May 1973, was a national sample of 1,455 households. Families in the survey answered questions about their dwellings, heating systems,

study considered not only energy used by families in their homes, but also the energy used directly by American households for private transportation. These two areas of consumption are jointly described here as "household energy."

The Washington Center's study was a pioneering effort in that it related actual records of electricity and natural gas usage to the social and economic characteristics of the households surveyed. As defined in this study, households include both families and single people living alone. The terms "household" and "family" are used interchangeably here.

The Washington Center also estimated consumption levels for gasoline, based on survey respondents' data on car ownership, the amount of driving done and the gasoline requirements of vehicles. Using the Washington Center's findings on household energy use in 1972–73, it should be possible for other researchers to continue monitoring energy consumption patterns in American households, thereby establishing a picture of trends over time, as conditions such as prices and fuel availability change. The energy situation is extremely fluid, making continued studies a necessary adjunct to policy making.

It should be stressed that the study is concerned with direct, primary energy use—that is, the total use of raw fuels. In the case of electricity, it includes the total amount of energy required to generate and distribute electric power to homes. In the generation and transmission of electricity, about 70 percent of the energy content of the original fuel is lost—or, in other words, electricity is 30 percent efficient. Therefore, total household energy use for electricity is more than three times the amount that is delivered to the user and registered on a home meter. Electrical usage data collected in the study's utility survey is converted to primary energy data[b] in order to reflect this fact.

The Washington Center study does not include the

energy-using appliances, and vehicles. Data about these energy related items were correlated with responses to questions about socio-economic status, living and transportation habits. The Response Analysis Corporation of Princeton, New Jersey, selected the sample, conducted the interviews, and collated the data for analysis by WCMS.

The second survey, conducted in summer 1973, was directed at electric and gas utility companies serving the sample households. With the permission of the surveyed families, WCMS asked the utilities how much electricity and natural gas the households used and how much they had paid for it during the preceding twelve months.

[b] The conversion rate used was 10,910 Btu's of primary energy for each kilowatt hour of electricity (kwhe).

energy used to manufacture the goods or perform the services which are ultimately consumed by householders. Some discussion of this indirect energy consumption, based on Energy Policy Project staff research, is given later in the chapter.

What is "typical"?

The American dream may be a vision of several luxury cars and a split-level house loaded with labor saving appliances, but the reality is a bit more modest. The "typical" American family bears little resemblance to television's famous "American Family," the Louds of Santa Barbara, with their sprawling air conditioned contemporary house, four cars, swimming pool, and jet traveling children. Most American families live much more modestly.

The "typical" American family lives in a five-room, single family house. The house structure—some 1,200 square feet in size—usually contains some insulation, but chances are just about even that it has neither storm windows nor a basement. Only 15 percent have central air conditioning, although almost half have at least one air conditioning unit.

Inside, most American homes contain at least six essential energy-using items: central heat, electric lights, hot water heater, stove, refrigerator, and washing machine. A television is present in almost every home, but it uses relatively little energy. Only half include clothes dryers, and one-quarter have dishwashers.

The automobile is also a feature of most households (about 80 percent), and 44 percent have two or more. The typical American family drives about 14,000 miles each year, and in 1972–73 got about 14 miles per gallon of gasoline in local driving. Almost nine out of ten Americans use automobiles (theirs or others') to get to and from their jobs, and almost three-quarters drive to work alone. The majority of heads of families (60 percent) take at least one car trip of 100 miles or more each year; one-quarter take at least one plane trip of 500 miles or more during the year.

The average American household, according to the survey, consumes a total of 341 million Btu's of primary energy each year. That is the equivalent of 848 gallons of gasoline plus over 8,000 kilowatt hours (kwh) of electricity and 142,000 cubic feet of natural gas per household. The average American family spends 6 percent of its income paying gas, electricity, and gasoline bills.

Many significant facts are hidden behind these "typical" energy images. In a country so large and diverse as the United States, a description of the "typical" family can only be suggestive of the ways people actually live and is not sufficient for policy formulation. Thus, we must dissect the averages and look at the variations by income groups.

Even within the United States, poverty—like wealth—actually covers a broad range of living circumstances. Some poor American families never achieve a reasonable level of economic well-being, compared to other Americans. Others are very young families struggling upward; for them, poverty may be temporary. Many are old people whose financial status has deteriorated because their incomes fail to keep pace with inflation. The poor often combine poverty with other disadvantages such as racial discrimination, lack of education or training, and physical handicaps. Any energy policy involving price changes and conservation requirements must take the special problems of the poor into account.

Energy use and income

The Washington Center's household survey looked closely at the relationship between the families' energy use and their incomes.[c] Families in different income groups have markedly different patterns of energy use as might be expected. The differences in their consumption levels for natural gas, electricity, and gasoline are shown in Figure 23.

Differences in consumption are smallest for natural gas, which is used almost entirely for essential functions such as heating, cooking, and water heating. The gap be-

---

[c] The Washington Center used the following income group classifications, based on respondents' 1972 household incomes:

Poor Households, defined as:
- 1 or 2 persons with annual household incomes of $3,000 or below
- 3 or 4 persons with annual household incomes of $5,000 or below
- 5 or 6 persons with annual household incomes of $7,000 or below
- 7 or 8 persons with annual household incomes of $9,000 or below

Lower Middle Households, defined as those with annual incomes between the poverty level and $11,999.

Upper Middle Households, defined as those with annual incomes between $12,000 and $15,999.

Well Off Households, defined as those with annual incomes of $16,000 or more.

The results of the survey were weighted to reflect the income distribution within all American households in 1972. The weighted distribution was 18 percent poor, 42 percent lower middle, 19 percent upper middle, and 20 percent well off. The median family income in 1972 was $11,116, according to census figures.[1]

Figure 23–**Household energy use by income group**

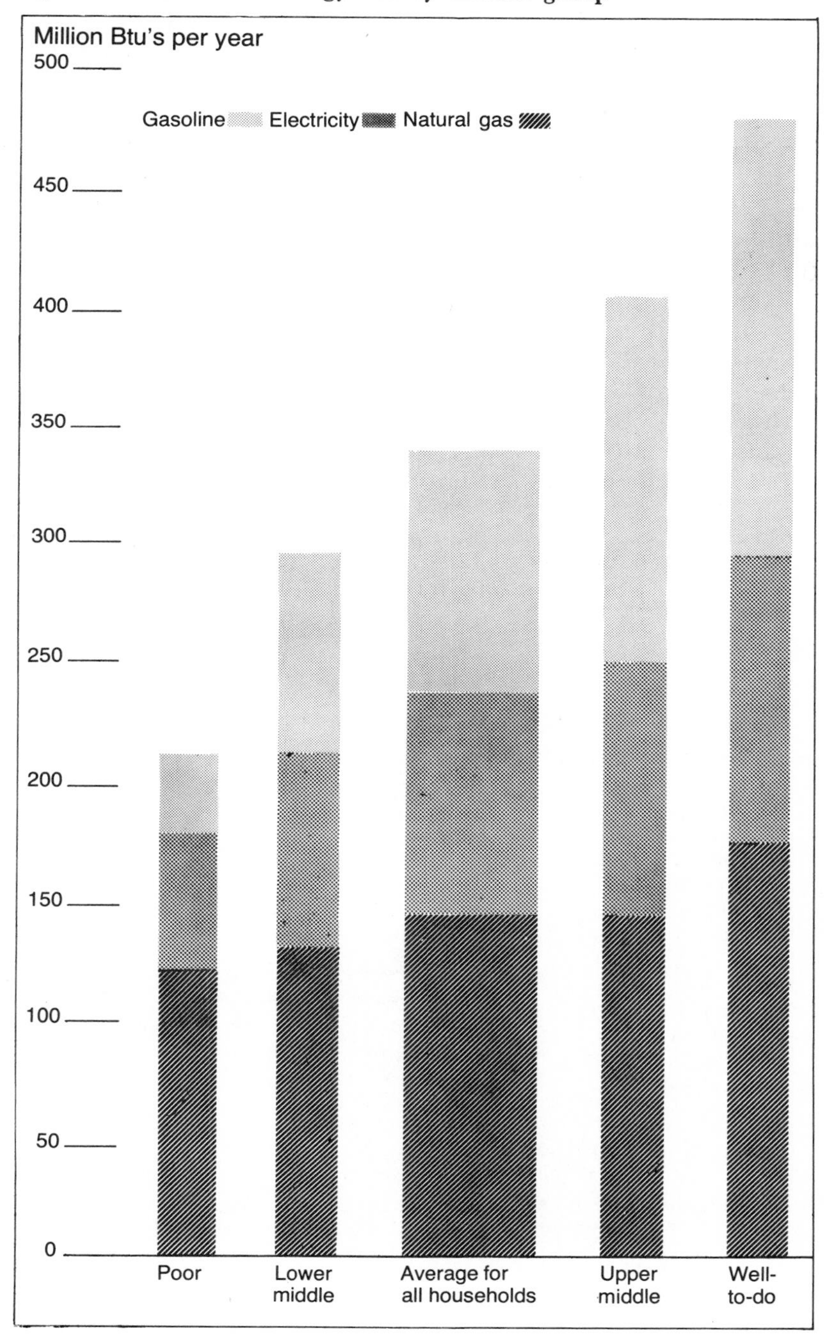

Note: Includes only natural gas, electricity, and gasoline
Source: Washington Center for Metropolitan Studies

tween income groups is slightly greater for electricity, which serves basic needs such as lighting and other purposes such as air conditioning and dishwashing, and greatest of all for gasoline, which covers a wide range of needs and wants.

The variations in energy use levels among the different income groups are less pronounced than the differences in income. While the average household income among the poor is about one-tenth as large as that of the well off, poor families use almost half as much energy as families who are well off. This means that energy, like other necessities such as food and housing, eats up a considerably larger share of a poor family's budget than an affluent family's. Table 26 shows the relative portions of family income spent on energy by different income groups in 1972–73. This table shows that poor families spent roughly 15 percent of their income on natural gas, electricity and gasoline during that twelve-month period, while the middle income groups

Table 26–**The percentage of family income spent on energy declines as income increases**

Income Status	*Average Income*	*Average Annual Btus (Millions per Household)*	*Average Annual Cost per Household*	*Percent of Total Annual Income Spent on Energy*
Poor: Total	$2,500[a]	207	$379	15.2
Natural gas		118	147	5.9
Electricity		55	131	5.2
Gasoline		34	101	4.0
Lower middle: Total	$8,000	294	572	7.2
Natural gas		129	153	1.9
Electricity		80	167	2.1
Gasoline		85	252	3.2
Upper middle: Total	$14,000[b]	403	832	5.9
Natural gas		142	166	1.2
Electricity		108	213	1.5
Gasoline		153	453	3.2
Well off: Total	$24,500[b]	478	994	4.1
Natural gas		174	200	.8
Electricity		124	261	1.1
Gasoline		180	533	2.2

Note: Electricity and natural gas expenditures based on billing data received from utilities. Gasoline expenditures estimated from respondents' quantitative information and the average 1972–73 price of 37¢ per gallon.

[a] 77 percent of the poor had incomes less than $3,000.

[b] Calculated from unpublished census data.

Source: Washington Center for Metropolitan Studies Lifestyle and Energy Surveys, 1972–73.

spent 7 percent and 6 percent respectively, and the well off spent about 4 percent of family income on energy. In the cases of natural gas and electricity, the poor also spend more for each unit of energy than do other income groups, because unit rates are highest for low volume consumers and decline as consumption increases. (Utility pricing is discussed in Chapter 10.)

Energy use for heating: variations by income groups

Much of the energy required at home is determined by the size and structure of the dwelling itself. Detached single family houses tend to use more heating fuel than either row houses or apartments since they are exposed to the elements on all sides. The bigger the house, generally, the more energy is needed to heat it. But the presence of insulation and storm windows can radically reduce the heating load of a given size house.

The WCMS survey reveals several facts about people's homes. A majority of all homes, regardless of their occupants' income status, are detached single family houses. But there is a general decline in the percentage of single family houses by income class, ranging from 83 percent of all well-to-do households to 58 percent of poor households. The number of families living in apartments, on the other hand, decreases as income rises from 32 percent of poor households to only 8 percent of well-to-do families. The size of a family's dwelling, as indicated by the number of rooms, is also related to income. While 77 percent of all poor families live in homes of five rooms or fewer, only 26 percent of the well off live in an equivalent amount of space.

Since poor families usually occupy smaller dwellings, and are the most likely of all the income groups to live in apartments or attached houses, one might expect their heating fuel consumption to be significantly lower than that of other income groups. But the relatively small difference in consumption of natural gas—the heating fuel used in 60 percent of all American homes—may be traced to the condition of homes, particularly the presence or absence of insulation and storm windows. The survey's results on insulation and storm windows (shown in Figures 24 and 25) are illuminating in this regard.

Figure 24–**Percent of households with no insulation**

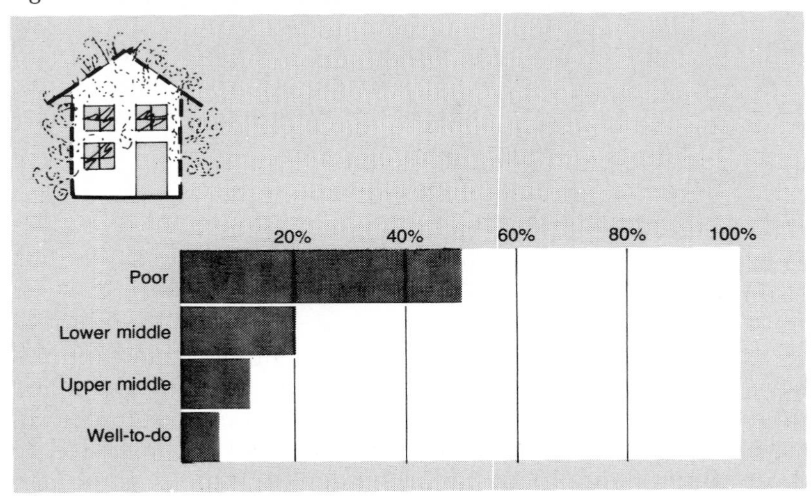

Note: Single family houses only
Source: Washington Center for Metropolitan Studies

Figure 25–**Percent of households with no storm windows**

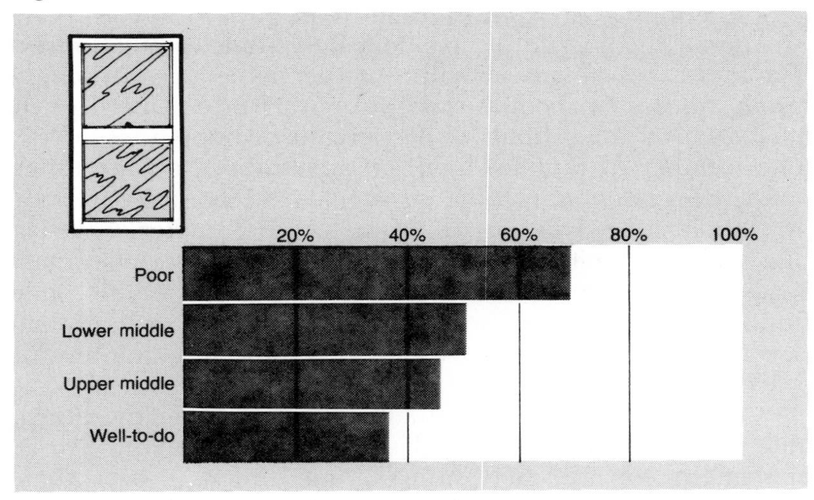

Source: Washington Center for Metropolitan Studies

Almost three-quarters of all American single family homes have some insulation.[d] However, over half the single family houses of the poor have no insulation at all, compared to a mere 5 percent of the houses of the well-to-do.

---

[d] This discussion is based upon information given by survey respondents who stated that they knew whether or not their homes were insulated. The "don't know" category (9 percent of respondents) was excluded.

The poor also lag behind other groups with regard to the existence of storm windows in their houses; only 31 percent of poor homes have them, compared to 63 percent of the well-to-do. These facts explain why the poor use more fuel per square foot of housing than the lower middle and well-to-do, and why their overall consumption of natural gas is only slightly below the levels used by the middle groups.

The simple function of home insulation and storm windows is to plug the leaks through which heat escapes. Without storm windows and insulation, much of the energy is used to heat the outdoors. A layer of insulation in the walls, floor and roof can substantially reduce fuel bills. This fact of life has been known to home builders for many years. As early as 1935, an experiment with twin houses in Detroit revealed that an insulated house used 38 percent less fuel than an uninsulated house over the course of one heating season; the insulated house's internal temperature fluctuated within only two degrees of the desired temperature, while the uninsulated house ranged up to 18 degrees from the norm.[2]

Low energy prices have served in the past to discourage investments in insulation. Unfortunately the poor, who can least afford to waste heating energy, suffer the most from the lack of insulation in houses and apartments which they own or rent. For the poor who own their own homes, as seven out of ten poor single family house dwellers do, there is a formidable barrier to installing insulation in their houses. Even if the fuel savings would pay for the initial cost of insulation within only a few years, poor families usually do not have the initial capital needed to purchase and install insulation. According to the Michigan Consolidated Gas Company, materials to insulate the attic roof of an existing house usually cost about $90, and installation might add $50 more.[3] A special program of loans offered by government or utilities for home insulation, with easy repayment as part of the fuel bill, is one way the poor could afford to insulate their homes. An insulation loan program is currently being tried by two companies in Michigan. The companies predict that in many cases, the dollar savings on fuel would cover the loan payments in a very short time.

Lighting

Lighting was the first and only function of electricity in the early days of home electricity. It remains an essential use, and it consumes about one-seventh of a family's residential electricity.[4]

The survey of households found that families at the lower end of the income scale are more thrifty about using lighting energy than are the more affluent families. The lower a family's income is, the more likely that family is to keep only one or two rooms lit during the evening hours; conversely, families at the upper end of the scale are more likely to light three or more rooms. This means that in lighting, as in heating, lower income families have little room to cut back on their energy consumption.

Appliances

The Washington Center data on appliance ownership (see Table 27) reflects income differences much as one might expect. The differences are far less pronounced for the most necessary appliances—stoves, refrigerators and washing machines—than for amenities such as air conditioners and dishwashers. Yet it is striking to note that

Table 27–**Percentage of households owning major appliances, by income group**

Appliance	*Poor*	*Lower Middle*	*Upper Middle*	*Well Off*
Stove	95	97	99	98
Refrigerator				
Manual defrost	74	51	39	30
Frost-free	24	48	60	69
Total	98	100	100	100
Separate food freezer	23	30	38	47
Dishwasher	3	13	39	55
Clothes washer				
Wringer	19	11	7	3
Automatic	44	64	84	90
Total	62	75	89	91
Clothes dryer	24	46	70	81
Television				
Black and white	74	63	57	64
Color	27	48	63	74
Total	94	96	98	98
Air conditioning				
Window	17	34	39	33
Central	4	10	19	32
Total	22	45	58	64

Source: Washington Center for Metropolitan Studies, Lifestyle and Energy Household Survey, 1972–73.

several appliances that go well beyond the bare necessities, including air conditioners, clothes dryers, color TVs and frost-free refrigerators, are owned by almost one-quarter of all poor households. Poor households thus own the most necessary appliances, and a minority own these additional items which save labor and provide enjoyment. Poverty is a relative condition; the poor American family appears well off in energy use compared to the poor in other parts of the world, yet, relative to other Americans, the poor family has very little extra energy to conserve.

The Washington Center computed an "appliance index" for households in its national surveys. The appliance index represents the average number of Btu's used annually by major household appliances, based on information from such sources as the Edison Electric Institute and the American Gas Association. An appliance index of 50 means that a family consumes 50 million Btu's of energy to run its appliances each year. While two-thirds of the poor have appliance indexes of less than 40, the same fraction of the well-to-do have indexes of over 60, indicating that appliance usage, like energy use generally, is mostly a function of income.

The energy efficiency of appliances on the market today varies widely. Consumers rarely have all the information they need to estimate the lifetime cost of operating appliances. There is no way to know, for example, that two air conditioners with the same purchase price and the same cooling power (rated as Btu's of output per hour) may require quite different amounts of energy—and money—to run. Nor is it made clear to buyers that frost-free refrigerators require as much as two-thirds more energy than standard models, and that one frost-free model can be far more efficient than another. Appliance manufacturers possess the information consumers need, and in some areas, including New York City, sellers must now display Energy Efficiency Ratings (EER's) for major appliances on sale. A federal "truth-in-energy" law, with a requirement to label appliances, giving the annual cost of operation in dollars at current prices would help consumers choose energy-using items that save both energy and money.

Energy on the road: poor, middle-income and well-to-do families

Gasoline, the major transportation fuel for individual Americans, requires an average annual household expenditure of $101, or 26 percent of total average annual energy cost, among the poor. Gasoline is a much larger part of a well off family's energy budget, averaging $533, or 38 percent of total energy expenditure, annually. It is in transportation that the greatest gap between rich and poor appears, according to the Washington Center surveys. Figure 26 illustrates the discrepancy between poor and higher income groups in gasoline consumption, and also shows how sharply gasoline consumption increases in families with more than one car, regardless of income group. Among the poor, there are fewer cars and drivers than there are households. In contrast, among the upper middle and well off there are more than two cars for every household, and almost one car for every driver. Almost half of all poor households own no car whatsoever, while 79 percent of all well-to-do own two or more. This means that while the poor represent 17 percent of all American households, they use only 5 percent of the nation's gasoline in 9 percent of its cars. The median distance driven by poor households each year in each car (when they own one) is only 5,000 miles compared to the 10,000 miles for the upper middle and well-to-do. Thus the poor not only have fewer cars, but also drive each car less than others do.

Among the households surveyed by the Washington Center, the poor reported consistently better gas mileage

Figure 26–**Gasoline consumption by income group**

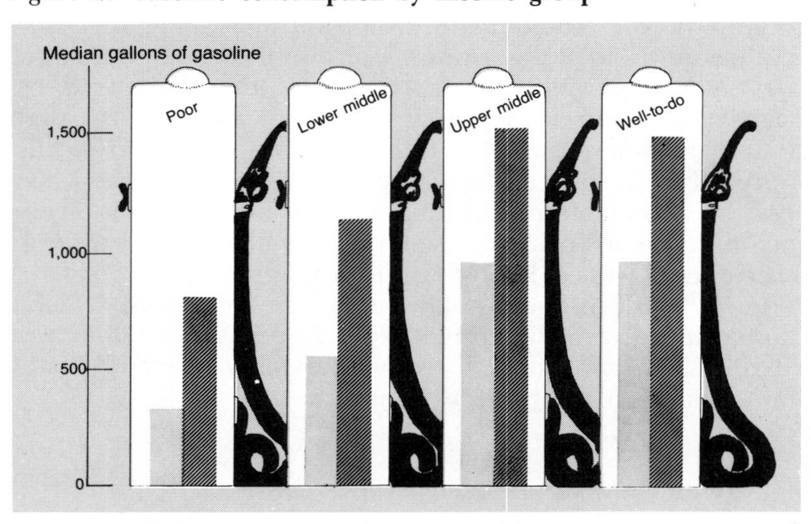

Note: Median gallons of gasoline used annually by households with one or more cars
Source: Washington Center for Metropolitan Studies

than other income groups. This response was based simply on the respondent's personal judgment. If this is in fact the case, the reason probably lies in the nature of the cars the poor drive. Over three-quarters of all cars owned by poor families were bought used, while a majority of the cars of the other income groups were new when bought. Poor people also keep their cars longer, according to the Washington Center's statistics. Three-quarters of poor families' cars are 1968 models or earlier, while most other cars—ranging from 56 percent for the lower middle group to 85 percent for the well off—are newer than that.

The average fuel economy of new cars declined every year between 1968 and 1974, partly because of the increase in average car weight over the same period. (Fuel economy declines in inverse proportion to auto weight.) Other factors contributing to the decline in fuel economy include the increased use of automatic transmissions, power steering, air conditioning and other accessories in newer cars. Emission control devices designed to mitigate auto related air pollution (first required in 1968) have also contributed to the decline in fuel economy, but their contribution has been no greater than that of weight increases.[5]

If fuel economy is one area in which the poor get a break, it is likely to be a short-lived advantage. If the poor continue to buy used cars—and it seems probable that they will, especially in times of inflation—the cars they buy in the next five years will be the gas-guzzling big cars made in the early 1970s. Poor families will need more gasoline, probably at higher prices, for the same amount of driving. New car sales in 1974 have shown a marked consumer trend toward smaller cars. But the poor will be making do in the next several years with the rejected large cars previously owned by the more affluent, and they will be paying dearly to operate them.

Indirect energy use: Btu's outside the household

While the preceding section provides a picture of direct household energy use as it relates to people's lives, it does not illustrate the relationship between the day-to-day activities of individual Americans and energy used elsewhere in the economy. People's way of life at home is supported by energy consumed in the industrial, commercial and transportation sectors to make the goods and perform the services they consume. The energy that goes into building automobiles, houses and furniture, and running federal, state, and local governments is, in a sense, part of a consumer's piece of the national energy pie.

Tracing the flow of energy through society and relating it to people's lives is a complex task, and there have been only a few pioneering efforts to describe and quantify small segments of indirect energy use. The Energy Policy Project has developed a rough estimate of certain categories of indirect household energy use, broken down by income groups. These estimates of indirect household energy use are shown in Table 28; and Figure 27 illustrates both direct and indirect uses. The processes used in arriving at the estimates are described briefly in the footnotes to Table 28.

Table 28–**Annual indirect energy use per household by income groups** (million Btu's)

Income Group	*Food*	*Autos*	*Housing*	*Appliances*	*Government Services*	*Other*	*Total*
Poor	38	35	10	6	65	199	353
Lower middle	65	82	11	7	65	319	549
Upper middle	79	121	13	9	65	544	843
Well off	94	147	16	10	65	763	1,095

Notes:

Income Group is defined in note c on page 116 above.

Food: Includes tractor fuel, fertilizer manufacture, container manufacture, food processing, trade and transport. EPP estimates derived from references 8, 9, 10, and 11. Allocated among income groups according to patterns of household expenditure on food.[12]

Autos: Includes auto manufacture and support industries such as highway construction, repairs, tires, operation of service stations and insurance companies.[13] Allocated among the income groups according to auto ownership patterns and miles driven, from WCMS data.

Housing Materials and Construction: Based on estimated energy consumption per square foot of housing,[14] amortized over 50-year lifetime for house. Allocated among income groups according to square footage of homes, estimated from WCMS data on number of rooms.

Appliances: Total energy use for manufacturing appliances[15] amortized over 8 years for water heaters, 14 years for all other appliances. Allocated among the income groups according to ownership data from WCMS (see Table 27).

Government Services: Based on estimated energy use in national defense[16] and on 15 percent of all energy used in the commerical sector, consistent with percentage of persons in commercial sector who are employed by government outside military. Assumed to be equal for all income groups.

Other: Includes such things as construction and operation of stores, office buildings, theatres and sports arenas, as well as personal consumption of nondurable goods —clothing, toys and books. Total "other" energy is the remainder when all energy previously accounted for is subtracted from 1972 total U.S. energy consumption of 72 quadrillion Btu's.[17] Allocated among the income groups by average household income.

Figure 27–**Direct and indirect household energy use, by income group**

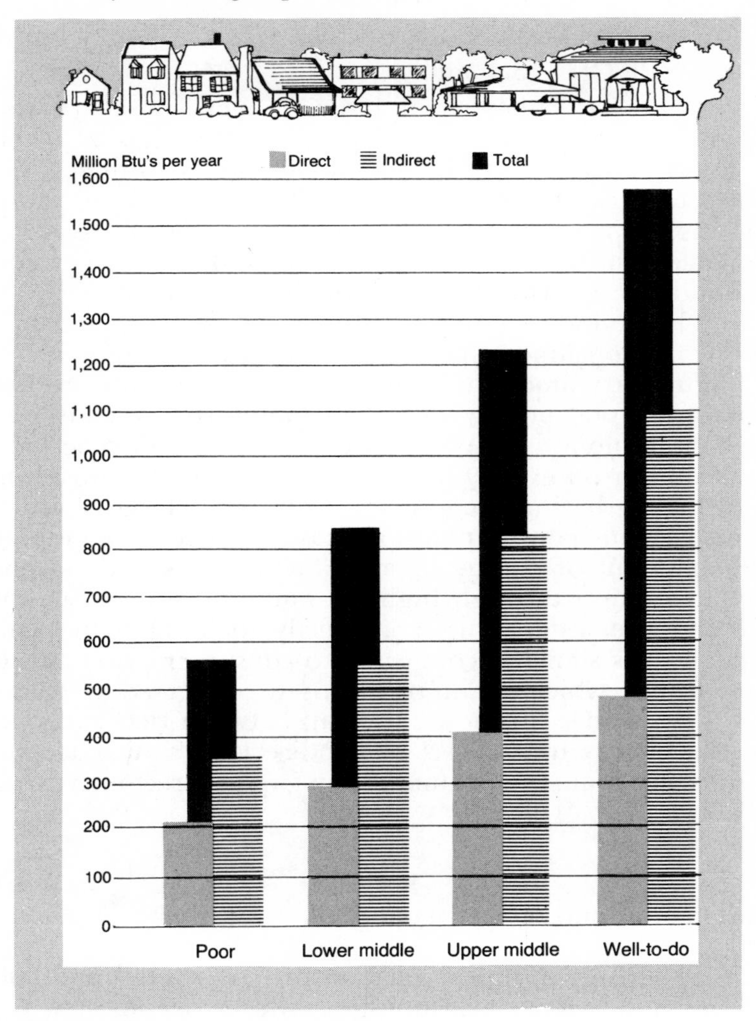

Note: Direct includes only natural gas, electricity, and gasoline
Source: Washington Center for Metropolitan Studies

Estimates of indirect energy use in this chapter are based on the overall national average of 3.2 persons per household.[e] The number of households in each income group, according to the Washington Center, is:

---

[e] According to the Census Bureau, the average number of people per household does not vary greatly by income group, although the distribution does. Poor households tend to be either very small or very large, while the other income groups tend to cluster in the middle.[8]

Poor: 11.8 million
Lower Middle: 27.6 million
Upper Middle: 12.6 million
Well Off: 13.4 million

These calculations of indirect energy use are very rough. A substantial part of indirect use, categorized here as "other," remains to be identified and quantified. The indirect household consumption allocated to food, autos, housing, appliances, and government services are based on available data, which is not abundant. Indirect energy consumption is a fertile field for further investigations.

These indirect energy estimates show how the ripple effect of consumer spending patterns, throughout the economy, vary among the different income groups in their impact on total energy use. The family that spends relatively little money on food because of limited income has a small impact on energy use in the food system. Those who do a lot of driving account not only for a large share of direct gasoline consumption, but also for a correspondingly great amount of energy to manufacture cars and support their operation through highway building, service stations and the like. Energy used indirectly to build homes and appliances is similarly connected to consumer purchases of these things, which are, in turn, closely connected to income levels. As Figure 26 shows, the gap between rich and poor in total energy use—direct and indirect—is similar to, and even larger than, the income group gap in direct energy use alone.

Conclusions and recommendations

There is a direct relationship between household income and energy consumption—the more money you earn, the more energy you burn. But the relationship between energy consumption and the percentage of family income spent for that energy is reversed. Thus, the poorer family uses less, but a bigger slice of its income goes to paying for that energy. Information about housing characteristics, appliance ownership, living habits, and automobile ownership and use suggest that the lower income groups use most of the energy they purchase for functions closely associated with their basic well-being. To cut back from current levels of energy use would be difficult for the lower income groups; similarly, retaining current levels or moving to higher consumption in the face of escalating prices could also cause difficulty.

This does not mean, however, that conservation policies will not aid lower income households. On the contrary, conservation made possible through more efficient cars and appliances and through better home insulation is a basic way of ensuring that families of all income groups can afford to buy the energy they need for such essentials as heating, driving to work, and doing laundry and cooking. Changing the capital stock and achieving greater efficiency require time, money, and carefully drafted policies. Such policies should include

- Mandatory labelling of major appliances, in terms easily understood by consumers, to inform buyers of the energy and dollar costs of operating them.
- Financing arrangements to permit lower income families to install insulation, storm windows and weather stripping in existing houses. A low interest loan program repayable as part of the family utility bill is one possible approach.
- Performance standards on automobiles, requiring new cars to meet minimum levels of fuel economy. The benefits of this policy will take several years to filter down to the lower income groups, who tend to buy used cars. This fact makes it all the more urgent to enact a gradually rising auto performance standard as rapidly as possible. The technology to build cars capable of getting twenty miles per gallon exists. (Auto performance standards are discussed in Chapter 3.)
- Development of federal contingency plans to assure low income people an adequate supply of energy in times of emergency. Fuel shortages are likely to recur in the next few years, although the severity, frequency and distribution of such shortages cannot be determined at present. A system of "energy stamps," emergency grants, or special allocations of fuel to low income consumers who demonstrate potential hardship could prevent poorer families from having to do without minimum necessary levels of heating fuels and gasoline.

Finally, the most fundamental answers to the energy distribution problem lie in solutions to the larger problem of poverty itself. If families were not poor, they could afford not only the energy they need for fundamental uses, but a comfortable margin of amenities as well. We cannot solve the poverty problem by proposing energy policies.

The Energy Policy Project urges that improving the economic status of America's low income citizens become a top priority item on the nation's, and especially the federal government's, agenda. Like soaring food prices, the rising price of energy is having a regressive impact upon the incomes of those who can least afford it, and makes the considerations of social reforms such as a program of income maintenance that much more urgent—along with effective policies to curb inflation and provide full employment. This is the time for refocusing our attention on the part of our population that author Michael Harrington once called "the other America."

CHAPTER 6

Energy, employment and economic growth

The *Technical Fix* and *Zero Energy Growth* scenarios describe how reduced growth in energy use in the United States can yield many economic, social and environmental benefits. Nonetheless, a crucial question must be answered. Can energy growth be reduced while economic activity grows at historical rates? Our research indicates that energy growth could be reduced while growth continues in the output of goods and services—without sacrificing national economic goals. This appears possible, particularly in the industrial sector of the U.S. economy, where most of our energy is consumed. Current studies indicate that the same is likely to be true for the household, commercial, and transportation sectors.[1]

What about the impact of slower energy growth on employment? Would it prevent the growth of new job opportunities? We think not. The energy and manufacturing industries that are the most energy-intensive activities in the economy employ relatively few people. Slower growth in these sectors could be more than offset by more jobs in the service sector, which does not depend heavily on energy, and in industries that make and install energy-saving equipment.

We believe that the fear of the ripple effect of economic disruption and lost jobs, if we do not continue high rates of energy growth, is unfounded. This fear confuses the impact of sudden supply disruptions with the quite different longer term effects of a slowdown in the growth of energy demand by way of economically efficient energy conservation.

Regardless of our energy future, major changes will occur in the U.S. economy before the year 2000. Some trends now discernible, especially population trends, may have the effect of slowing down growth in energy demand. If we anticipate such changes and their energy implications, public decisions on transportation, housing, and similar social programs can help to shape future patterns of personal consumption to be less energy demanding.

Our conclusions are based upon two quite different analytical approaches. We looked broadly at the energy-economy connection, dividing the economy into nine separate sectors.[a] We also looked at individual energy consuming activities, projected their normal growth, and then worked out the effects of applying available technologies to improve the energy efficiency in each activity.[b] The results of both analytical techniques are remarkably similar. The general conclusion of the analyses, stated simply, is that neither the economy nor employment necessarily suffers from lower growth in overall energy use.

Periods of rapid economic expansion in this country have generally coincided with rapid growth in energy consumption.[2] In order to decide whether this relationship is necessary to the nation's economic well-being, we must examine the underlying question: what are the principal determinants of economic growth, and how do they relate to growth in energy consumption?

Economic growth

Economic growth in the consumption of goods and services occurs in four broad areas: personal consumption

[a] See Appendix F, which describes the Data Resources, Inc. model.

[b] See Appendix A, Energy Requirements for Scenarios.

expenditures by consumers; gross private domestic investment by business; government purchases of goods and services; and the trade balance—the difference between exports of U.S. goods and services and imports from abroad.

Because personal consumption expenditures are larger than the other three combined, changes in their pattern have a major influence on the pattern of economic growth. Advancing production technology is a key element in changing consumer buying patterns over time by the introduction of new products. For any given year, consumer spending patterns determine to a large extent the mix of goods and services that make up the gross national product. But over a period of years, technical advances in production can and do change consumer spending patterns. The advent of the automobile is an example.

Many other factors interact to determine the growth of the economy. Productive capacity grows as a result of new investments in capital stock, and technical change makes that capacity steadily more efficient; better education and training improves the quality of the labor force. These kinds of changes combine to increase the productivity of both capital and labor, permitting Americans to enjoy a rising standard of living.

Other elements are growth in population and the labor force which create potential for greater output and provide the demand to absorb it. Economic growth is also influenced by countless government actions and changing policies, and by the availability of raw materials from domestic and foreign sources.[3]

Energy demand

The five factors that chiefly affect energy demand are: population growth, price, personal income, technical efficiency, and the growth and mix of the stock of capital goods. In general, as income and population increase, the demand for energy also increases. Energy demand tends to decline as price or efficiency increases. The impact of changing growth and mix of the capital stock upon energy use is variable. In the past, energy demand grew as capital stock increased, but more efficient equipment, or energy-saving substitutes for capital equipment, can both change the pattern of the past.

Consumer, or final demand for energy (such as electricity for air conditioning or gasoline for cars) is influenced by price, consumer preferences, and income. Relative price—the price of energy in relation to prices for other goods—is especially important. But energy demand may increase while prices are rising, if strong preferences for such services as air conditioning or auto trips offset the dampening effect of higher prices. (This is referred to as "inelastic," or nonresponsive, consumer demand.) Rising personal income can also counteract the effect of rising prices. Falling real prices for energy-using goods, such as freezers and air conditioners, also tend to increase the demand for energy. This is true for most goods whose demand is "derived," and is based on the demand for intermediate goods (electricity demand by the aluminum industry), or is complementary to the demand for different goods (gasoline demand and auto demand).

Industrial and commercial demand for energy is indirect. Energy is an input needed to produce goods and services. If the demand for the products grows, then demand for the energy it contains also grows. For example, the aluminum industry's demand for electricity is based on the economy's demand for aluminum. If aluminum demand is up, then, all other things being equal, electricity demand also will be up.

In the United States, most energy is used in or by machines, equipment, buildings, and appliances, which in general constitute the capital stock of the economy. Energy cannot be eaten or worn. It is used through the medium of capital goods, from complicated industrial machinery to the lowly household iron. The rate of growth in this capital stock, its changing mix, and changes in the efficiency with which it transforms energy into goods and services are fundamental determinants of the growth in U.S. energy demand. Higher prices and uncertainty about availability have compelled closer attention to the ways in which energy is used. Industry's efforts to reduce energy requirements by conservation measures and by developing and applying engineering devices are increasing the efficiency of energy use.[4]

A decline in future U.S. energy growth could occur if we increase the energy efficiency of capital goods, that is, add to or replace existing capital stock with energy efficient substitutes, or change consumption patterns to conserve energy.

Capital investment is crucial to economic growth because it leads to productivity increases and greater efficiency in resource use. It plays an equally important role in determining energy demand. Public policies such as investment tax credits that accelerate the rate of capital spending have in the past contributed to growth in energy demand, especially when, as in the past two decades, energy prices were low relative to the prices of other resources.[5] When it becomes economically attractive to conserve energy, then the same new capital investment that spurs economic growth may slow down energy growth rates because it uses energy more efficiently.[c]

Economics of energy conservation

Our scenarios concentrate on technically efficient changes in specific activities to reduce energy demand. For these changes to become probable, they must also be economically attractive or "cost effective": the money value of the energy savings must exceed the cost required to achieve it. (Both benefits and costs properly discounted over the life of the activity.)

To determine what is economically efficient is not easy. Calculations may be based on incomplete information concerning future prices for energy and uncertainty concerning the availability of certain fuels. Our expectations about future prices of energy are now dramatically different from those of the past two decades. When energy prices were constant, or falling, relative to the general price level, machinery was purchased and plants built in order to use "money saving" processes, even if these were technically inefficient and wasteful of "cheap" energy. Given current energy price levels, the installation of an energy-wasting manufacturing process is apt to be economically inefficient as well.

Some findings

Given these rather straightforward relationships, what can we say about the future? An economic model developed for the Project by Data Resources, Inc. provides a broad based measure of the impact of reduced energy growth, and concludes that a transition to slower growth —even zero energy growth—can indeed be accomplished

[c] An engineer might say that what is technically or thermodynamically efficient is now economically efficient.

without major economic cost or upheaval.[d] The study indicates that it is economically efficient, as well as technically possible, over the next 25 years, to cut rates of energy growth at least in half. Energy consumption levels could be 40 to 50 percent lower than continued historical growth rates would produce, at a very moderate cost in GNP —scarcely 4 percent below the cumulative total under historical growth in the year 2000, but still more than twice the level of 1975.

The following are the chief conclusions of the Data Resources report.

- Substantial economies are possible in U.S. energy input with the present structure of the economy, without sacrificing the continued growth of real incomes.
- Such energy conservation does have a "nontrivial" economic cost in real income; in 2000, both under *Technical Fix* and *Zero Energy Growth,* real income would be about 4 percent below the *Historical Growth* figure.
- Our adaptation to a less energy-intensive economy would not reduce employment; in fact, it would result in a slight increase in demand for labor.

Other Project-sponsored studies also support the conclusion that we can safely uncouple energy and economic growth rates.[6]

For example, the Conference Board and the Thermo Electron Corporation have completed Project studies showing that energy/output ratios for U.S. manufacturing are expected to fall rapidly in the future.[7] Some energy intensive industries such as steel could maintain current levels of output with one-third less energy than now used. The Conference Board study included actual field surveys, and a detailed mail questionnaire completed prior to the 1973–74 oil embargo, and was thus based on normal trends in prices and costs. It found that

> Significant savings in energy use have been realized by the manufacturing sector in the past. Energy use per unit of product declined at a 1.6 percent rate from 1954 to 1967. As a result, while total manufacturing output rose 87 percent, total energy use rose only 53 percent. This was achieved in a period of stable or declining relative prices of energy.

[d] Appendix F contains a report on this work by Data Resources, Inc., with an explanation of its methods, results, and conclusions. The model is based upon the past (1947–1971) relative costs of labor, capital, materials, fuels and electricity, and relates energy and economic growth. It provides an understanding of how much one of these economic inputs can be substituted for another. A more complete account of the work is to be found in Dale Jorgenson and H. S. Houthakker, "Energy Resources and Economic Growth," a draft report to the Energy Policy Project, 1973.

> Recent sharp increases in energy prices, together with present and expected interruptions in supplies of energy, will result in an acceleration in energy savings in manufacturing. The projections presented in this report indicate that energy use per unit of output will decline at an average rate of 2.0 percent from 1967 to 1980.
>
> . . . The existing stock of productive equipment and likely rate of replacement in the industries studied were taken into account, together with known technology. Because of this, the projections represent economically probable developments, [at pre-embargo prices] rather than technically possible optimums.[8] (Words in brackets added.)

The dynamics of employment change in the United States

Each era of U.S. economic growth has been accompanied by increasing investment and employment in new industries: in the last third of the nineteenth century, railroads, agricultural equipment, steel, and oil industries; in the first two decades of the twentieth century, the public utilities—electric, gas, and telephone—and urban transit systems; in the 1920s, the auto and radio industries; after World War II, television, aircraft and air travel, electronics; and in the 1963–73 period, housing, defense, and space programs.

Labor-saving technical change that required large capital investment and high energy consumption have dominated productivity trends during the postwar period.[9] GNP grew three times as fast as the labor force, yet unemployment was persistent, and even on the rise. Much of this unemployment resulted because people looking for work lacked the skills that available jobs demanded. Persistent unemployment has arisen from such causes as defense and space program buildups and cutbacks, regional shifts of industries, increasing foreign production of some imported consumer goods, shifts in jobs from one economic sector to another, and job declines in some activities because of changing productivity trends and consumption patterns.[10]

U.S. unemployment will never decline to zero. Workers continually quit, change residences and occupations, and move in and out of the labor force. Simultaneously, businesses and industrial firms come and go, expand and contract, change location, and shift their type of production. Even in a full, growing economy a 2 to 3 percent "frictional" unemployment rate can be expected.

From 1950 to 1971, the fastest growing sector of the

economy in terms of U.S. employment was trade, finance and services (+10 percent), followed by government (+6.4 percent) (See Table 29).

Though agriculture and manufacturing grew during this period, their relative shares of total economic activity declined by about 9 percent in production of goods and services; about 10 percent in national income; and more than 13 percent in employment. Jobs in agriculture declined from 7.2 million in 1950 to 3.4 million in 1971. While the proportion of workers in manufacturing declined, the absolute number of jobs in the manufacturing sector increased by about 3.3 million.

Table 29–**Relative changes in national income, total employment, and gross national product by major sectors of the U.S. economy, 1950–1971**

	1950–1971 Percentage Change in Shares of		
Industry Group	*National Income*	*Total Employment*	*GNP*
Agriculture & mining	−5.5	−10.6	−5.6
Manufacturing & construction	−5.0	4.0	−7.0
Trade, finance & services	+4.6	+10.0	+5.9
Transport, communication & utilities	− .8	− 1.5	− .3
Government	+6.3	+ 6.4	+5.1
Total economic growth[a]	+255%	+41%	+269%

Notes:

All figures here reflect absolute percentage changes for each sector between 1950 and 1971. Thus, in 1950, Agriculture and Mining generated 9.5 percent of total U.S. national income; in 1971 it generated 4.0 percent.

In 1971, total national income amounted to $854.8 billion in current dollars. The absolute percentage shares of this total were: Agriculture and Mining, 4 percent; Manufacturing and Construction, 31.5 percent; Trade Finance and Services, 39.7 percent; Transportation, Communications and Utilities, 7.8 percent and Government, 17 percent. The make-up of GNP is similar to the above make-up of National Income. In 1971, total U.S. employment was 74 million persons. Agriculture and Mining employed 5.4 percent of this total with Manufacturing and Construction employing 29.6 percent; Trade, Services and Finance 41.9 percent; Transportation, Communication and Utilities 6.0 percent and Government 17.4 percent.

[a] GNP grew 109% in constant dollar terms during 1950–1971. The above percentage changes in shares and economic growth percentages are based on current dollar amounts.

Such figures show the dominance of the service sector (services, trade, finance, communications, government, transportation) in the growth of the U.S. economy over the past twenty years. The manufacturing and agricultural sectors taken together showed a net loss of a half-million workers, precisely because these sectors have experienced high productivity gains. The value of output in these two sectors increased 250 percent between 1950 and 1971. (In constant dollars, the increase was 109 percent.)

Consumer spending also changed during this period, although the share spent for goods decreased. This may seem surprising, given our national reputation as gadget consumers; but the fact is that our expenditures for services, including education, insurance, recreation, and health, have increased more than our spending for goods. At the same time, increased labor productivity in industry added to the increased demand for services, helped shift employment growth to services, trade, and government.

Changing consumer preferences have also affected the energy industries. For example, the development of large diameter natural gas pipelines, and growing consumer preferences for natural gas enabled the Gulf Coast natural gas industry to enter into traditional coal markets. These factors, together with the changeover to diesel fuel locomotives and the rapid automation in mining, led to a drastic decline of more than 300,000 jobs in the coal industry since 1947, mostly in Appalachia. This deep regional pocket of unemployment has proved to be stubborn, in spite of the national economic growth of the past two decades.

Employment effects of supply disruptions versus demand conservation

During the 1973–74 Arab embargo, we experienced a new set of employment problems that were directly linked to energy availability. Layoffs by auto plants; airlines; boat, airplane, and recreational vehicle plants; and tourist and resort businesses reportedly increased the number of unemployed by about 200,000 persons between October 1973 and January 1974. This was one-third of the total increase in unemployment of 600,000 workers in that period.[11] Clearly, the nation's economic problems during this period stemmed only partly from energy shortages.[12]

The disruptive economic and employment effects of short-run energy supply problems can be mitigated by good planning. We can attribute the sudden shutdown of auto plants and travel related businesses as a result of the Arab oil embargo to a near total inability to adjust to the problem in the short run. But many businesses were little affected, either because they had fuel storage capacity or were able to reduce their energy use through belt-tightening or substitution of other inputs for energy. Most firms and households lie somewhere between these extremes. But our success or failure in adjusting to an unexpected energy shortage tells almost nothing about how we could adjust over a longer period, if we anticipated the problem.

Although a sound energy policy seeks to avoid shortages, our energy destiny is not entirely in our own hands. Unforeseen, uncontrollable events upset the best laid plans, and shortages will no doubt occur in the future.[e] The United States has yet to adapt fully to tighter supplies, higher costs, greater environmental awareness, and foreign policy problems. If shortages do occur, foresight and sensible planning can help to keep disruption in employment to a minimum.

Obviously, the government's emergency planning must concentrate on avoiding shortages that will cause unemployment. This means being sure that alternative fuels are available to industries now dependent on a single source. We must recognize that oil is not the only fuel that can be interrupted. The next really cold winter could bring severe shortages of natural gas for industrial customers. And even supposedly reliable coal supplies could become as scarce as oil during the embargo, if the coal miners call a strike and stay out for weeks or months.

The potential employment impact of energy conservation in the energy producing and energy intensive industries

The energy industry itself and the economy's energy intensive industries[f] have continuously become more capital intensive and, in the process, have provided a smaller share

[e] For a full discussion of the energy problems the country faces for the next few years, see Energy Policy Project, Exploring Energy Choices (P.O. Box 1919, New York, N. Y. 10001) (75¢).

[f] An energy intensive industry is defined here as one whose ratio of total energy consumed to total output produced significantly (generally by at least a factor of four) exceeds the average energy/output ratio for total U.S. manufacturing. See Table 31 for a listing of broad industry groups that are considered energy intensive, and Table 33 for a more detailed listing of such industries. A similar criterion holds for classifying industries as capital or labor intensive.

of the nation's jobs. This is easily seen by comparing the shares of energy, labor, and capital in these industries. They consume about one-third of total U.S. energy; they account for about 45 percent of U.S. industrial production (or about 15 percent of GNP); and they provide only about 10 percent of total employment. But as a group they account for one-half of the new capital requirements (including construction) in the industrial sector (see Tables 30 and 31).

The energy industries of the U.S. economy consist of seven major industry groups: coal mining; oil and natural gas extraction; refined petroleum and coal products; pipeline transportation; electric, gas, and combination utilities; petroleum bulk stations, terminals, and other fuel wholesale merchants; and gasoline service stations and fuel oil dealers.

Total employment in the United States increased 41 percent between 1950–1971, while employment in the energy industries increased only 5.5 percent, or 115,000 workers. The net increase in energy industry employment was due entirely to growth in the utility and fuel marketing industries, with gasoline service station employment dominating the trend. During this period, employment actually decreased by 341,000 in mining, refining and pipeline transportation industries (see Table 32).

Forty-three percent of the jobs in the U.S. energy industry are in fuel distribution; mining and refining accounts for 27 percent; and utilities account for 30 percent. The significant decline in coal mining employment (300,000) and the large increase in service station and fuel dealer employment (248,000) tell the story of the changes over the past two decades.

Five manufacturing groups are clearly the largest industrial energy consumers: primary metals; stone clay and glass; food and kindred products; chemicals and allied products; and paper and allied products.[g] They consume two-thirds of the energy used annually by U.S. manufacturing.

In 1971 these five groups employed 4.8 million workers—7.3 percent of total U.S. employment and 26 percent of U.S. manufacturing employment. While total U.S. employment increased 41 percent between 1950 and 1971, total employment in these five energy intensive industries was static. Employment within two of these groups

[g] Refined petroleum and coal products are included in the energy producing industry group, rather than in manufacturing.

Table 30–**Relative importance of U.S. energy industries by selected economic measures, 1971** (Percent)

Economic Measure	*All U.S. Energy Industries*	*Coal & Oil Production*	*Oil & Coal Refining*	*Pipeline Transport*	*Electric and Natural Gas Utilities*	*Wholesale Dealers*	*Retail Dealers*
Total employment (employees)	3.0	.54	.26	.02	.89	.29	.97
Total employee compensation ($)	3.2	.75	.51	.03	1.13	.23	.60
Total national income ($)	4.0	.54	.92	.06	1.63	.22	.59
Total industrial production (value added in $)	12.0	5.12	1.80	n.a.	5.07	n.a.	n.a.
Total wholesale trade ($)	3.8	—	—	—	—	3.8	—
Total retail trade ($)	6.0	—	—	—	—	—	6.0
Total new P & E investment ($)	27.0	2.2	7.2	.10	17.3	.4	.3
Total construction investment ($)	15.2	n.a.	n.a.	.33	14.50	n.a.	n.a.

Table 31–**Relative importance of major U.S. energy intensive industry groups by selected economic measures, 1971** (Percent)

Economic Measure	*U.S. Energy Intensive Industry Group*	*Primary Metals*	*Chemicals & Allied Products*	*Stone, Clay and Glass*	*Paper & Allied Products*	*Food & Kindred Products*
Total employment (employees)	7.3	1.6	1.1	.8	.9	2.1
Industrial production (Federal Reserve Index)	31.8	6.3	9.4	2.9	3.6	9.6
New plant and equipment investment ($)	13.6	3.4	4.2	1.1	1.5	3.3
Share of manufacturing's gross energy consumption (Trillion Btu's)	67.8	26.8	17.0	7.3	7.6	6.3

actually declined (primary metals went down 5.5 percent, food and kindred products 11.6 percent). Two advanced faster than total manufacturing, but slower than the overall economy: chemicals increased 32 percent, and paper 30.9 percent. (Table 31 shows the energy consumed by these five groups.) Primary metals and chemicals clearly are dominant.

Table 33 ranks the top fifteen energy intensive industries (as measured by their energy–value added ratios) from the standpoint of energy consumption and value added.[h] They consume almost 45 percent of the energy used by U.S. manufacturing, produce less than 9 percent of manufacturing's value added, and account for only 6 percent of manufacturing jobs. The great bulk of U.S. energy consumption is tied to industries that have generated almost no net new jobs in the past two decades, and that employ less than 10 percent of the current work force.

In these industries, the efficiency of energy use has increased even as energy prices have fallen. This increase in efficiency is likely to accelerate in the future. Much of the decline in energy growth in a future following the *Technical Fix* scenario will occur in this sector—through technical

[h] Value added is value of shipments less costs of intermediate materials; essentially, that is the value of capital and labor entering the final product. It serves as an indicator of actual output or production of industries by preventing double counting of products sold from one industry to another.

Table 32–**Total employees in U.S. energy industries, and U.S. energy intensive manufacturing industries, 1950–1971**

Industry Classification	*1950*	*1971*	*1950–1971 (Percent Change)*	*Percent of 1971 Total Employment*	
Total employment in U.S. energy industries	2,075,000	2,190,000	+5.5	2.96	
Anthracite, bituminous coal & lignite mining	441,000	138,000	−69.0	.19	
Oil & natural gas extraction	266,000	261,000	−2.0	.35	
Petroleum & coal products	218,000	191,000	−12.0	.26	
Pipeline transportation	24,000	18,000	−25.0	.02	
Electric companies & systems	239,000	296,000	+24.0	.40	
Gas companies & systems	118,000	168,000	+42.0	.23	.89
Combination companies & systems	169,000	190,000	+12.0	.26	
Petroleum bulk stations, terminals & other wholesale	131,000	212,000	+61.0	.29	
Gasoline service stations	469,000	618,000	+53.0	.84	
Fuel & ice dealers	n.a.	99,000	n.a.	.13	
Total employment in energy intensive manufacturing	4,709,000	4,828,000	+15.0	7.31	
Primary metals	1,247,000	1,178,000	−5.5	1.6	
Stone, clay and glass	547,000	588,000	+7.5	.8	
Food & kindred products	1,790,000	1,582,000	−11.6	2.1	
Paper & allied products	485,000	635,000	+30.9	.9	
Chemicals & allied products	640,000	845,000	+32.0	1.1	
Total employment in U.S. energy industries and energy intensive industries	6,784,000	7,603,000	+12.1	10.27	

change in the long term, and through simple belt-tightening austerity in the short term. But the effects of energy conservation efforts in these industries on employment would be minimal.[13]

Future U.S. economic growth: some energy implications

In the past, technical change affecting energy efficiency occurred almost entirely on the supply side—in the production, transmission, and conversion of fuels to electric power and petroleum products. Very little improvement took place on the consuming side, in such goods as autos, furnaces, and houses (fluorescent lamps and heat pumps are two exceptions).

Higher costs and prices now provide an economic incentive for technical improvements in consumption, thereby decreasing the amount of energy needed for the same level of service or activity and making future economic growth less energy intensive.[14] Other developments may contribute to less energy intensive economic growth. A projection of economic and consumer expenditure trends over the next two decades lends support to this prospect.

Population, labor force, and employment

The United States has recently reached a fertility rate which, if continued, would achieve zero population growth early in the next century. Latest estimates of the fertility level in the United States now show a rate of 1.9 or less, which is below the replacement rate of 2.1 children per female. A few years ago, the best population projections were 283 million in 2000; today 263 million, or perhaps lower, is more likely. This declining forecast has certain predictable economic and social consequences.[15]

Some 25 million new workers will be seeking their first job during the 1970s, seven million more than in the 1960s; the bulge is the legacy of the postwar baby boom. This growing number of job applicants will put an extra burden on full employment policy.

But the pressure should be off in the 1980s, as a result of the slowdown in births during the mid sixties. The Bureau of Labor Statistics projects annual labor force

Table 33–**Top fifteen U.S. energy intensive industries, 1967**

SIC[a] *Number*	*Industry Title*	*Energy–Value Added Ratio (1,000 Btu's $)*	*Gross Energy Consumed (Trillion Btu's)*	*Total Manufacturing Energy Consumption (Percent)*	*Value Added (Million 67 $)*	*Total Manufacturing Value Added (Percent)*
—	All manufacturing	59.22	15,515.4	100.0	261.984	100.0
3274	Lime	818.18	81.9	.5	100.	—
3334	Primary aluminum (5)	726.04	589.4	3.8	812.	.3
3313	Electrometallurgical products (12)	678.05	131.0	.8	193.	.1
2812	Alkalies & chlorine (8)	636.69	266.9	1.7	419.	.2
3241	Hydraulic cement (6)	634.25	515.2	3.3	812.	.3
2819	Industrial inorganic (2) chemicals, n.e.c.	423.15	971.3	6.3	2,295.	.9
3251	Brick structural tile	404.62	101.6	.7	251.	.1
2631	Paperboard mills (7)	316.08	476.9	3.1	1,509.	.6
2611	Pulp mills	293.68	98.0	.6	334.	.1
2813	Industrial gases (14)	280.12	112.3	.7	401.	.2
	Subtotal (ten industries)	—	3,344.5	21.5%	7,126	2.7

[a] The Department of Commerce's Standard Industrial Classification numbers for all of U.S. industry.

Source: The Conference Board report, *Energy Consumption in U.S. Manufacturing* (Ballinger Publishing Co., Cambridge, Mass., October 1974).

growth at 1.8 percent for 1968–80, declining to 1.1 percent for 1980–85 (see reference cited in Note 18).

The labor force is projected to grow from 91 million in 1973 to 108 million in 1985, and to at least 127 million in 2000. Total employment was 84 million in 1974; it is projected to reach 103 million in 1985. Table 34 summarizes

Table 33 (continued)–**Top fifteen U.S. energy intensive industries, 1967**

SIC[a] *Number*	*Industry Title*	*Energy–Value Added Ratio (1,000 Btu's $)*	*Gross Energy Consumed (Trillion Btu's)*	*Total Manufacturing Energy Consumption (Percent)*	*Value Added (Million 67 $)*	*Total Manufacturing Value Added (Percent)*
2661	Building paper and building board	271.64	49.9	.3	184.	.1
2818	Industrial organic chemicals, n.e.c. (3)	266.30	952.1	6.1	3,575.	1.4
2621	Paper mills ex. building paper (4)	255.99	603.2	3.9	2,356.	.9
2815	Cyclic intermediates and crudes (10)	205.35	149.8	1.0	730.	.3
2911	Petroleum Refining	307.52	1,459.2	9.4	4,745.	1.8
3312	Blast furnaces and steel mills (1)	203.21	1,810.6	11.7	8,910.	3.4
	Grand Total	—	6,910.1	44.5	22,881.	8.7

Source: Table 4 Conference Board Report

the estimates for population and the labor force and employment totals that are consistent with the Census Bureau's Series E projections.[16] While growth in the labor force will slow after the early 1980s, it will not necessarily inhibit growth in general economic activity if participation rates and productivity increases are slightly higher than past trends. And it will lessen the pressure for high growth rates to maintain full employment.

Income and consumption patterns

In the year 2000, GNP may be two to three times what it was in 1970. Average household income, now about

Table 34–**Population, labor force, employment levels—selected past years and projections for 1985 and 2000**

	1950	*1960*	*1970*	*1973*	*1985*	*2000*
Population:[a] (millions)	152.3	180.7	204.9	210.4	235.3	264.4
Labor Force (millions)	63.9	72.1	85.9	91.0	108.0	127.0
Participation rate (%)	42.0	40.0	41.9	43.3	45.8	48.0
Employment[b] (millions)	58.9	65.8	78.6	84.4	103.0	121.2
Unemployment (millions)	3.3	3.9	4.1	4.3	5.0	5.8
Unemployment rates (%)	5.3	5.5	4.8	5.1	4.6	4.6
Population, below 20[a] over 65 (millions)	64.1	86.2	97.3	97.6	100.0	109.6

[a] Series E: fertility = 2.1
[b] Civilian and military
Sources:
Social Indicators, Office of Management and Budget
1985: Bureau of Labor Statistics (reference 19)
2000: Population, Resources, and Environment, p. 41 (reference 16)
1985: Bureau of Labor Statistics (reference 19)
2000: Calculated based on 4.6 percent unemployment rates
Data Resources, Inc. (1950, 1960, 1970, 1973 figures from Economic Report of the President, February 1974.)

$12,000, might exceed $21,000 (in 1972 dollars),[i] even if the work week were reduced to 30 hours.[17]

Disposable personal income per capita is likely to grow from $3,816 in 1972 to $6,346 in 1980, and to $8,400 by 1985.[18] This is a 3.1 percent annual increase in the face of a projected slowdown in economic growth during the period 1980–85. In general, economic analyses consistently project significant growth in every form of personal income measure, including average household income, per capita personal income, and per capita disposable income. Personal income is either spent, saved, or taxed away. Taxes are expected to take an increasing share of personal income, but the increase is small—less than 2 percent more by 1985.[19] Saving is also expected to increase.

With lower birth rates, the population will become more middle-aged. With more work-age adults and few

[i] These results are derived from projected increases in productivity of the labor force, the slower growth in population and labor force, and moderate increases in labor force participation rates. See the reference in Note 17.

dependent children in the population, private saving is likely to increase. Governments will not have to spend as much on education and welfare. When higher savings are invested, capital formation will increase, which will contribute to a rising GNP and per capita income. Current alarmist projections of enormous capital needs, based on historic growth in the development of new energy supplies, do not take into account these demographic trends. Nor do they consider the likelihood that the capital requirements for energy conservation will be substantially less than those needed to increase supply at historical growth rates (see Chapter 3).

As income increases, people show more preference for services: education, health, entertainment, travel and recreation. While demand for appliances, autos and furniture will continue to grow because of replacement needs, the lower rate of household formation will slow this growth. This effect is aside from the likelihood of saturation of household appliances in U.S. homes by 1985 (see Chapter 3). Finally, judging from past experience, high personal and family incomes can be expected to increase the demand for higher quality, more expensive goods. Thus, while there may be a slowdown in units produced, sales revenues can be maintained.[20]

Public investments will shift away from child oriented programs, such as building schools, towards services such as crime prevention, pollution control, health care, housing, recreation, and transportation. While such shifts have important consequences for future energy growth rates and mixes, those who forecast energy resources and investment needs usually have not taken adequate account of them.

Energy implications

As noted, an older and wealthier U. S. population should display a pattern of consumption favoring personal services and higher quality goods. A shift in demand toward higher quality durable goods—autos, stoves, air conditioners—coupled with higher energy prices should lead to energy-efficient design improvements, particularly if consumers are better informed about life cycle costs. Generally, a shift in consumption patterns toward more services means less demand for energy growth. Some services, however, such as the resort industry, are heavily dependent upon transportation. Growth in those services does not automatically imply that energy growth will slow down, unless the transportation supporting them becomes more efficient.

We do not expect that people will change their consumption patterns primarily, or even significantly, merely for the joy of saving energy. We do expect consumers, businesses, and governments to react to higher energy prices by conserving energy. This reaction will have a major impact on the energy efficiency of future products and the energy costs of future services.

Summary

Future energy growth may decline without disrupting the U.S. economy, through greater energy efficiency. This may mean higher capital, materials, and labor costs, but it also means benefits: less pollution, lower operating costs, and reduced risk of social and economic disruption.

Our principal finding—that energy growth and economic growth and employment can safely be uncoupled—may puzzle the gas station owner or local retailer, to whom the 1973–74 winter oil supply disruption meant less business. As noted, however, there is a great difference between a sudden disruption in energy supply and a carefully planned, long term energy conservation program. The kind of economic growth we expect, and the way it differs from past growth, will have major implications for our future energy requirements.

These are our general conclusions.

- Government policy measures to stimulate capital investments by businesses, such as investment tax credits, should recognize that they may stimulate growth in energy demand as well. On the other hand, tax incentives may be useful to encourage investments that increase energy use efficiency.
- U.S. manufacturing has realized significant energy savings in the past, during a period of stable or declining relative prices of energy. The past rate of improvement was a 1.6 percent decline in the energy/output ratio for total U.S. manufacturing, and a pre-embargo study indicates that a 2.0 percent rate of decline is probable out to 1980. Given much higher energy prices, fear of future shortages, and explicit government actions, even greater energy savings are likely.

• Energy saving is economically attractive today. Energy conservation measures should pass the test of economic efficiency as well as thermal efficiency. Our studies indicate that the conservation measures in the *Technical Fix* scenario (see Chapter 3) meet this test.

• It is reasonable to expect that energy conservation in the most energy intensive manufacturing industries will have little, if any, adverse effect on employment in these industries.

• Energy conservation will not disrupt the nonmanufacturing sector of the economy, if long term government policies toward housing, transportation, R&D and environment are consistent with the objective of conservation, and allow time for adjustment.

• Because of the slowdown in population growth, growth in the labor force is expected to slacken after 1980. A slower population growth could bring about economic growth requiring less energy. An older and wealthier population probably will consume more services and high quality goods, which tend to be energy conserving.

Energy, employment, and economic growth are interdependent—but they are in no way linked inevitably to the patterns of the past. The United States can grow and prosper and have plenty of jobs—and still conserve energy.

CHAPTER 7

U.S. energy policy in the world context

The world oil market, an artery to Western European, Japanese and American prosperity, has undergone an extraordinary metamorphosis in just four years. During that period the oil exporting nations, which represent a small fraction of the world's population, income, and military might, gained almost unlimited control of world oil prices. They were able to force the importing nations, rich and poor alike, to pay literally tens of billions of dollars more for oil imports.

Since 1970 the members of the Organization of Petroleum Exporting Countries (OPEC), or the cartel, as it has come to be known, raised taxes almost at will on the oil

Note: This chapter draws on some of the material developed by members of the staff of the Brookings Institution in a forthcoming study of energy and U.S. foreign policy. This chapter is not, however, a summary of the Brookings study. It represents the Energy Policy Project's independent analysis.

produced within their borders. In the four years ending January 1, 1974, the price of oil rose about 515 percent (see Fig. 28). As a consequence, the total oil revenues of the OPEC members soared from $7.8 billion in 1970 to $23 billion in 1973. They could reach $90 billion in 1974. To put these numbers in perspective, the total revenues earned by all developing countries from exports in 1972, including oil, amounted to about $19.6 billion.[1]

If our analysis is correct, the world has experienced more than just a temporary aberration on the supply and demand charts; we have witnessed a fundamental shift in

Figure 28–**Estimated market price of Persian Gulf oil**

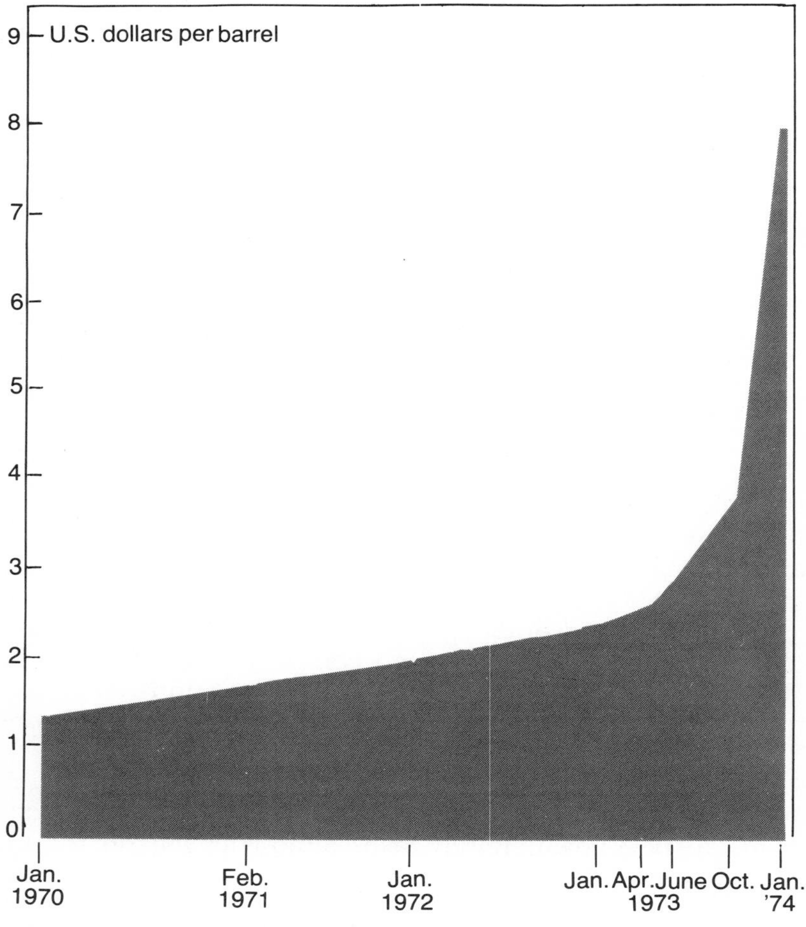

Note: Saudi Arabia Light Oil (34°), FOB Ras Tanura

Source: World Bank

the power relationships between the world's industrial powers, including the United States, and the major oil exporting nations such as Saudi Arabia, Iran, and Kuwait. Its implications are far-reaching to future international oil companies' operations as well as to American diplomacy; we will discuss those implications in some detail later in the chapter. Let us begin our analysis by asking how such an unprecedented turn of fortunes came to pass.

As might be expected with a development of this magnitude, its origins lie deeply rooted in the past, are complex, and are open to differing interpretations.

Middle Eastern journalist Leonard Mosley takes one controversial point of view. He argues that the OPEC countries were able to negotiate higher prices from the international oil companies during the early 1970s because the companies were not concerned with higher oil prices. "They would, after all, simply pass the increases over to their consumers," Mosley reports. What did concern the oil companies during these negotiations was the OPEC nations' demand for "participation"—that is, a share in the ownership of oil company operations inside their countries. Their resistance was, of course, for naught and the OPEC nations are gaining "participation" quite rapidly.[2]

But events have weakened the force of Mosley's argument. It should be noted that the last giant price hike by OPEC of 120 percent between October 16 and January 1, 1974 was done without any negotiations with the international oil companies. The Middle Eastern OPEC nations simply announced increased prices, they became the world prices, and the importing nations paid them.

From a different perspective, economist M. A. Adelman believes that the U.S. government's failure to oppose escalating OPEC demands in the early 1970s put power and motive into the producers' hands."[3] Adelman also contends that the international oil companies played a key role:

> It is essential for the cartel that the oil companies continue as *crude oil marketers,* paying the excise tax before selling the crude or refining it to sell as products. Were the producing nations the sellers of crude, paying the companies in cash or oil for their services, the cartel would crumble. The floor to price would then be not the tax-plus-cost, but only bare cost. The producing nations would need to set and obey production quotas. Otherwise, they would inevitably chisel and bring prices down by selling incremental amounts at discount prices. . . . Every cartel

> has in time been destroyed by one, then some other, member chiselling and cheating; without the instrument of the multinational companies and the cooperation of the consuming countries OPEC would be an ordinary cartel.

He proposes that the only way the oil importing countries will obtain cheap and secure oil is by "a simple and elegant maneuver—get the multinational oil companies out of crude oil marketing; let them remain as producers under contract and as buyers of crude to transport, refine, and sell as products."[4] It is an intriguing proposal. But it would seem that over time, the idea will be tested in any case as more and more oil company assets are nationalized by the exporting countries. No one knows whether this structural change in the world oil market will produce the far-ranging change which Adelman envisions. However, there is reason to remain skeptical.

The marketing role of the international oil industry may have hastened OPEC's ascendency to power as did the disunity among the governments of the oil importing nations, but they were not decisive. What could be described as the "oil dependency factor" was decisive. Oil is vital to modern industrial economies, and oil is not readily replaceable with other energy sources in a short time at almost any price. Unplanned interruptions in its flow—even relatively minor ones—cause some nasty short term side effects in employment and incomes, as the United States learned in the winter of 1973–74.

From 1970 on, as the United States, Western Europe and Japan grew increasingly dependent on oil imports to fuel their expanding energy consumption, they were willing to pay more and more for that next barrel of imported oil rather than not have it, especially since it still cost far less than alternative sources. (Curiously, the United States did not at the same time build up strategic reserves as insurance against a cutoff.) Hence, while oil prices were rising, the United States was increasing its oil import demands by 78 percent—from 3.4 million barrels a day in 1970 to 6.2 million in 1973; Western Europe was increasing its oil imports by 19 percent, and Japan's were rising by 33 percent[5] (See Fig. 29). Clearly, the pull of oil dependency is powerful enough to override price considerations—to a point. It could not last indefinitely of course—oil is not heroin. Eventually oil prices did level off in 1974 as the OPEC nations began to realize that further increases could "kill the goose's" ability to pay higher prices.

Figure 29–**Oil imports, USA, Japan, Western Europe, 1950–1973**

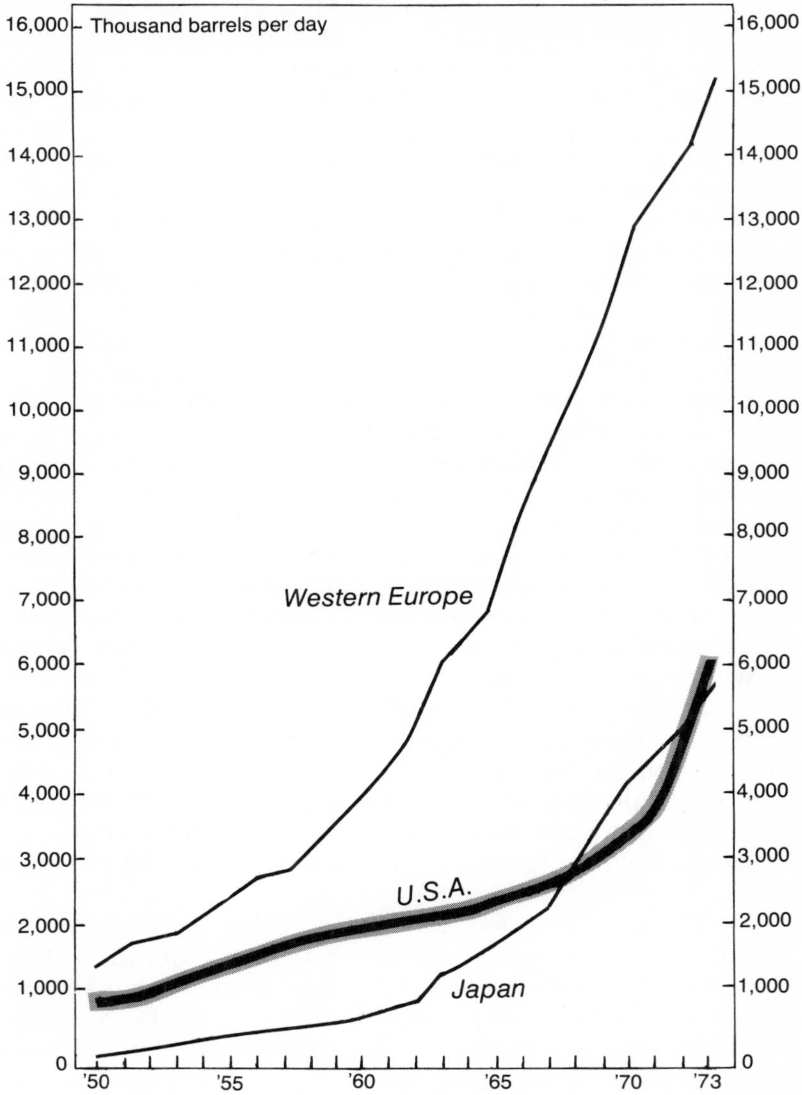

Note: Western Europe includes U.K., Germany, Italy, Scandinavia, Benelux
Source: BP Statistical Review of World Oil Industry

Future oil prices

Can OPEC freeze prices at something like the current level, with producing nations earning about $7 a barrel?[a] It is impossible to say because the world has no practical experience either with high oil prices or with a cartel such as OPEC—that is, a cartel of financially solvent governments who control a nonagricultural, nonrenewable commodity for which there is no substitute, at least in the short run.

A series of facts, however, suggest that a drastic reduction in price—is unlikely.

First, there is the fact that a handful of nations control such a large share of the world's readily available oil and seem committed to a common policy of maintaining these price levels. The size of the Middle Eastern resource base is enormous. At the end of 1973, the Middle East possessed 55.4 percent of the world's proved oil reserves and accounted for 36.8 percent of the world's oil production but only 2.3 percent of its consumption. Contrast these figures with the United States' 6.3 percent of the world's proved reserves, 18.3 percent of the production, and 29.5 percent of the consumption; or Western Europe's 2.6 percent of the reserves, less than 1 percent of the production, and 27 percent of the consumption.[6] These resource facts mean that if a few key producing nations are of the same mind on oil pricing policy, they possess the market power to make the prices stick.

Second, there are of course alternative energy sources to oil, but their rapid expansion is restrained, in the short term, by human or environmental problems. The inherent limitations of coal, nuclear power, and oil shale are explored elsewhere in this book. Oil's competitive edge over the alternative energy sources, it appears, will hold for some while.

Third, the steady growth in world demand for oil imports (10.8 percent a year over the past decade), combined with the price record of the past four years, has been convincing evidence to OPEC of the high market value of this finite resource they now control. Consider the shrewd comments of the Shah of Iran:

> In 1947 the posted price for a barrel of oil in the Persian Gulf was $2.17. Then it was brought down to $1.79, and that lasted until 1969. So there were 22 years of cheap fuel that made Europe what it is, that made Japan what it is. Then the price of wheat jumped 300 percent, vegetables the same, and sugar in the past six years increased 16 times. So we charged experts to study what prices we should put on oil. Do you know that from

---

[a] In 1974, host government revenues on contract oil sold to oil companies were about $7 a barrel; on oil sold at auction, their revenues were about $11 a barrel.

> oil you have today 70,000 derivates? When we empty our wells, then you will be denied what I call this noble product. It will take you $8 to extract your shale or tar sands. So I said let us start with the bottom price of $7; that is the government intake.[7]

History seems to indicate that in the international oil market the exercise of monopoly-like power is more the rule than the exception.

Before World War II the few companies that produced Middle Eastern oil operated as a cartel, with written rules and regulations for dividing the market, limiting output, and supporting prices. After the war, the cartel's power was temporarily broken by the U.S. government, which insisted on a "lowest competitive price" in its massive Marshall Plan purchases for Europe. But the Marshall Plan ended, the Korean War began, and the so-called "seven sisters"[b] (plus one smaller French company) reestablished market control. Prices stabilized, and even increased twice in the 1950s.[8]

During the 1960s, however, the entrance into the market of independent oil companies and state owned companies, and the rising tide of cheap Middle Eastern oil, loosened the cartel's hold on production and prices. Oil prices actually dropped. Most oilmen, be they in Riyadh or at 30 Rockefeller Plaza, refer to this particular period as one of "a low return on investment" and generally they do not look back upon it with fondness. By the beginning of the 1970s, the major international oil companies' share of crude oil production outside the United States and the communist bloc countries was down to about 75 percent.[9] (See Chapter 9 for oil companies' control of U.S., domestic market.)

The downward trend in prices continued to mid-1970. Up until that time, long term contracts had been at lower prices than short term ones, indicating that the industry expected still lower prices in the future, even as far as 10 years ahead. But then, in a series of negotiations with the oil companies, in Tehran, Tripoli, and Geneva, OPEC demanded and got increased taxes on its oil. The decline in the oil companies' monopoly control had created something of a power vacuum in the world oil market—and OPEC filled it.

One key to future oil prices is the production plans of OPEC nations, especially Saudi Arabia and, to a lesser extent, Iran. Some countries such as Libya and Kuwait are

---

[b] The seven include five American companies—Exxon, Gulf, Mobil, Texaco, Standard of California—plus British Petroleum and the Anglo-Dutch Shell group.

already restricting output at current prices. They feel current oil earnings are sufficient to meet their internal needs for revenue. Perhaps the extreme example is tiny Abu Dhabi, whose oil revenues in 1974 could come to $76,923 per capita (at $7 a barrel government take). However, other countries such as Indonesia (oil revenues of $43 per capita), Nigeria ($198 per capita), and Iraq ($635 per capita) have indicated that they will maximize their oil production. But this increase in supply will have little impact upon price if import demands grow even moderately from today's level of 34.1 million barrels a day.[10]

Saudi Arabia, on the other hand, could swamp the oil market. Possessing 21 percent of the world's petroleum reserves, Saudi Arabia is capable of raising its production from 7.3 million barrels a day in 1973 to 20 million barrels a day by 1980. With a population of only 8 million, Saudi Arabia has a limited ability to spend its oil revenues internally, and in 1974 those revenues could reach $3,125 per capita. Yet the Saudis have expressed the desire for considerable military and technological assistance from the United States and other oil importing nations. But whether the Saudis would go it alone from the rest of OPEC and increase production, to the extent of sharply reducing world oil prices, in return for guns and engineers is doubtful.

Iran could possibly increase its production from 5.9 million barrels a day in 1973 to 10 million in 1985. With a population of about 32 million, with one of the most ambitious development programs in the Third World, and with a growing appetite for expensive military hardware, Iran needs increasing oil revenues. But Iran is one of the most militant advocates within OPEC for higher prices, for the fruits of cartelization (shown in Figs. 30 and 31) have not escaped the Iranians' notice.

World demand for oil probably will grow more slowly in the future due to the higher prices. But the market for oil will continue to be an expanding one. The OPEC nations such as Iraq, Nigeria and Iran, which have the economic need and resources to expand their production, will share whatever growth takes place along with Saudi Arabia. Some nations may well wish to sell more oil than their share, but the interest of each OPEC nation in keeping prices in the current range, or higher, is far more important to them financially than any gain from marginal increases in production.

The sharp price reductions required to expand greatly the market for oil would seem counterproductive

Figure 30–**Crude oil production, selected OPEC countries, 1963–1973**

Figure 31–**Government oil revenues, selected OPEC countries, 1963–1974**

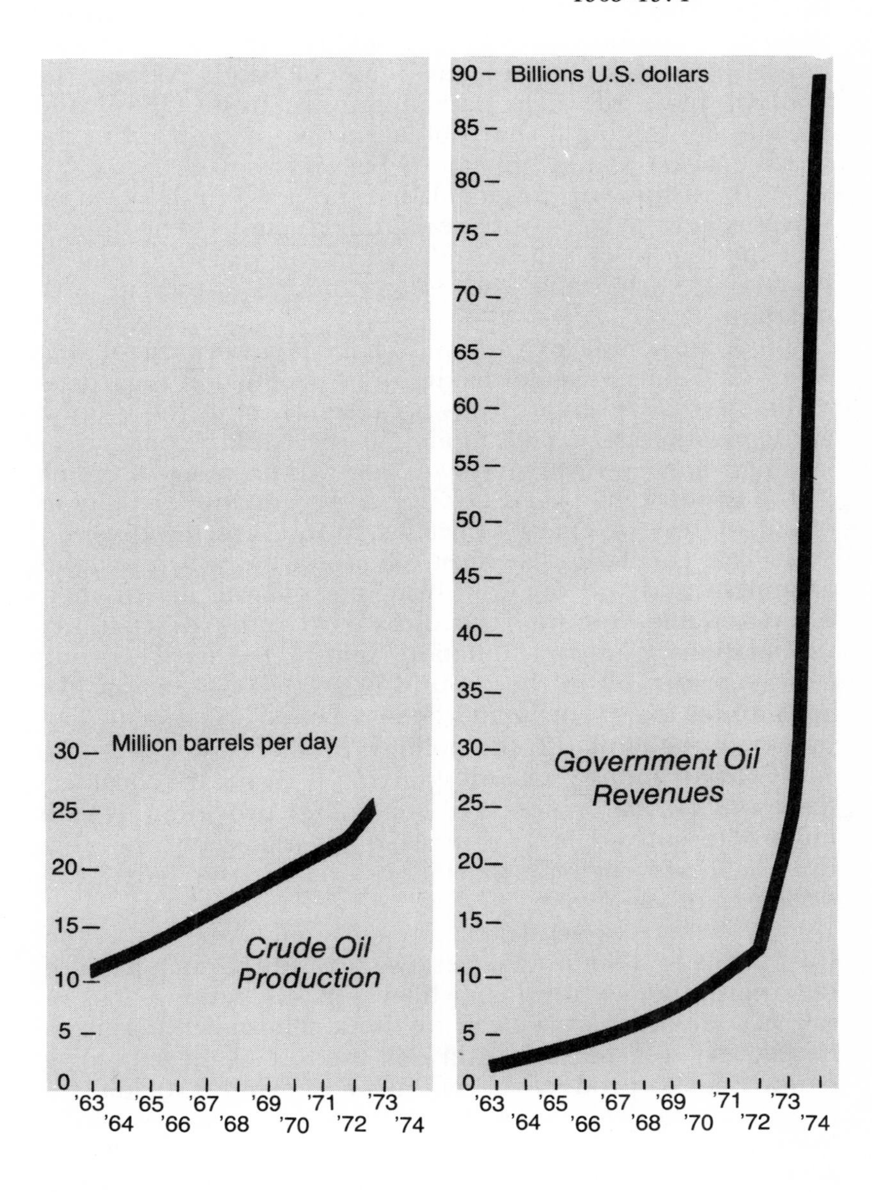

Note: Includes Abu Dhabi, Algeria, Iran, Iraq, Kuwait, Libya, Qatar, Saudi Arabia, Venezuela
Source: BP Statistical Review of World Oil Industry

Notes: Includes Abu Dhabi, Algeria, Iran, Iraq, Kuwait, Libya, Qatar, Saudi Arabia, Venezuela

Figure for 1973 is an estimate; for 1974, a projection.

Source: The Middle East and North Africa, *London: Europa Publications, Ltd.*

for any of the OPEC nations—except perhaps Saudi Arabia, which may be in the best position to maximize revenues through price reductions. But the Saudis show no interest in rolling back last year's doubling of oil prices. Putting aside the question of whether selling more barrels at a much lower price makes any sense to Saudi Arabia, the political price of their breaking away from OPEC and unilaterally forcing a sharp break in the price of oil in the world market would appear to be very high.

It is thus our judgment that the new era in world oil prices is here to stay—with certain fluctuations. The hope is that the real price will be reduced in the years to come by remaining fairly stable and not continuing to escalate with inflation.

A final note of caution. There is an argument that every oil country should maximize its production regardless of its internal revenue needs because (1) oil in the ground earns no interest, unlike money in the bank; (2) oil prices are high now and will go down; and (3) the value of the oil in the ground may even decline in the coming 25 years as other sources of energy, such as fusion, are developed.

We conclude that most oil producers will reject this argument and will opt for stretching out the life of their nonrenewable resource because (1) in times of currency devaluations, rampant inflation, and a trend of soaring energy prices, oil in the ground is to a certain extent like gold under the Frenchman's bed—a better investment than money in the bank; (2) the political problems of what to do with surplus money, including giving it away, are considerable; and (3) the prospects of technology providing us with major alternative energy sources to oil—especially ones that can be used in transportation—within the next two decades are minimal.

There is a tendency among some Americans and Europeans to assume that because a resource such as oil is out there and we need it, whoever has it must sell us as much as we demand. This is, we think, a mistaken notion. It might very well be within the self-interest of oil exporting nations not to produce all the oil we demand. Canada, for instance, concerned about its own domestic energy needs, has put the United States on notice not to expect increased oil exports. Countries as diverse in interests as Venezuela and Libya have declared that they will conserve some of their oil in order to power their own industrialization in the future. It would not be at all surprising if other oil exporting nations adopted a similar point of view as the years roll on and they see their oil reserves decline.

The conservation connection

The radical change in the world energy market poses a number of foreign policy problems for the United States. The most pressing appear to be vulnerability to politically motivated oil cutoffs, international economic disorder, United States relations with Japan and Western Europe, and the poverty of developing countries. The following is not intended to be a definitive survey of energy-related foreign policy problems. Rather, we hope it to be a brief introduction to the salient issues.

The United States can attempt to mitigate these problems through a variety of multilateral actions, some of which will be discussed later. But the most effective unilateral action the nation can take for coping with these international problems is energy conservation.

Vulnerability: Before the outbreak of the October 1973 Arab-Israeli War, about 17 percent of the oil consumed in the United States originated from the Arab countries. After the lifting of the Arab embargo, the United States seemed headed towards a resumption of that level of dependence. At this writing, U.S.-Arab relations have improved dramatically.

How much dependence on Arab oil is too much? There is no definitive answer. A certain amount of trade with the Arab world is probably useful to sharpen and balance the United States interest in the area. But so long as the United States supports Israel with military equipment and the Arab-Israeli dispute continues, Arab oil may be used as a political weapon against us. It is clearly within the interests of the United States to encourage vigorously, as it is now doing, a peaceful settlement in the Middle East.

But there has been no settlement as yet of the deep differences between the Arabs and Israelis. A pattern of rapidly increasing imports from the Arab nations to the United States would inhibit our freedom of action during the time—which could extend over many years—before a permanent peace is negotiated. For now, the only prudent course is to plan for the worst and be prepared to do without any Arab oil. Stockpiling at levels to cover at least 90 days' supply of Arab oil would help. Obviously, however, the less energy we consume, the less vulnerable we are to interruption. In the case of the vulnerability problem, energy conservation is an insurance policy taken out against the possibility that diplomacy can fail.

International economic disorder: The sudden oil price increase has caused massive transfers of funds from oil importing nations to oil exporting nations. The account surpluses of the exporters are expected to be on the order of $50 billion in 1974.[11] Fortunately, the dire predictions by some that the resulting deficits in other nations' accounts would capsize the world's monetary and trade systems have not yet materialized. But the oil price hikes certainly have aggravated already existent economic woes—from inflation to wobbly currencies—in the world.

The full shock of the oil price-induced imbalances on the economies of the financially weakest consuming nations will apparently be felt in the next year or two. No one can be sure what will happen. A round of competitive exchange rate devaluations and restrictive trade measures, which in themselves would create further disruptions, are feared. The world's financial experts are struggling with this unprecedented problem and are trying to devise ingenious measures to cope with it. Certainly, energy conservation actions that slow the oil import demands of Western Europe, Japan and the United States will help to cushion the full blow from increased oil price when it hits.

Poverty in developing countries: Increased oil prices have dealt a specially cruel blow to developing countries which import oil. These price hikes come on the heels of rising food grain prices in recent years. The price of wheat and rice have doubled and tripled since 1968 (although they have declined modestly of late). Fertilizers, so vital to the continuing of the Green Revolution, have doubled in price too, and are now in extremely short supply.

In just one year, 1974, developing countries which must import oil will have to pay $15 billion more for that oil and $5.5 billion more for food and fertilizer than they did the previous year.[12] Some countries, such as Malaysia, Zambia, Morocco, or Zaire, should be able to cope because they can dip into their accumulated reserves, they have access to capital markets, and they are getting continued high prices for their exports. But there is another group of developing countries which cannot afford to buy the oil, food, and fertilizer which they need. Perhaps one-quarter of the world's people live in these, the poorest states: the whole of the Indian subcontinent—Bangladesh, India, Sri Lanka—west and east Africa, and parts of tropical Latin America.

So dire is their economic condition, so distinct is their deprivation from the rest of the developing countries, that at the 1974 Special Session of the United Nations General Assembly on problems of raw materials and development, the term "Fourth World" became common in describing their plight. The World Bank reckons that these countries will need additional financial aid of about $6.8 million in 1975, which is roughly a 45 percent increase over the aid they will receive in 1974.[13] The situation presents a real challenge to the United States to enlarge its direct aid and to set an example for the oil producers by offering the food we export to the "Fourth World" at discount prices.

Relations with Western Europe and Japan: The entry of the mammoth U.S. energy market into heavy demand for oil from the world market (U.S. oil imports rose 31 percent in 1973) has raised fears in import-dependent Western Europe and Japan. They are paying more and perhaps will be obtaining less oil because the United States seems determined to buy all it can get while preaching consultation and multilateral cooperation. Indeed, a self-serving tendency in matters of oil appears to be growing among the United States and its allies. Some of that was manifest during the selective Arab oil embargo of 1973–74, in sharp contrast to the close cooperation between Americans and Western Europeans during the 1956 Suez Crisis when the flow of oil was disrupted. The oil scramble has spilled over to strain our political-economic relations with Western Europe and Japan. Efforts are underway to achieve better coordination, but the tension is apt to grow if the United States competes with its allies in the purchase of more and more Middle Eastern oil. Inflation is a clear and present danger. Any policy of "beggar your neighbor" is a very dangerous game for any nation to play.

The salutary effects of energy conservation upon U.S. relations with Japan and Western Europe are important, but equally important is agreeing on a system to share available energy resources in the event of an emergency. Japan and Western Europe will have to continue to depend upon oil imports to supply more than three-quarters of their total oil demand.

More upward pressure on world oil prices from increased U.S. imports will immediately ricochet into the

Western European and Japanese economies and, as the Brookings Institution observes, "An oil-induced recession in Western Europe and Japan would have severe economic repercussions in the United States."[14] One need only recall that European financial instability (the collapse of the Kreditanstalt of Vienna, and so on) was a factor in the making of the Great Depression in the United States. Cooperation in sharing available resources, in conducting research and development of new sources, and in conserving energy are certainly essential actions for Americans and their Western European and Japanese allies.

International implications of the scenarios

Let us turn now to the three scenarios discussed earlier in the book and probe their foreign policy costs and benefits. Our purpose here is to examine the interplay between our energy policy options and what seem to be existing U.S. foreign policy aims. We do not intend to suggest that energy concerns necessarily supersede foreign policy, or vice versa. The salient point, it seems to us, is that whatever energy policy Americans ultimately choose, it will be shaped in part by events in the world which are beyond U.S. control.

Historical Growth: In the Domestic Oil and Gas option of this scenario, U.S. oil imports would decline from the 1973 level of 6.2 million barrels a day to 4 million in 1985, with gas imports of about 2 trillion cubic feet a day. At this level of oil imports, the stress on the world oil market from U.S. demands would ease, thereby moderating somewhat the attendant foreign policy problems of relations with Europe and Japan, international economic instability, and poverty in the developing countries. In addition, oil imports of 4 million barrels a day are sufficiently small that the United States would have the option of buying only from the more secure sources if it so chose.

At least 1 million barrels a day under this approach could probably come from Canada (official Canadian estimates peg oil exports to the United States in 1985 at 2 to 2.5 million barrels a day) with an additional 0.5 million coming from Latin America. The remainder could be purchased outside the Western Hemisphere, probably without serious political consequences.

Obviously, this supply option offers the advantage of greater flexibility in foreign policy. Its principal drawback is that domestic expansion may not take place without protecting domestic energy against the prospect of being undercut by less costly foreign sources. We cannot dismiss the fact that huge increases in oil production in the Middle East could take place at *costs* of 20 cents per barrel. While we believe prices will stay in the range of $7 a barrel, an investor cannot ignore the risk of price cuts large enough to ruin investments in domestic oil production that actually cost $7 a barrel. American oil buyers will want to buy oil at a lower price regardless of where it comes from. And they will buy it if the U.S. government does not prevent them from doing so through trade restrictions.

The government could hope that industry would assume the risk and allow embryonic domestic industries such as oil shale and coal gasification, which are essential to achieving this scenario and which are likely to produce higher cost energy (in the range of $7–10 a barrel), to suffer the economic consequences of being no longer competitive.

But the greater likelihood is that the domestic investments simply will not take place. The government could try to provide protection with tarrifs, subsidies (the "Lockheed alternative"), or guaranteeing minimum prices for specific production. But all these actions are protectionism by one name or another, and protectionism is the bane of the multilateral, international trading system from which our economy draws so much sustenance. In the end, the foreign policy benefits of this supply option must be weighed against the potential risks of protectionism, and against its environmental and consumer costs.

The High Import option in the *Historical Growth* scenario is, from the foreign policy vantage point, a high stress option. Interestingly, the High Import option under historical growth conditions is the course which the United States was on before the outbreak of the October War.

In this option, U.S. oil imports would climb to 11 million barrels a day by 1985 and account for almost half the nation's total oil consumption. Gas imports would increase sixfold from their current levels and by 1985 would reach 6 trillion cubic feet (TCF), which many experts consider the upper limits of what is feasible for imported gas from Canada and LNG from abroad.

U.S. oil import demands of this magnitude would put intense upward pressure on world oil prices, thus worsening the plight of the developing countries considerably, unless very generous increases in aid were forthcoming.

It would also strain U.S. relations with Western Europe and Japan and severely test the world monetary system. This option would appear to require great diplomatic restraint and finesse on the part of the United States and its allies, as well as exceptionally close consultation and timely cooperative action to insure that the international economic system stays afloat.

It is, of course, difficult to foresee the international scene or the world economy a decade from now. But by far the most immediate foreign policy problem under the High Import option would be trying to cope with our vulnerability to politically motivated oil cutoffs, especially if the fundamental Arab-Israeli conflict were still unresolved. U.S. imports would steadily increase over the next decade and by 1985 we would need about 6 million barrels of oil a day from Arab nations under this option.

There is a basic question of whether exports in such quantities would be economically available over the next 25 years. But if they were, in order to maintain secure oil supplies the United States would appear to have four alternative courses.

One, it could become necessary to withdraw U.S. support of Israel as the political price for growing volumes of secure oil. We assume that as other, less drastic options exist—and they do—the American people will not pay this price for more oil. Second, in theory the United States could, by military force, invade and occupy an Arab oil producing nation. In practice, such an action is out of the question for a nation concerned with peace in the world; furthermore it risks a war with the Soviet Union and is impractical even without Soviet intervention because of the vulnerability of the oil fields to guerilla action. The United States would probably end up with less oil than it had at the beginning.

A third approach, which the United States seems to be pursuing today, is for this country unilaterally to seek special arrangements with one or more major oil exporter, say Saudi Arabia and Iran. In return for technical and military assistance and preferential treatment for other exports, as well as some form of barter agreement, the United States might get long-term commitments for specified quantities of oil—perhaps 4 to 6 million barrels a day from Saudi Arabia and another 2 to 3 million barrels a day from Iran.

Even if the agreements could be reached, they would signal a major departure by the United States from the principle of multilateralism in international economic relations. Other oil importers could be expected to seek similar agreements, possibly touching off cut-throat competition among Western Europe, Japan, and the United States to tie up the oil output of the Persian Gulf and certainly straining relations among allies. The United States might also find itself deeply committed to the survival of the specific governments of the countries with whom the agreements are signed. Moreover, no such agreements can guarantee prevention of an embargo if there is an impasse in the Arab-Israeli settlement and the United States fails to support the Arab position.

The last alternative would be for the United States to work in close conjunction with the Western Europeans and Japanese—probably through the Organization of Economic Cooperation and Development (OECD) or the energy coordinating group that was the outgrowth of the 1974 Washington conference of oil importing industrial nations—and seek multilateral agreements with the OPEC countries concerning oil supply and price guarantees. However, such agreements are difficult to reach and their track record in other international commodities trade has been very uneven.

It makes sense, under any scenario or growth pattern, to develop an emergency stockpile of oil commensurate with the level of imports from insecure sources. Under the high import option, oil stockpiling is an absolute necessity if the United States is to be assured of a secure oil supply.

According to Brookings: "Storing crude oil in steel tanks is the simplest form of oil stockpiling[c] and the form that can be fitted most easily into the existing supply system." Assuming a world price of $6 a barrel, Brookings reckons that it would cost the United States about $12 billion (at $1.2 billion per year) to create a stockpile that by 1985 could replace 6 million barrels a day of imports for *one year.* (A stockpile to replace Arab imports for 180 days would cost $6 billion, and for 90 days $3 billion.)[15] This certainly is not a cheap alternative. But in the absence of a dramatic change in the Middle Eastern picture, such a stockpile would seem prudent for the United States if it

---

[c] An emergency stockpile would be in addition to stocks normally maintained for commercial purposes. In 1973 such stocks represented about 30 days of normal consumption.

chooses the High Import road of *Historical Growth.* A logical way to pay for it would be through a tariff levied on oil imports.

In the High Nuclear option under *Historical Growth,* the United States would import 6 million barrels of oil a day through the end of the century and gas imports would be 5 TCF by 1985. As we have seen, the nation could be selective up to a point in choosing its sources, if necessary, at this level of imports. A mixture of oil imports from secure sources could account for 4 million barrels a day. The other 2 million could be backed up by stockpiles at roughly one-third the costs discussed above.

To this point we have discussed the myriad problems of constantly adjusting to the perturbations of the world oil market. But limiting oil imports by pushing the rapid expansion of nuclear power raises another energy-related foreign policy concern—the proliferation of nuclear weapons.

The nuclear genie is out and ready to spread. The commercial application of nuclear power for electricity is on the verge of rapid expansion throughout the world. In 1973 there were 67 nuclear power plants operating outside of the United States in 29 different countries, with a total capacity of 161,000 megawatts.[16] The United Arab Republic and Israel now also seem destined to join the club. As the number of nuclear plants grows, the traffic in nuclear materials will grow too. The problem, in brief, is that it is a relatively simple matter for a government to take nuclear material from a civilian nuclear program and make crude nuclear weapons.

The time has come for a concentrated effort to reduce this awesome threat to humanity. In a first step toward trying to stem the further proliferation of nuclear weapons, we urge that the nuclear exporting nations (United States, Russia, Canada, France, Great Britain, India, et al.) seek agreement not to export additional nuclear fuels and equipment, either civilian or military, to the strife-torn Middle East, which is so rich in other energy resources. Also agreements should be sought to cover other regions of the world that are as yet untouched by nuclear power. As part of this agreement, the nuclear exporters should provide poor countries with technical and financial assistance so that they can develop more economical, domestic resources —solar, hydroelectric, organic wastes, geothermal, and the like. These multinational efforts to restrict nuclear exports should, of course, be part of a broader effort within the nuclear community of nations to disarm their arsenals and to create adequate safeguard systems (see Chapter 8) to keep their own civilian nuclear materials out of the hands of hijackers and terrorists.

The United States also has the option of unilaterally cutting off its export of nuclear reactor and uranium enrichment, which is substantial; but this action would be of questionable effectiveness because the United States does not have a nuclear monopoly in the world and would-be importers can go elsewhere.

We have raised the nuclear weapons proliferation problem here, under the High Nuclear option, because it seems to us that it is a most pressing problem for humankind and that the United States, rather than being the world's leading nuclear salesman, should take the lead in stopping the spread of the ingredients for making bombs—which includes power plants—before it is too late.

The risk of nuclear annihilation increases in a world where energy demands escalate at historical rates. Thus, reduction of the nuclear weapons proliferation risk, which grows inexorably with the expansion of nuclear power, is a most urgent priority under this option. Of course, the problem arises in every option under all three scenarios and should be addressed by the United States regardless of the energy course it takes.

Technical Fix: The main benefit of this scenario in the foreign policy area is that it moderates all the energy-induced international problems just discussed and gives the United States more room to maneuver in trying to cope with them. The foreign policy costs of this scenario seem to be low but they are difficult to evaluate because we have no real experience with the combination of the 1970s international realities and a slower energy consumption growth rate.

Under the *Technical Fix* Self-Sufficiency option, the United States would import only 2 million barrels of oil a day by 1985. If need be, all of it could come from relatively secure Western Hemisphere sources, and special trade arrangements would probably be unnecessary. The key question to ask about this option is, why reduce imports to only 2 million barrels a day? Is this a case of oil security overkill? The foreign policy benefits of not importing an additional 2 to 3 million barrels seem slight. Hence, the environmental and consumer costs of producing those additional 2 to 3 million barrels at home rather than buying them abroad should be carefully weighed.

Another question arises under the Self-Sufficiency option of *Technical Fix.* Having virtually withdrawn from the world oil market, will the United States tend to go it alone in other areas of international concern that yield only to multilateral solutions—currency, trade, and capital flows? If it does, the tendency must be resisted, for the economic consequences of this approach would be profound. Finally, the problem of that low *cost* Middle Eastern oil recurs. If they decide to compete against some of our more costly sources, how will the United States respond?

Under the Environmental Protection option of *Technical Fix,* the United States would import 5 million barrels of oil a day by 1985 and a fairly small amount of gas—3 TCF. The foreign policy problems appear to be routine. The oil imports would account for roughly one-quarter of the nation's total oil requirements. Imports from Arab nations need not exceed current levels. It would be prudent to develop modest stockpiles. The United States would be free to buy oil from a diverse group of relatively friendly nations under this option.

Zero Energy Growth: Between the present and 1985, *Zero Energy Growth* (*ZEG*) takes the same track as the *Technical Fix* scenario. Peering ahead into the near future of international political and economic events is, at the very best, a speculative and murky endeavor. And in the more distant future, the world context becomes extremely opaque. For this reason, we shall not try to evaluate the international implications of *ZEG* except to suggest that the U.S. example of achieving stability in energy consumption is bound to have major implications for the rest of the world. One potential benefit looms very large—the United States will be consuming a smaller share of the world's finite resources under *ZEG,* thereby improving the odds that there will be something left for the majority of the world's peoples who have not yet partaken of the feast.

Project Interdependence

It has been suggested that the world's oil importers should band together into a kind of counter-OPEC and bargain collectively with OPEC in an enfort to bring down prices. We think this proposal impractical for two fundamental reasons. One, major oil importing nations such as France, and possibly Japan, do not seem ready to join such an organization. Second, it could engender an air of confrontation between exporters and importers that would solve nothing, except for making OPEC more unified, and perhaps lead to further supply interruptions. Indeed, supply interruptions or slowdowns could become the producers' bargaining answer to the strike in labor negotiations. And that would be unfortunate.

Instead, a less confrontational and more conciliatory approach would be mutually beneficial. The suddenness of the world oil market's turnabout has obscured, we think, the real points of shared interest between oil exporters and importers.

Take, for example, the matter of the oil companies' operations within the exporting nations. Many of the governments of these countries see their self-interest served by the nationalization of those assets. As for the governments of the oil importing nations, it is important not to permit the takeover of the oil company assets to intrude upon real concerns—oil prices and supply—either by occurring in a way that disrupts production, or by becoming a bargaining chip for the exporters during multinational discussions.

Also, both exporters and importers alike have expressed concern for the plight of the "Fourth World." At the Special Session of the U.N. General Assembly, U.S. Secretary of State Henry Kissinger spoke of "a collective decision to elevate our concern for man's elementary well-being to the highest level" and suggested that the "unprecedented agenda of global consultations in 1974" was a measure of that collective decision. The Shah of Iran has proposed that a fund of $3 billion be created, with the oil countries contributing $1 billion and the industrialized nations the rest, to aid the poorest countries next year. Both sides have acknowledged that if massive and immediate action is not taken, the thousands dying of famine today in west and east Africa and on the Indian subcontinent will become millions. In an age of instant, global communications, it is hard to be an innocent bystander to widespread famine.

The oil exporters and importers could translate these concerns into specific mutual actions on two fronts. They could create a special emergency fund. Also the oil importers, especially the United States, could agree to sell food and fertilizer to Fourth World countries at greatly reduced prices, and the oil exporters could agree to sell them oil at comparably discounted prices. The lines of credit they granted could be coordinated as well.

Another area of shared interest lies in market stability. The governments of the importing nations have a vital interest in seeing that their nations get a dependable supply of oil. Rapid, unplanned gyrations in the level of supply mean disruptions, not the least of which is unemployment. On the other side, of course, the oil exporting nations have a perfectly legitimate interest in getting enough for their oil so that they can develop their still basically poor countries before this nonrenewable resource of theirs is gone. But excessive fluctuations in the flow of oil revenues to them makes efficient, planned development impossible. Also, they need the technological assistance that only the industrialized world can provide.

In addition, both sides are genuinely troubled by international monetary fluctuations. Already stung by Western currency devaluations in recent years, the OPEC nations have a strong interest in currency stability. The large oil riches they now possess have given them an enormous stake in the world's economic well-being. The producing nations want to invest as much as they can in their own countries, but many are building up rather large reserves of money they cannot spend at home. The orderly flow of some of that money back into the industrialized oil importing countries in the form of long term capital investments makes eminently good sense for both sides. For the importing nations, it would help to cushion the currency, balance of payments, and recessionary impacts of the increased oil prices. The exporters would of course earn money from these investments over an extended period of time. They would also want to use some of their growing reserves to buy goods manufactured in the industrial countries—tractors, computers, generators, and so on.

The flood of money into the bank accounts of the oil exporters has caused considerable consternation in Western Europe and the United States. The idea of "them" (which usually means the Arabs) having billions of dollars or pounds, francs, or whatever at their disposal is for some a profoundly unsettling idea. It seems to us, however, that the recent bonanza of the oil exporting countries has given them an enormous stake in the world economy. They suffer right along with everyone else from runaway inflation, currency declines, and shaky capital markets. It makes no sense for the exporting nations—now that they have their enormously increased oil revenues—to try to create economic disorder, because then their reserves will be worth less, their investment opportunities will be diminished and, if things really get bad, something even worse might happen—people might not be able to buy their oil.

In our judgment, the many shared interests of oil exporting and importing nations can best be advanced through systematic, government-to-government, multilateral discussions. The forum or framework—whether it be a new organization or under the auspices of an existing one, such as the United Nations—is far less relevant than that the discussions actually occur.

It should be apparent by now that the international trade in oil can no longer be left entirely to the so-called marketplace—a handful of international oil companies dealing with the governments of the oil exporting nations. The changes that world oil has undergone in four years are profound and represent a fundamental power realignment in the international community.

To put it bluntly, the major oil-rich nations, which formerly were weak and, therefore, did not need to be reckoned with politically, are now world powers by virtue of the financial reserves they have accumulated and the control they exercise over the oil artery to the industrialized West. Only through government-to-government discussions do adjustments to this new, political reality seem possible.

Western Europe, Japan, and the United States need to ease their growing import demands by trimming the fat from their energy budgets, and to begin developing adequate stockpiles of oil. At the same time it would clearly make sense for the oil exporting nations to enter into multilateral discussions with the importing nations. The underlying purpose of such discussions would be to seek agreements that would stabilize the impact of world oil trade.

The agreements envisioned here could include several elements.

As a start, a mutual agreement of general principle on the matter of nationalization could be in order, with the oil importing nations pledging not to interfere in the internal affairs of the exporting nations, and the exporters agreeing to provide reasonable compensation to oil companies for nationalized assets—taking into account both the enormous contribution made by the companies in discovering and developing the oil and the profits earned by the companies from that oil over the years.

In addition, both exporters and importers could agree not to impose artificial barriers to the flow of investment and trade among them. They could also seek agreement on levels of supply that the producers would guarantee to sell at price ceilings which at least did not exceed the January 1, 1974 level. Whether or not the consuming nations could obtain modest reductions in price probably would depend most upon the extent to which they pursue energy conservation at home.

In return, the producing nations could receive guarantees of systematic technological assistance and assurance of a market for the agreed-on volumes of oil exports at prices that would not drop below an agreed floor. The idea would be to reach agreements on the general range of quantities required and prices charged, leaving room for fluctuations as market conditions change. An adjustment mechanism would have to be agreed upon to take into account changing demand patterns, currency devaluations, and inflation.

Obviously such agreements could not happen all at once. Perhaps they would begin with discussions which explore the practical range of price, demand, and supply levels. Then, perhaps informal agreements within very broad limits could be tried, and, if found useful, formal agreements for short periods could be tested. If the approach still proved viable, further adjustments could then be made in the agreement.

The whole process would certainly take a number of years and the discussions between exporters and importers would no doubt grow tedious at times. But even if specific price, supply, and demand agreements were never reached, the exchange of views and identification of common interests in other areas—perhaps international investment and trade flows—could prove to be important to both sides.

Epilogue

While focusing on a subject such as the international aspects of energy policy, it is all too easy to lose perspective. Let us step back for a moment and make a few last observations.

None of the energy policy options laid out in this chapter, except for the High Import case under *Historical Growth,* seems to create foreign policy problems that can be described as insurmountable.

Second, in weighing the international costs and benefits of any policy option, it would be wise to take into account the unpredictability of the world context, and build in as much room for foreign policy maneuver as possible. Events move swiftly and can render obsolete the narrow strictures of yesterday. For example, just 22 years ago the President's Materials Policy Commission framed the world context in terms which sound strangely medieval today.

> The United States . . . is today throwing its might into the task of keeping alive the spirit of Man and helping beat back from the frontiers of the free world everywhere the threats of force and of a new Dark Age which rises from the Communist nations. In defeating this barbarian violence moral values will count most, but they must be supported by an ample material base. . . . The world has seen the narrowness of its escape from the now dead Nazi tyranny and has yet to know the breadth by which it will escape the live Communist one—both materialist threats aimed to destroy moral and spiritual man.[17]

Four years ago any of the following events would have been described by us as wildly improbable.

- A selective Arab embargo that succeeded in keeping oil out of the United States while still getting it into most other countries.
- An American President visiting Damascus and receiving a warm welcome.
- An American President selling a nuclear reactor and material to the United Arab Republic and to Israel, all in the same trip.
- A 515 percent increase in the price of oil that bore no relationship to cost.
- A withdrawal of all Soviet forces from the United Arab Republic upon the orders of the President of the United Arab Republic.
- An American Secretary of State, who is Jewish, negotiating a truce between Arabs and Israelis at their joint request.

Yet they all happened and, one way or another, affected U.S. energy policy.

The cardinal virtues in energy-related international matters boil down to just one: flexibility. And that appears to be attainable through determined moderation in oil imports and imaginative multilateralism.

CHAPTER 8

Energy and the environment

The recognition of environmental damage as a serious consequence of industrial activity, and of energy use in particular, is a relatively recent phenomenon. Air pollution was the first major environmental issue to capture public attention. General awareness of it in this country can probably be traced to the appearance of smog in Los Angeles in the 1940s and the 1948 air pollution episode in Donora, Pennsylvania, in which half the town's population of 14,000 became ill and twenty died. National goals of protecting the public health and welfare were set out in the Clean Air Act of 1963, and made specific in the Clean Air Amendments of 1970. While much progress has been made, clean air is not yet a reality.

The public debate about nuclear power began to capture public attention in the late 1960s when commercial atomic power plants began to become operational. The

United States Atomic Energy Commission, in response to public pressures, has tightened the standards on low level radiation routinely released and that controversy has subsided. More recently, public attention has turned to the questions of reactor safety and nuclear theft—issues that are still unresolved. Moreover, several long term issues, including the question of whether our institutions are capable of managing nuclear power, have not yet been squarely faced.

Land use is also likely to be a major concern in the years ahead. Population growth, urbanization, industrial and commercial expansion, and resource exploitation are forces that are rapidly changing the American landscape —upsetting ecological balances in natural areas, bringing rapid and radical changes to the social fabric in sparsely settled areas, and generally reducing diversity in the land environment.

The problems associated with land use are of prime concern in the undeveloped areas in the Rocky Mountains and in the Atlantic and Pacific coastal zones that face pressure for the development of energy resources. There is a growing consensus that systematic land use planning policies are needed to ensure that valuable land resources are not irretrievably lost to poorly planned development. Unfortunately, as the recent Executive Branch–Congressional struggles on proposed land use legislation indicate, the nation is slow in meeting this challenge.

The use of energy poses an inevitable energy-environment conflict, whatever the supply option. These environmental problems can be remedied to a certain extent with technological controls—at a price. It must be recognized, however, that some risks are inherent in any particular energy technology, because complete environmental control either is too expensive or is impossible to implement. In many areas, however, a great deal of environmental protection can be achieved at a price most people should find acceptable. The critical energy-related environmental questions are: What environmental risks do we as a society feel we can live with? And, what energy sources, growth policies, and environmental protection strategies are likely to be the most effective in avoiding unacceptable risks?

To a certain extent, we in the United States have the luxury of being able to ask what kinds of environmental risks we choose to live with, because, as we have seen in Chapters 2, 3, and 4, we do have distinct supply options for the future. With policies that result in lower growth in the demand for energy, those options become more numerous. We have identified air pollution, nuclear power risks, and land use problems as the central environmental issues for energy policy today. The intensity of problems in each of these areas varies significantly with both the future growth pattern the country chooses, and with the supply mix the country chooses for a particular growth pattern.

In this chapter, we discuss what we consider to be the central environmental issues of energy policy for both fossil fuel and nuclear energy sources. We focus on the major concerns and the need to make comparative judgments. The risks and the benefits of alternative energy supply options ultimately must depend upon value judgments and public decisions. A decision on oil drilling off the Atlantic coast or shale development in Colorado, for example, cannot be made solely on the basis of any calculation of costs and benefits which, at best, reflect only a fraction of the values that people hold dear. A comparative overview of what is happening to the environment, together with some rough estimates of what it will cost to pursue present and possible future environmental protection goals, can help the citizen make these judgments.

Fossil fuels

Fossil fuels have been the mainstay of our energy system, accounting for 95 percent of total U.S. energy consumption in 1973. And even with the most aggressive growth in nuclear power, fossil fuels will dominate energy supply, at least until the turn of the century.

Oil and gas account today for over three-fourths of total U.S. energy consumption. Natural gas is our cleanest major energy source; oil is the most versatile. Petroleum products can be used as fuel in all sectors: transportation, the home, industry, and commerce. But oil and gas may be our scarcest energy sources over the long term.

Coal resources are much more abundant, large enough to meet the present level of total U.S. energy use for several hundred years. Coal is today usually regarded as our dirtiest energy resource. But there is good reason to be hopeful that with an adequate R&D "cleanup" effort, coal could meet a substantial fraction of U.S. energy needs for centuries. A principal advantage of slower energy growth (as in the *Technical Fix* scenario) is that the country can buy time to clean up coal. In stark contrast to a recent U.S. Department of Interior target of tripling coal production by 1985,[1] the *Technical Fix* scenario allows us the option of increasing coal production gradually until 1985—after which the fruits of our R&D effort would enable "clean" coal to account for a major share of energy growth.

Oil shale is one of the nation's more abundant energy sources, with oil in high grade shale (at least 25 gallons per ton) estimated to be the equivalent of the present levels of U.S. oil consumption for 100 years.[2] But the environmental problems of obtaining it with present technology are more difficult and far-reaching than for coal.

Coal mining and oil shale recovery

Coal mining requires a lot of land, whether the coal is mined with surface methods or underground. Up to now, most coal has been mined in Appalachia and the Midwest. But shortages of oil and gas and the rapidly growing demand for low sulfur coal to meet air quality standards have created a rush to develop western coal. Safety and environmental problems of coal mining are qualitatively different in each of these regions.

Underground mining–coal miners' health and safety: There are problems of acid mine drainage and subsidence of the surface in underground coal mining,[3,4] but the principal concern is the health and safety of the men who go down in the mines. In 1973, a total of 131 fatal, and several thousand nonfatal, injuries were reported in U.S. bituminous coal mines, which employ roughly 150,000 miners.[5] The workers in coal mining and preparation industries have suffered more than double the days of disability per million employee hours worked than have such high-injury industries as metal mining and milling, construction, lumber and wood products, and primary metals. Coal mining is thus among the most hazardous occupations in the nation.

How much safer can coal mining be made? A partial answer can be gleaned from a look at the safety records of some major coal producing companies. The range in the safety records for companies in this group is remarkable. As shown in Table 35, the injury rate for underground mining in 1972–73 ranged from a high of about 115 injuries per million man hours to a low of 5.3, with a 1973 industrywide average of 50.2.[6]

Table 35–**Underground mine injury rates for major coal producers**
(Nonfatal injuries per million man hours)

Company	Underground Coal Mine Nonfatal Injury Rate	
	1972	1973
Peabody	53.7	50.2
Consolidation	34.5	19.0
Island Creek	33.1	22.2
Pittston	85.1	73.5
U.S. Steel	5.3	7.0
Bethlehem	9.0	7.9
Eastern Associated	107.6	115.1
Old Ben	49.7	44.3
North American	96.4	89.3

Source: U.S. Department of the Interior, Mining Enforcement and Safety Administration.

U.S. coal mine fatality rates 1966–1973
(Fatalities per million man hours)

Type of Mine	*1966*	*1967*	*1968*	*1969*	*1970*	*1971*	*1972*	*1973*
Surface Mines	0.61	0.55	0.67	0.64	0.59	0.40	0.33	0.30
Underground Mines	1.14	1.04	1.66	0.95	1.21	0.87	0.60	0.50

Sources: National Coal Association, *Bituminous Coal Data 1972*, Washington, 1972. *Coal Age*, Vol. 79, No. 2, February 1974.

Mines owned by the steel companies appear to have the best safety records. It is noteworthy that the accident rate at the low end of this range is low for any occupation. It is less than the 1970 rate for such industries as wholesale and retail trade (11.3), real estate (11.4), institutions for higher education (7.4), and the federal government (6.6).[7] The underground fatality and injury rates for the safer coal companies are frequently much lower than for many surface mines. There is no inherent reason why the underground mining of coal cannot be made a reasonably safe occupation.

There has been a significant improvement in the accident record since the passage of the Federal Coal Mine Health and Safety Act, as shown in Table 35, but the fatal injury rate for underground mining in the United States is still two to four times that of Great Britain.[8] Wide variation among U.S. companies and the much better record of British coal mining strongly suggests the high human cost being suffered by U.S. coal miners is not necessary.

Black lung, a respiratory disease caused by breathing excessive amounts of coal dust, is another occupational hazard of underground coal miners. There are currently about 125,000 cases of this disease in the United States, with an estimated three to four thousand deaths each year in which black lung is the underlying or contributory cause of death.[9] There is no cure. Prevention is the only alternative.

The landmark Federal Coal Mine Health and Safety Act of 1969 provides benefits to black lung victims and makes provisions for the protection of miners from black lung. The total cost of the black lung benefits program is estimated at $8 billion. The Act set dust standards to limit the amount of dust a miner could be exposed to. Any miner whose X-rays disclose signs of black lung (but not a totally disabling amount) is given the right under the Act to transfer to a less dusty job.

If the dust standards of the 1969 Act are strictly enforced, black lung disease can be drastically reduced for new miners beginning employment after implementation of the standards.[10]

Surface mining and the land: About 50 percent of all coal produced today comes from surface mines. The feasibility of reclamation depends greatly on the geology and climate of the region.

In the past, surface mining or "strip mining" has ravaged large areas in the Appalachian region, where more than half the country's total surface mining has occurred. Unreclaimed mine sites have polluted thousands of miles of streams with both acid mine drainage and sediments from continuous erosion. Reclamation efforts have been limited and generally ineffective. It is particularly difficult to reclaim the land in hilly or mountainous areas.

Damage from strip mining leads to economic losses from declining recreational and agricultural opportunities for people in the local area.[11] Even more painful is the damage to neighboring property owners—for example, mudslides from strip mine spoil banks burying their land and houses. Kentucky and West Virginia—the states where most of the damage from strip mining has occurred—both enacted legislation several years ago requiring proper reclamation, grading and revegetation of strip mined land. But a law on the books does not guarantee effective reclamation, since enforcement is in many cases sadly lacking. Pennsylvania is the principal exception; it has implemented a program regulating strip mining that appears to have satisfied the interests of both environmentalists and coal companies.

The Midwest: The coalfields of Illinois, Indiana, and western Kentucky, which accounted for about 30 percent of surface coal production in 1971, are in an area of relatively flat terrain and rainfall of 30 to 50 inches per year. These geographic facts permit reclamation at moderate cost—if it is preplanned in the mining operations. However, coal in the Midwest has a higher sulfur content, so that coal development in this area must be limited until sulfur oxide emission controls are put into use.

Much of this midwestern land that has been strip mined has not been reclaimed, even though it could be good farm land. Illinois, for example, has about 40,000 acres of surface mined lands which have not been put back into productive use.[12] Land having a "moonscape" appearance is all too common. What was once some of the country's best farm land is now often capable of only one agricultural use—pasture. If strong reclamation laws are adopted that require replacing topsoil in mined areas, as successfully demonstrated in Europe,[13] the sharp decline in productivity of reclaimed land can be avoided.

The West: In the western states the problem of reclamation is more serious. In many areas with light rainfall, where soil cannot retain moisture, reclamation is not feasible at all, according to a study conducted for this Project by the National Academy of Sciences.[14] But if the best available technologies were applied, stable revegetation might be established in western areas favored with good soil and adequate rainfall. Favorable conditions appear to exist in the mixed grass region of the Northern Great Plains and the Ponderosa pine and mountain shrub zones of the Rockies. These areas contain about 60 percent of the surface mineable coal reserves of the western United States. Reclamation, where feasible, would add only a few cents a ton to the price of coal, but its success would depend on an intensive, coordinated effort over many years. No such efforts have as yet been undertaken in this country. Successful reclamation requires strong new laws and enforcement mechanisms that work.

The problems posed by surface mining in the West also affect water supply. In most of the near-surface coal beds in the West, the coal seams trap underground water. Removal of the coal seam would disrupt this aquifer and diminish the region's water supply.

Development of western coal would also intensify existing competition for scarce water resources. Most coal deposits in the western United States that can be mined by surface methods are located in water-short areas. While the amount of water required directly for mining is relatively small, the water required to support secondary development, including the influx of people from the industrialization of the area, is not small. Coal burning power plants and coal gasification or liquefaction plants would be substantial users of water.[14] This water demand would compete with other uses, including agriculture, other industries, and recreation. Water requirements for large coal conversion operations in the arid West may well exceed the limits of available water supplies in the area.

One method of supplying large quantities of water to coal conversion would be to import water by huge aqueducts over hundreds of miles, as suggested in the U.S. Department of Interior's Montana–Wyoming Aqueduct Study.[15] But the transport of large quantities of water into any western regions might lead to substantial population growth in areas that are today committed to low density uses such as ranching, wildlife preservation, and outdoor recreation. Once fixed, a system of interbasin water transport preempts other options for water use and limits land use choices as well. Areas distinguished by scenic, recreational and cultural amenities should be preserved whenever possible, and water resource planning should include careful protection of unique water and land values from harmful development.

An alternative is to separate the mining from the conversion activities by exporting the coal to power plants or coal synthetics plants in other regions with adequate water supplies. The social and demographic changes from mining and reclamation activities alone are much less than in the case where coal conversion technologies are also developed. Modern technology has made it possible to mine 5 million tons of coal per year with about 120 men.

Because oil shale is located primarily in Colorado, Utah and Wyoming, its development problems are similar to those of western coal. But the disposal of the spent shale is an additional serious concern because of the large volumes of waste and the difficulty of reclaiming the spent shale banks. Oil shale recovery may be also limited by water resources. The requirements are greater than for mining coal, but are only one-sixth to one-third as great as those for producing synthetic coal liquids. Water pollution is also a problem.[2] Recovering oil by fracturing and heating shale underground (the *in-situ* method) offers the prospect of alleviating some, but not all, of these problems, the most far-reaching of which is the impact of industrialization on a sparsely settled region.

In the case of coal, the problems of regional development and water scarcity can be, in large measure, avoided by restricting coal operations to resource extraction, with the coal shipped elsewhere for conversion. But this option does not exist for oil shale, since the oil must be extracted from the shale at the mine site. Thus the pressures for regional development are likely to be much greater with oil shale than with coal.

Western coal or shale oil development cannot be considered in isolation from water resource management policy and, indeed, economic growth policies for that region. A framework for dealing with such regional development problems is described Chapter 11.

Offshore oil and gas recovery

Offshore production has been an oil and gas source of growing importance for many years, accounting for about 18 percent of domestic production in 1972. Offshore oil and natural gas have been produced primarily from wells off Texas and Louisiana in the Gulf of Mexico, with limited production off California and Alaska. Possible future Outer Continental Shelf (OCS) development areas are shown in Figure 32.[16]

Oil spills resulting from both accidents and chronic routine discharges are a major environmental concern arising from offshore oil and gas operations, and are the subject of an Energy Policy Project sponsored study.[17] Large accidental spills receive the most publicity, but the chronic low level pollution may have more serious long term consequences. Oil spills from offshore operations frequently affect coastal waters, which are particularly productive marine habitats.

Salt marshes are among the most biologically productive environments on earth and are particularly sensitive to the effects of chronic oil pollution. Estuaries, which serve as valuable marine breeding grounds, are also particularly vulnerable to damage from oil spills.

Unfortunately, the present body of scientific knowledge concerning ecological impacts of oil pollution is not

Figure 32–**Potential oil basins, Outer Continental Shelf**

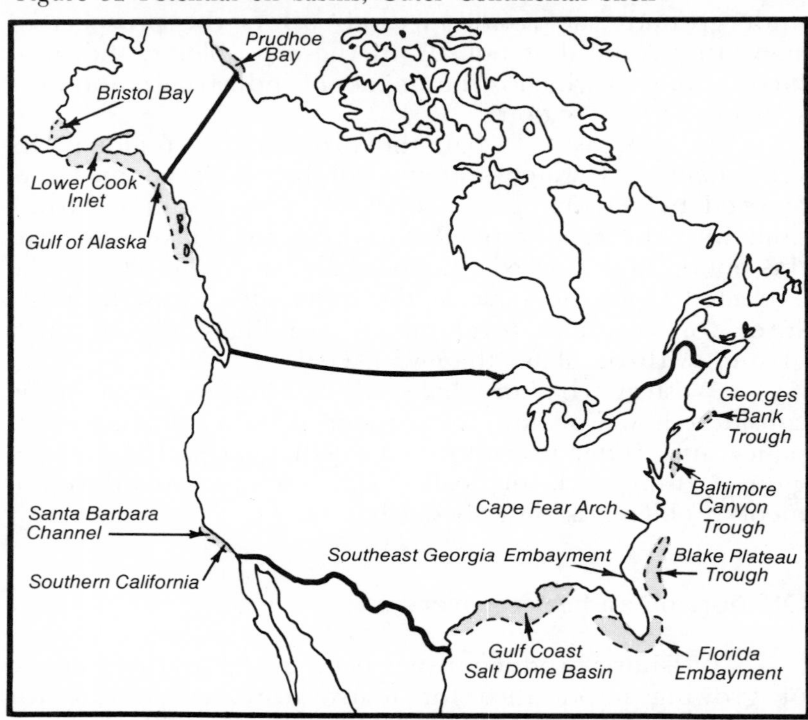

Source: D. E. Kash et al., *Energy Under the Ocean* [16]

adequate to assess the long term environmental impact of oil spills near such potentially vulnerable areas. As oil spill effects may be irreversible, it would be prudent to avoid development in sensitive areas, at least until scientists understand the long term impacts much better. The slower growth of the *Technical Fix* scenario allows a breathing period to wait.

Successful offshore oil and gas development leads inevitably to the need for supporting facilities somewhere onshore. The exploration phase of offshore operations in virgin areas requires sites for construction of drilling rigs and platforms, and the development or improvement of ports to serve oil related vessels. Once production of oil and gas begins, the storage tanks, pipelines, terminals, refineries, and gas processing plants must be ready. In addition, petrochemical complexes and electric power plants may be attracted to the region by the green light for industrialization that offshore drilling signals, and by the availability of oil and gas.

The impact that these onshore developments have on

the fragile coastal environment and economy depends on the character of the region before offshore development, the pace at which development takes place, and the planning process used to accomplish development. Offshore operations can dramatically change the natural character of a coastal region and preempt traditional coastal activities such as fishing, tourism, and recreation. Since 75 percent of the American people live in the 30 coastal states, offshore oil and gas development can have a direct impact on a large segment of the population.

As with western coal and shale oil, offshore oil and gas development must be carried out cautiously, taking into account national, regional, state, and local needs. A balancing of needs and resolution of conflicts requires broad public participation at all planning levels, as well as meaningful communication between the public and the private companies seeking coastal facilities.

Air pollution

The quality of the air we breathe directly affects health and well-being. It is well established that episodes of high air pollution have caused hospital admissions to increase and deaths to occur in especially susceptible groups.[18] Air pollutants can increase either the frequency or severity of respiratory illness such as colds, sore throats, bronchitis and pneumonia, even among the healthiest people. Air pollution has especially aggravating effects on people with asthma, and chronic lung and heart diseases. It can also damage crops, animals and property.

It is impossible to measure precisely the cost of air pollution to the nation's health and welfare, because knowledge of pollution effects is incomplete, and not all adverse effects can be given dollar values. The range of estimates provided by EPA in 1974 of $6 to $18 billion in 1970 alone shows that the problem is indeed serious;[19] this, combined with the available medical evidence, makes the air pollution cleanup effort an urgent national priority.

Recent cleanup efforts appear to have improved air quality, with measurable reductions in the concentration of particles and sulfur dioxide in urban air in recent years.[20] This trend may be reversed if there is a substantial shift in urban power plants from the use of oil to the use of high sulfur coal without adequate pollution controls. Moreover, small particles—a class of pollutants for which there are as yet no adequate controls or standards—may prove to be much more damaging to health than those pollutants for which standards have been set. Data being collected on small particle sulfate pollution over wide regions of the country also suggest that air pollution can no longer be considered solely an urban problem.

Air pollution is a major concern in energy policy decisions because the combustion of fossil fuels is responsible for almost all man-made air pollution, contributing (by weight) in 1969:[21]

- 98 percent of the sulfur dioxide (SO_2)
- 80 percent of the carbon monoxide
- 60 percent of the hydrocarbons
- 40 percent of the particle pollution

Two air pollution control problems of particular importance to energy policy are automotive pollution and pollution from fossil fuel burning electric power plants.

Automobiles in urban areas produce hydrocarbons, carbon monoxide, oxides of nitrogen and lead particles. Problems posed by direct emissions are compounded by the formation of secondary pollutants in atmospheric chemical reactions. Photochemical smog, the product of such reactions, gives rise to the familiar urban haze and to vegetation damage, eye irritation, and respiratory distress.[22]

How bad is automotive air pollution in urban air? Consider carbon monoxide. Studies have shown that at carbon monoxide levels two to three times higher than the level permitted by the air quality standards, heart disease patients have lower tolerance of exercise, and healthy adults show decreased physical and mental performance.[23] A recent study shows that none of the nation's largest cities meets the carbon monoxide standards set by EPA to protect public health.[24]

Beginning with 1968 models, new cars have been required to meet increasingly stringent federal emission standards, as required by the Clean Air Act and its Amendments. However, in areas such as Los Angeles, emission control devices can't do the whole job, and auto travel itself would have to be reduced to meet air quality standards.[25] Improved urban public transit systems could help achieve this goal.

Many automobiles built in 1975 and later will include a pollution control device called the catalytic converter. Besides reducing pollution, catalytic converters improve fuel economy over 1974 models. Unfortunately, automobiles equipped with this device emit sulfate particles through reactions in the converter with sulfur in the fuel.[26] Such sulfates are now thought to be hazardous to health even in low concentrations. While the total amounts of sulfates emitted by automobiles with converters may not be large, the sulfate pollution would be concentrated along heavily populated roadways and could cause a serious hazard in a few years.

Within ten years the widespread use of alternative engines such as the diesel and stratified charge engines could permit automobiles to meet stringent emission standards without the catalytic converter.[27] The diesel, in particular, combines low emissions with a 20 to 30 percent improvement in fuel economy.[28] Such alternatives, combined with the use of smaller, lighter cars, could do much to improve urban air quality and slow the growth in petroleum demand.

For the foreseeable future, one of the most hopeful approaches to controlling air pollution in the cities is to burn less gasoline, since each control device brings problems of its own. Programs to build urban bike paths, to provide better bus service, and to encourage car pooling can be effective in helping to clean urban air—with the added bonus of conserving petroleum.

The emission of sulfur dioxide (SO_2) and particles from fossil fuel burning electric power plants is another major source of harmful pollution. The health effects of SO_2—which range from moderate breathing problems to premature death, among people suffering from heart and lung diseases—appear to occur when SO_2 gas is associated with particles in the air or with other pollutants.[29]

A recent report of the Department of Health, Education and Welfare concluded that adverse health effects occur where SO_2 concentrations are twice the level permitted under the present standards.[30] Moreover, the report cites considerable scientific evidence of discernible health effects at the level of the current standards. The standards thus allow only a slim margin of safety, if any, for protecting human health.

The HEW report also points out that almost all the SO_2 released to the atmosphere is eventually converted to sulfate particles. This is significant because evidence suggests that the adverse health effects of sulfates, for which there are no federal standards, occur at much lower atmospheric concentrations than do SO_2 effects. Thus, in order to deal adequately with the dangers of sulfate pollution,

SO_2 emissions must be controlled much more stringently than even the present standards call for.

Such considerations lead us to adopt the HEW report view that there is no scientific basis for relaxing SO_2 standards. In fact, there is considerable evidence that the standards should be tighter. The real issue is how to enforce the current standards and tighter ones if necessary.

Unfortunately, low sulfur fuels are in short supply. The scarcity of natural gas and the desire to reduce dependence on Mideast oil are leading to power plant conversions to coal—which often means high sulfur coal. While there is a great deal of coal with a sulfur content below 0.7 percent in the ground, it is distant from markets, and its availability over the next few years is limited because coal transport facilities are inadequate and time is needed to open new mines. Plants already burning eastern coal cannot switch to lower sulfur western coals to meet standards without substantial plant modifications.

Some electric power companies believe the best way to meet air quality standards for SO_2 is to use tall stacks to disperse the pollution, except in periods of an atmospheric inversion when a low sulfur fuel would be used. But this approach can spread sulfate pollution—which as we said is harmful at much lower concentrations than SO_2—over a wide area. Data being collected indicate that the entire Northeast, bounded by Boston, Washington, Cincinnati, and Chicago, is already significantly polluted with sulfate particles from fossil fuel combustion.[31] This spread of the sulfur pollution problem also creates acid rain, which can lead, in turn, to changes in leaching rates of soil nutrients, acid buildups in lakes and rivers, corrosion of structures, and adverse effects on plant growth.[32] Evidence grows that dispersion of sulfur from tall stacks is an ineffective pollution control strategy.

The most viable option for meeting air quality goals for SO_2 in this decade, in our opinion, is the installation of flue gas desulfurization (commonly called stack gas scrubbers) in coal fired power plants.

This view, however, is hotly contested by some utility companies. The Chairman of the Tennessee Valley Authority stated in early 1974, "The country's knowledge of scrubbers has not yet progressed to the point where TVA can have any degree of assurance that it is not buying a billion-dollar pig in a poke."[33] This view is supported by Federal

Power Commission officials, who argue that scrubbers fail to meet the reliability requirements of the utility industry.

The Environmental Protection Agency, in an effort to get an up-to-date assessment of the status of SO_2 control technology, held two weeks of hearings on this matter in October 1973. Witnesses included representatives from utilities, trade associations, and manufacturers of pollution control equipment as well as environmental and public interest groups. Some of the principal findings of the EPA hearings panel were as follows:

> • Flue gas desulfurization (FGD) technology must be installed on large numbers of power plants if SO_2 emission requirements adopted pursuant to the Clean Air Act are to be met in the 1970s. . . .
>
> • With several noteworthy exceptions, the electric utility industry has not aggressively sought out solutions to the problems they argue exist with FGD technology. . . .
>
> • Although most utility witnesses testified that FGD technology was unreliable, that it created a difficult sludge disposal problem, and that it cost too much, the hearing panel finds, on the basis of utility and FGD vendor testimony, that the alleged problems can be, and have been, solved at a reasonable cost. The reliability of both throwaway-product and saleable-product FGD systems has been sufficiently demonstrated on full scale units to warrant widespread commitments to FGD systems for SO_2 control at coal and oil fired power plants. . . .[34]

This assessment suggests that there are no fundamental obstacles to the implementation of scrubbers. The problems that utilities are encountering are not different in kind from the difficulties in reliable operation of new power plants. The difference is that utility officials are reluctant to buy this expensive equipment as long as they can avoid it.

Meeting air quality standards will require a widespread commitment to the use of scrubbers. Variances and delays in compliance with the standards may be necessary in some cases, because of the lead times for building and installing scrubber systems. In our view, if such variances are needed, they should be coupled with a firm commitment to install scrubbers at the earliest feasible date.

For the longer term, there are several promising clean coal technologies on the horizon to remove the sulfur and ash from the coal before or during combustion. Options such as solvent refining of coal, coal gasification, and fluidized bed combustion may be commercially available, beginning in the 1980s, if adequate funds are devoted to their development.

Small particle air pollution

Substantial evidence is accumulating that particles too small to see directly (smaller than a couple of microns) are an especially hazardous component of air pollution.[35] Yet there are no air quality standards requiring that small particles be controlled, nor has the technology to control them adequately yet been developed.

When people inhale air, many of these small particles pass through the natural filters of the respiratory system and become lodged in the deep recesses of the lung.[36] Small particles can carry harmful materials such as adsorbed gases, sulfates, and trace metals deep into the lung. Insoluble particles deposited there can remain for weeks, months, or even years. In contrast, larger particles are trapped by the filters in the upper respiratory zone and are rather quickly removed.

Small particles in the air exacerbate the sulfur dioxide (SO_2) pollution problem in two ways. First, in humid air, soluble salt particles form droplets, and SO_2 in the air dissolves in these droplets. Small droplets inhaled can carry SO_2 deep into the lung, where it has an irritant effect on lung tissue. Second, when these particles contain metallic compounds (including lead, vanadium, manganese, iron, and copper) the dissolved SO_2 can be rapidly converted to much more irritating sulfates and sulfuric acid.[30]

Small particle sulfates produced from SO_2 appear to cause adverse effects at concentrations much lower than the level considered harmful for SO_2 itself. Average annual sulfate concentrations in some rural areas appear to be close to the level which is harmful to human health. The toxicity of sulfates depends on the chemical form and the total sulfate concentration. Good information on sulfates, just now beginning to become available, is needed to fully assess the public health implications of this class of pollutants.[37]

Particles bearing trace elements not only facilitate the formation of sulfates from SO_2 but often present serious hazards themselves. Recent studies have found that many of these trace elements are found especially concentrated in small particles.[38] Living organisms possess little or no tolerance for many trace elements. When carried into the deep lung by small particles, between 50 and 80 percent of these trace elements are usually absorbed into the bloodstream. The lung is thus the major gateway to the bloodstream for toxic elements in airborne particles.

Coal and oil combustion are both major sources of

small particles, but the chemical composition—and the toxicity to man—may vary significantly between coal and oil, as well as from one source to another. Fossil fueled power plants and automobiles both emit particles bearing toxic trace elements.[a]

A substantial fraction of the small particle pollution is formed in chemical reactions among airborne pollutants. In Los Angeles about 35 percent of the particle matter in the air is produced in the atmosphere.[35] Different primary pollutants can work together to form worse secondary pollution. Sulfates can be produced from SO_2 in the atmosphere through a photochemical process involving oxides of nitrogen and hydrocarbons, derived primarily from automotive emissions and other mobile sources.[30]

The present air quality standards for particles, which are stated in terms of total particle weight, do not effectively control small particle pollution. In fact it is quite possible for air quality to decline in a particular geographical region despite a drop in total particle mass, if the mass of material in the small particle range increases. Commonly used electrostatic precipitators and scrubbers may be as much as 99 percent efficient at removing particles by weight, yet remain inadequate for removing small particles.

This kind of air pollution stands out as a major unsolved environmental health hazard relating to fossil fuels. It is a matter of urgency that the Environmental Protection Agency develop and implement standards for small particles and promote R & D on their control. An effective control program must contend with the diverse sources of small particles and the variations in toxicity from one type of particle pollution to another.

There are several hopeful control technologies that are already advanced to the prototype stage, where experiments are being conducted to determine their efficiency

---

[a] Residual fuel oil burned in power plants often produces small particles containing vanadium, chromium, nickel, iron, and copper. Most of the trace metals in gasoline are associated with fuel additives or are introduced during the distribution process by pipelines and storage tanks. Lead is the most concentrated trace element in gasoline, and the EPA has ordered that it be phased out of gasoline over the next several years.

Several trace metals (lead, antimony, zinc, mercury, arsenic, cadmuim, nickel) appear to be associated with the inorganic sulfur in coal, so possible health hazards arising from these trace elements would tend to be greater for high sulfur coal, like much of that found in the Midwest. Two other trace elements, beryllium and selenium, generally occur in greatest concentrations with Appalachian coal, and in lowest concentrations with western coal.[39] Regional variations of some important trace elements (like fluorine) have not yet been assessed. Such regional differences may be important for coal development planning.

and economic feasibility. High efficiency electrostatic precipitators, fabric filters, and improved scrubbers are especially promising for the control of particle emissions. Research is also underway on methods to reduce metallic particle pollution by removing these metals from coal prior to burning.[40]

To limit sulfate formation, SO_2 standards much stricter than those currently required will be needed.[41] Modified scrubbers look promising for this purpose in the short run, and precombustion sulfur removal from fuel is a hopeful approach for the longer term. Another approach to limiting sulfate formation is to control the automotive pollutants and metallic particles that promote sulfate formation in the atmosphere.

Research and development on all aspects of small particle control must be pursued aggressively to bring adequate control technologies quickly into commercial use. New standards for small particles are urgently needed along with these control technologies to protect public health and welfare.

Global limits to energy use

Man's energy use can affect the global climate. These matters are speculative but they are a source of concern because, if the problems are real, they may limit the quantity of fossil fuels that can be safely consumed on this planet.The most immediate global concerns stem from the atmospheric buildup of carbon dioxide (CO_2) and small particles, both of which are linked to the burning of fossil fuels.

Carbon dioxide in the atmosphere strongly absorbs the heat radiation that the earth returns to space, balancing incoming solar radiation. This absorption process, known as the "greenhouse effect," helps to keep us warm on the earth's surface. Additional CO_2 in the global atmosphere appears to have the effect of heating the earth's surface and cooling the upper atmosphere.[42] Heating and cooling effects which are rather small may have far-reaching climatic consequences. Over the last century, the CO_2 level in the atmosphere has risen 10 percent. This buildup is generally believed to be due to the burning of fossil fuels. If recent trends in global fossil fuel consumption persist, the CO_2 level could go up another 20 percent by 2000.[43]

Particles in the atmosphere may have a contrary

effect on climate. A dust layer of particles in the atmosphere tends to cool the earth's surface by blocking out sunlight; it may also have the effect of warming the air where the dust layer is suspended. The surface cooling effect is likely to be greater near the poles than near the equator, since at high latitudes sunlight must travel through more dust laden air before reaching the ground.[43]

Major sources of particles in the air include sea salt, volcanic activity, windblown dust from natural processes, agricultural activity and construction, and combustion processes—especially the burning of fossil fuels. Particles arising in combustion are especially worrisome because many are small and tend to remain suspended in the air a long time. Unless adequate controls are developed, their contribution will grow with the worldwide growth in fossil fuel use. One estimate is that by the year 2000 the global burden of particles less than five microns in diameter may increase by 25 percent from fossil fuel combustion alone.[44]

Recent research indicates that additional CO_2 and particles in the atmosphere, acting in concert, could be factors in trends toward drought in lands near the equator, where survival is dependent on the monsoon rains. One climatologist finds that the decrease in rainfall in the Sahelian zone south of the Sahara desert, where drought and famine are now in the fifth year, correlates with what could be expected from the increases in atmospheric CO_2 and particles.[45] This "Sahelian effect" indicates that climate may be changing even today as a consequence of man's energy use.

A slight global heating or cooling, by way of its effect on the relatively thin Arctic sea ice pack, could trigger large scale climatic change. The cooling effect of particles in the atmosphere might possibly lead to another ice age, if initial cooling causes the Arctic sea ice to grow.[43] The white ice absorbs less sunlight than the dark water displaced, so that the initial growth of the ice pack would lead to further cooling, then more growth, and so on.

Some scientists fear the opposite effect: that the increased surface heating from CO_2 would prove more important than the cooling from particles. In that case, a sequence of events leading to a complete melting of Arctic sea ice could develop. The complete melting of Arctic sea ice could bring about climatic changes that would result in widespread disruption of agriculture. On a longer time scale it could cause a complete melting of the Greenland ice cap, which contains enough water to raise the world's mean sea level more than twenty feet.[43] While such melting may take a hundred years or more, it could be difficult to stop, once it started.

This climatic problem is coupled to the public health problems caused by small particles. If technology brings small particle pollution under control, then CO_2 may emerge as more important than particles in affecting climatic change. The burning of fossil fuels inevitably places an extra load of CO_2 in the earth's atmosphere. It is difficult to conceive of a technical solution to this problem.

Correlations between energy use and climatic change are not well established. However, for the first time in history, man's activities appear related to substantial changes in important indicators of global climatic change. By the year 2000, further increases of 20 to 25 percent in the atmospheric load of CO_2 and particles may arise from fossil fuel use. Some crude but reasonable models suggest that these increases could lead to major climatic changes. Vigorous support for research on climatic implications of energy use should be part of any national energy policy. Moreover, the potential for man's triggering climatic change provides one motive for seriously exploring at this time zero energy growth policies in this country and in other developed countries of the world.

Overview of fossil fuel environmental problems

Our analysis of specific fossil fuel problems leads us to the following conclusions and recommendations:

Coal Extraction: If the nation is to count on coal as an energy resource for the long term, we must solve critical health and safety problems for underground coal mining, and solve surface mining problems as well. A comparison of safety statistics among U.S. mines, and U.S. to British experience, suggests that coal mining can be made far safer than it is today. Moreover, strict enforcement of dust standards set in the 1969 Coal Mine Health and Safety Act should lead to a drastic reduction in the incidence of black lung disease among new miners. Making mines much safer is both possible and essential.

Coal surface mining reclamation problems are qualitatively different in each of the three major regions: Appalachia, the Midwest and the West; and the problems also differ within the regions. Water resource availability is a limiting factor for development of coal conversion industries in the West.

Surface mining should be limited to sites where reclamation is feasible, and highly valued natural areas should, in any case, be exempt. Coal mining should be carried out only if it is in accord with overall demographic and economic growth plans for a region. Chapter 11 discusses the institutional issues posed by these regional development problems in the context of decision making on federal resources.

Rehabilitation plans must be developed before any surface mining operations begin. Strong state and national reclamation laws and strict enforcement are needed to provide guidelines for those plans. Strict long term monitoring of reclamation attempts is also critical. This regulatory action should be combined with a severance tax to be paid by coal companies, in partial compensation for diminished land values from mining.

Offshore oil and gas recovery: Since very little is known about the long term ecological effects of chronic oil pollution, it is vital to identify and protect especially productive and potentially vulnerable estuaries, salt marshes, and fish and wildlife habitats. It should be the rule to avoid new development near these areas until it is evident that the biological systems of the areas are resilient, and that technical and legal protection of their character is assured.

Careful assessment and planning for onshore impacts of offshore oil and gas development are essential. Federal leases should require that offshore operators comply with state and local land use management plans; development of these plans must precede lease sales in virgin areas. States should be provided with federal funding and technical assistance, through the Coastal Zone Management Act of 1972, to assure that the planning job will be done and done well.

Air pollution: The best scientific evidence suggests that there is little or no margin of safety in the primary air quality standards that federal law has established. The standards must be maintained if the goals of protecting public health set out in the Clean Air Act are to be achieved.

One of the most pressing near term problems is the control of SO_2 emissions at fossil fuel burning power plants. A widespread commitment to stack gas scrubbers is the best available option for meeting air quality goals. A sulfur pollution tax may be effective in providing financial incentives to encourage industry to comply.[46]

Utility growth plans make the SO_2 control problem harder to handle. At the end of 1973, utilities had plans for 164,000 megawatts of new fossil fuel generating capacity,[47] which amounts to more than 35 percent of today's total generating capacity. Getting off the historical growth curve, with energy conservation strategies outlined in Chapters 3 and 4, would do much to alleviate the SO_2 control problem. A switch to peak power pricing policies would also help.

The problem of air pollution from autos requires a combined strategy: first, to develop emission control devices in accordance with the Clean Air Act standards; and second, as part of a national energy conservation effort, to encourage the use of small cars and to improve public transit. This is another area where a pollution tax, levied at the time of the auto purchase and at periodic inspections, could spur the industry to move quickly toward developing better controls.[46]

For small particles, which are an especially hazardous component of air pollution, there are no air quality standards, and adequate controls are not available. It is a matter of high priority to gain a better understanding of the health effects of small particles, to set standards, and to develop control strategies.

Research and development efforts on new fossil fuel technologies such as coal gasification and solvent refining of coal should be geared to making air pollution control a design feature, taking into account not only the goals established in present standards, but also potentially serious problems such as small particles.

Global climatic effects: There are several indications that man's use of fossil fuels is altering the global climate. Such changes may well dictate limits to man's use of fossil fuels. Energy policy considerations require intensive research on these problems in order to better understand the limiting factors involved. Concern about global limits is one motivating factor for zero energy growth in the developed nations of the world.

How much pollution control?

It is useful to remember that pollution control is an attempt by society to achieve benefits that are worth more to us than the cost of the controls. We must be sure that the public benefits—whether quantifiable or not—are worth the price. Air pollution controls are needed to protect human health as well as to reduce property damage. In passing the Clean Air Act, Congress decided that protecting human health is important enough to justify the cost of controls. This means that standards must include enough margin of safety so that sensitive population groups, as well as the population as a whole, will not suffer. The margins of safety in the actual standards are not large, but the law requires that they be satisfied whatever the cost.

Obviously it is important to know the cost, because there is a limit to what clean air is worth. As it turns out, the costs of clean air are not high. By far the most significant of these costs are for stack gas control of SO_2 and for reducing automotive pollution. Rough estimates suggest that controlling air pollution adds no more than 10 percent to the price of electricity or to the cost of owning and operating a car.

These two are the largest energy related pollution control costs. They constitute only a fractional increase in the cost of providing energy services and, in fact, are not really new costs to consumers at all. We are merely shifting societal costs to the price of energy instead of paying them in the form of lung disease and other health effects. By reflecting these costs in the price of energy, we encourage conservation, which in turn can lead to less pollution.

Translating a decision about pollution control to a cost vs. benefit formula is difficult, and often misleading, because in many respects the health and environmental benefits represent intangible values. The fact that benefits are not easily translatable into dollars does not make them any less valuable today or for generations to come. An example is a development that involves irreversible commitments and where the benefits of restricting development are intangible, such as preserving a wilderness area for outdoor recreation. Who can know how much such a wilderness area would be worth to future generations? Certainly on a discounted cash flow basis, very little would be worth saving.

As we have seen, there are only small margins of safety, if any, in the air pollution standards we now have. In the future, as we learn more about health effects of pollution, it is likely we will find adverse effects at even lower levels. To protect public health in the case of sulfate pollution, for example, may require controlling SO_2 to a much higher degree than the present standards call for. This may, of course, drive up the cost of pollution control, but it would be done to drive down the illnesses and even deaths of human beings. With energy conservation policies and aggressive pursuit of R&D on alternative environmental control methods, it should be possible to protect human health and lives and also reduce the cost of these controls.

Fossil fuels: limiting factors and energy growth

We have described a number of environmental problems associated with fossil fuels and have specified actions that will help alleviate them. It must be remembered, however, that there are limits to our ability to solve these problems, and that it will take time to do what can be done. The rate at which energy demand grows is thus very important.

The adverse impacts of resource development —whether it be coal or oil shale in the West, or offshore oil and gas on the Atlantic or Pacific coasts—can be mitigated through good planning, but they cannot be eliminated. With each of these fuels, substantial industrial development could follow resource extraction in areas that are now largely devoted to recreation, fishing, ranching, wildlife preservation, and the like. Fossil fuel development and related industrialization will bring inevitable conflicts between those who welcome the changes and those who prefer the status quo. With high growth in energy demand, these conflicts must be resolved, for the most part, in favor of development.

Unfortunately, development tends to be irreversible. When we opt for development of fuel resources, we lose another finite resource—undeveloped land. We also bring substantial changes into people's lives where the development takes place. With reduced energy growth, as in the *Technical Fix* scenario, there is enough leeway to consider the case for nondevelopment in particular areas. It is possible to avoid areas where wilderness, recreation, or traditional life styles are highly valued and instead choose areas where fossil fuel and industrial development is generally welcomed.

Wilderness areas are a diminishing resource. Their value cannot be assessed in traditional cost-benefit terms, but it is bound to increase. The demand for wilderness recreation will grow with population and incomes, while the wilderness itself remains the same or shrinks.

Air pollution problems appear to be controllable

through "technical fixes" of one kind or another. But we are, in a sense, chasing a moving target as we implement controls in the face of worsening pollution from greater energy use. Moreover, new knowledge tells us that present standards themselves are inadequate. A program to control small particles—a dominant unsolved air pollution problem—will be much more complicated than existing programs to control other pollutants. It will take years to accomplish. In the face of these difficulties, energy conservation becomes an important part of an environmental protection program. It makes the development of effective controls more manageable and less expensive.

Finally, we have seen that global climatic problems are closely tied to the use of fossil fuels. At least in the case of CO_2, there may be no way to avoid a buildup in the atmosphere, short of restricting the use of fossil fuels. Curtailed energy growth has an important role both in buying time to implement pollution controls and in avoiding some of the limiting problems posed by fossil fuel use.

As to the most serious land use concerns, the growth rate in overall energy consumption is crucial. If we continue historical growth trends, we can be sure a massive development of the coastal zones and the Rocky Mountain resources will be necessary. With the lower rate of growth in the *Technical Fix* scenario, we can do without growth in two of our four troublesome sources of energy (the other two are nuclear power and imports). Hence, the natural setting of the Rocky Mountains and the coastal zones will be candidates for preservation or at least very slow regional development. These are the essential trade-offs the nation will need to make; but without a policy of energy conservation, both these areas will inevitably be fully developed, and important environmental values will be lost in the next decade.

Nuclear energy

Nuclear energy offers the potential for meeting a significant fraction of our energy needs far into the future. If the breeder reactor is successfully developed, low cost U.S. uranium resources could meet electric energy needs for thousands of years.

Nuclear power offers a potential alternative to our present heavy reliance on oil and gas (now accounting for more than three quarters of total U.S. energy use). Over the longer term, nuclear power might even displace liquid fossil fuels in transportation, either through electrified transportation or through hydrogen fuel produced at the power plant.

There are short term foreign policy benefits in pursuing the nuclear electric course, since nuclear power can make up much of the growth in electric power that otherwise might require increased oil imports. Further, some of today's oil fired power plants could be replaced by nuclear plants, freeing some oil for other purposes.

Nuclear power has significant advantages in terms of air pollution and land use. Nuclear power produces no chemical air pollution, and in normal operations, radioactive emissions can be kept to very low levels. The clean air benefit of nuclear power plants is especially important for the next decade or so, while we learn to burn coal in an environmentally acceptable manner. The land use advantage with nuclear power is also significant, since coal surface mining requires about 85 times as much land as uranium ore mining for today's light water reactors[48] to produce an equivalent amount of electricity.[b] Nuclear energy also offers an escape from some of the global climatic problems of fossil fuels. These potential benefits have spurred the country to emphasize nuclear fission as the major candidate for long term energy growth.

Yet harnessing the atom involves the production of deadly radioactive materials. Because of the danger, nuclear power development has emphasized safety from the beginning. As a result, there has not been a single nuclear related fatality in the U.S. civilian nuclear power program. Yet in spite of this record, doubts remain about the adequacy of safeguards and controls built into nuclear systems to protect the public.

The critical questions are:

- Do nuclear control systems provide the needed safety?
- Will they withstand the test of time?
- Are nuclear facilities secure from acts of violence?

Nuclear control systems: design vs. performance

Nuclear systems contain many safeguards to prevent accidents. These safeguards provide multiple barriers be-

[b] But if the coal surface mined to run today's uranium enrichment plants is taken into account, the advantage is only 15-fold in favor of nuclear power. Future enrichment plants need not run on coal fired electricity, and new enrichment technologies may require substantially less electric energy input.

tween radioactive materials and the outside environment, and are designed to be sufficient to cope with the worst accidents considered at all plausible to occur (called "design basis" accidents). Other accidents are conceivable but are not safeguarded against, because they are judged by experts to have such a small likelihood of occurring (say, less than one chance in a million per plant per year) that they do not warrant the high cost of extra protection.

Concern about nuclear safety systems centers on the following questions: Will safety systems work as designed? If so, will they avoid a major accident? Do the "design basis" accidents represent all accidents that should be guarded against? These questions arise with regard to reactor safety, spent fuel transport, and quality control problems.

Reactor safety: Most of the reactor safety controversy has focused on whether light water reactors are adequately designed to prevent a loss of cooling water from the reactor core, followed by a meltdown of the fuel and a subsequent release of radioactivity. Such an accident could be initiated by the rupture of one of the main pipes carrying cooling water to or from the reactor core. These pipes, 2–3 feet in diameter and made of steel 2–3 inches thick, are designed to high standards.

However, a loss of coolant accident initiated by a major pipe rupture is not impossible. In the event of such a major pipe rupture, the emergency core cooling system (ECCS) is designed to deliver backup cooling water to stop the accident from becoming a hazard to the public. If a rupture in the pipe occurred, *and* the ECCS did not work, the heat generated would be great enough to melt the fuel within minutes. If the fuel melts, a chain of events could lead to radioactivity's escaping all through the containment structures with potentially catastrophic environmental consequences. Ultimately, the fuel would melt down through all man-made structures.

A critical question is how likely such a pipe rupture and the simultaneous failure of the emergency cooling system may be. No such accident has yet occurred, but we have only a few years' operating experience with a relatively small number of reactors. Failure probabilities for these pipes and other nuclear components are not well established, but the AEC has provided an estimate of one chance in twenty thousand per reactor per year, for the probability of such a pipe rupture.[49]

Because of the high standards set for nuclear piping, the AEC believes this estimate to be conservative. An estimate used in a Westinghouse risk assessment study is one chance in ten thousand per reactor per year for such a pipe rupture.[50] These numbers mean the emergency core cooling system might be called upon to perform its functions once every decade or two if there were 1,000 reactors, as projected for the year 2000 in the *Historical Growth* scenario.

As matters stand now, it is argued by critics of the program that the emergency system may not work and may not prevent a serious accident if it did.[51,52] Some nuclear critics argue that other serious accidents not considered likely enough by the AEC to be safeguarded against—such as a rupture of the reactor vessel itself—may not actually be so unlikely.[53] In case of reactor vessel rupture, the emergency cooling system would not prevent the escape of massive radioactivity.

The importance of effective safety systems is dramatized by showing what could happen if such systems fail when an accident occurs. One unlikely, but conceivable, accident sequence would involve a loss of cooling water, a partial meltdown, and a breach of the reactor containment structure.[c] If such an accident occurred near a metropolitan area, 100,000 or more people could be exposed to radiation at dangerous levels. The likely consequences would be:

- 1,200 deaths from acute exposure and cancer[50]
- 10,000 cases of thyroid cancer[50]
- 6,000 genetic defects over the next five generations[54]

Nuclear reactors are designed with great care to keep the chances of catastrophic accidents to very low levels. The nub of the reactor safety debate is whether such efforts are in fact good enough to ensure public safety. Estimates of the likelihood of a catastrophic accident vary. A forthcoming AEC study of reactor safety risks is reported to indicate that the AEC estimates the chances at less than one in a million per reactor per year.[55] But the report is already being criticized by scientists who doubt the value of such risk assessment[56] and those who believe that major accidents could be expected as often as once a decade.[52]

Spent fuel transport: After nuclear fuel is burned up in a reactor, the spent fuel is shipped to a fuel reprocessing

---

[c] In assessing a design basis loss of coolant accident, the AEC assumes that 50 percent of the radioactive iodine and 100 percent of the radioactive noble gases escape to the containment.[49] Here we assume that 25 percent of the iodine and 100 percent of the noble gases also escape the containment.

plant, generally located hundreds of miles away, where plutonium and uranium are separated from the fission product wastes. The spent fuel is shipped by truck or rail in massive shielded casks that are designed to prevent radiation exposure. But this transport link in the nuclear fuel cycle is especially vulnerable to accidents. There will be many shipments (10 to 60 shipments per reactor per year) of intensely radioactive spent fuel over long hauls in the years ahead. The nuclear power industry is growing in such a way that power reactors, fuel reprocessing plants, and long term radioactive waste disposal sites will generally be located considerable distances apart.

The AEC has formulated guidelines for spent fuel cask design to assure that casks can withstand serious accidents. As in the case of reactor safety, the AEC identifies a worst likely accident sequence, and safeguards are engineered so that in a design basis accident public risks are negligible.[57] A recent analysis points out that there may be serious public hazards in a design basis accident involving transport of spent fuel; it calls attention to the potential for the release of radioactive cesium, which the AEC has not considered significant.[58] It is by no means certain that this concern is valid, since it is based on a theoretical analysis. But it does point up the urgency of carrying out experiments with loaded casks under simulated accident conditions to provide a proper basis for risk assessment.

Quality control: In order for engineered safeguards to be effective, they must work properly; quality control in the manufacture and operation of nuclear systems is essential. No serious accident has yet occurred in the nuclear power industry. Nevertheless, general experience with quality control for nuclear power plants is not reassuring. For example, a recent AEC review shows that during 1973 there were 861 "abnormal occurrences" at operating nuclear power plants, about half of which had direct or potential safety significance.[59]

Opinions differ on the seriousness of the problem. The AEC and the industry are considering standardization of reactor design and changes in regulatory procedures to improve quality control.[60] These problems may be merely "growing pains" of a new technology being rapidly commercialized. But they could also be signals of a more basic difficulty. Managing nuclear power requires the highest standards of engineering design and construction. Strict discipline and meticulous attention to detail are required in operating power plants.[61] Some critics argue that such very high standards may not be achievable.[52,62]

Nuclear control systems: the test of time

Managing the deadly wastes generated in nuclear power plants is a task that extends to the indefinite future. Nuclear fission products[d] must be isolated from living organisms for several hundred years. Actinides[e] typically must be isolated for a million years. Both the wastes themselves and contaminated equipment and facilities must be managed with great care. Are we adequately planning for difficulties that lie ahead in this management effort? Are we capable, both technologically and institutionally, of carrying out the necessary long term programs? We consider these questions in several contexts: storing high level radioactive wastes for centuries or millenia, decommissioning obsolete nuclear facilities, and managing a plutonium energy economy.

Long term storage of high level radioactive wastes: At nuclear fuel reprocessing plants, unburned uranium and plutonium are recovered from spent fuel to be reused, and bulk high level radioactive wastes are reduced to a form suitable for long term storage. With present regulations, these wastes must be solidified within five years and the resulting solids must be shipped to a federal repository for disposal within ten years. The volumes of these wastes are not large. The difficulty is that they must be isolated for up to a million years.

The question is, how can we be sure that *any* storage "vault" will remain intact for so long? The wastes must be stored in a place that will be immune from floods, earthquakes, or man-caused intrusions (in case knowledge about the hazards involved is lost). It is very difficult, of course, even to list all potential future uncertainties because it is almost impossible to think in terms of a million-year future. In the final analysis, we will have to ask whether the best we can do in this regard is good enough.

The waste disposal scheme regarded by the AEC as most promising is burial in natural salt deposits. Indeed,

---

[d] Products of the nuclear combustion (fission) process like strontium-90 and cesium-137.

[e] Radioactive materials produced in nonfission processes in reactors, like plutonium-239, with a 24,000 year half life.

until recently, a salt mine in Lyons, Kansas, was the chosen site for the first waste repository; but complications due to prior human activities developed (old oil wells had been drilled through the salt) and now the AEC is looking for another salt mine disposal site. Alternative options for disposal are also being reconsidered.

The salt deposits look attractive because they are very stable formations. Salt deposits have remained free of circulating ground water since they were formed several hundred million years ago; and earthquakes are very infrequent in the areas where these deposits lie. But disposal in salt means that the wastes become irretrievable after a few years, at least with present technologies. Heat from the radioactive wastes would cause the salt to flow plastically around the steel canisters containing the waste, sealing them off completely. The steel canisters themselves would be "eaten away" by the action of the salt within a short time. Thus retrievability is lost, and with it the opportunity to cope with potential unforeseen accidents that might occur in the indefinite future.

The alternative is to maintain the wastes in a completely retrievable condition—in surface "mausolea" for instance. Retrievable wastes could be removed to other sites if circumstances arose that threatened to spill the wastes into the environment. But the price of retrievability is eternal vigilance over the waste stores. It is not easy to choose between these two options. Nontechnical considerations, such as the stability of the future societies and our responsibilities to them, are key factors that must be considered.

Various alternatives to disposal in salt have been proposed—burial in the icecaps of Antarctica,[64] removal to outer space,[65] and disposal in deep cavities created by nuclear explosions,[66] to name a few. Each option has both advantages and drawbacks.[63,67,68] Also, technology is being pursued to convert long lived actinide wastes into fission products that will be hazardous for only a few hundred years, thereby making waste storage a more manageable problem.[63]

For the near term, the AEC has decided to postpone a final choice and "buy time" by building a temporary retrievable surface storage facility where solid wastes would be stored until a long term solution is found.[69] This approach makes a great deal of sense if we can be reasonably sure that a permanent solution will be forthcoming. But this strategy poses a fundamental question: Is it desirable to pursue vigorously the development of nuclear power without a firm assurance that technology is at hand to satisfactorily manage long term hazards?

Decommissioning nuclear facilities: A second difficult issue is what to do with obsolete nuclear facilities such as old reactors and fuel reprocessing plants. The problem is this: Plants must be phased out of operation after twenty to thirty years of operation, primarily because they will have become technologically obsolete. Radioactive contamination makes this a hazardous task. Present plans call for industry to turn over reprocessing plants to the states in which they are located, after they are taken out of service. But the states have no idea of what they will do with them.

A basic difficulty is that reprocessing plants and nuclear reactors are not designed to be decommissioned. The radioactivity hazard is generally worse in reprocessing plants, but it can be a special problem in a reactor which has had an accident, as in the case of the Fermi breeder reactor in Detroit.

A full assessment of the decommissioning problem should be carried out—promptly—before the new reprocessing plants coming on line are fully contaminated, and before reactors proliferate throughout the country. Institutional and economic questions are at least as important as technical ones. Who should be responsible for decommissioning? How should decommissioning be paid for? How will decommissioning costs affect the economics of the nuclear fuel cycle?

Public health implications of a plutonium energy economy: One troublesome legacy of a commitment to nuclear power is that humanity will have to manage plutonium very carefully for the indefinite future. Plutonium, one of the most toxic substances known to man, is produced in the present generation of nuclear reactors and is expected to be the principal fuel for fast breeder reactors.

A small quantity of plutonium-239 smoke deposited in the lung—about one ten-thousandth of an ounce—would kill a person through radiological destruction of lung tissue. A quantity smaller than one-millionth of an ounce would give rise to a substantial risk of lung cancer. The lethal dose of plutonium-239 is smaller than that of most other poisons. It is at least 20,000 times more toxic than cobra venom or potassium cyanide, and 1,000 times more toxic than heroin or modern nerve gases.[70]

Plutonium, unlike many other hazardous radioactive substances, can be contained with rather simple shielding. A leakproof plastic bag is adequate to stop the radiation effects of plutonium from reaching man. But if plutonium enters the body, by breathing a fine (micron-sized) plutonium-bearing dust or smoke, it is very hazardous. Plutonium carried into the deep lung may be immobilized there for one to four years. Part of this is eventually cleared from the body, and part is transferred to the lymph nodes surrounding the lung, to the liver, and to the bones.

One major uncertainty is whether present standards for plutonium in the lung are stringent enough. The present standards (allowing a maximum lung burden of about ten-billionths of an ounce) fail to take into account the fact that the intense radioactivity of a plutonium particle in the lung is wholly absorbed by the tissue adjacent to it, rather than being uniformly distributed over the lung. It is presently uncertain whether this concentration of the exposure results in greater or lesser cancer risk, but if those scientists who argue that this concentration increases the risk are correct, then present standards could be too lax by a hundredfold, a thousandfold,[71] or even more.[72,73]

The potential for public exposures to plutonium occurs at various points in the fuel cycle.[74] Accidents in shipping plutonium or at plutonium processing facilities pose possible opportunities for plutonium releases.[75] One disturbing development concerning potentially serious risks to future generations is the present method of disposing of low level radioactive wastes contaminated with plutonium. Such wastes are now disposed of at shallow commerical radioactive waste "graveyards"—sometimes without even proper records being made. While the plutonium in these wastes is very dilute, the total quantities are significant; they are expected to amount to about six tons a year by the turn of the century.[74] Since these commercial graveyards are not well isolated, it is inevitable for some of the plutonium buried there to eventually reach the open environment. It is imperative to stop these disposal practices, and to give far more care to wastes contaminated with plutonium and other actinides.

Controls against acts of nuclear violence

Another broad class of nuclear problems involves the intentional disruption of nuclear power systems for malicious purposes. The problems range from diversion of nuclear materials to weapons purposes by nation states, to acts of sabotage against nuclear facilities by small terrorist groups.

The potential for acts of nuclear violence is difficult to assess; it is greatly dependent on the prevailing social climate of the nation and the political climate of the world. It has been argued, moreover, that "the toxicity and persistence of radioactive substances has radically altered the power balance between large and small units."[76] An individual acting alone, or in concert with very few others, has the power to do great harm to a whole society by sabotaging nuclear facilities, or using stolen nuclear materials to commit acts of terrorism. Recognizing the importance of this problem area, the Project commissioned a study on one major aspect of nuclear violence—theft of nuclear materials for the purpose of making weapons.[70]

Many of the same materials used for nuclear power generation are the stuff nuclear explosives are made of. Atomic weapons capable of causing great damage are relatively easy to make, once appropriate nuclear materials are at hand.[70] The resources of a world superpower are not needed, as India's acquisition of the bomb has shown. In fact, unclassified publications provide nearly all the technical information that is needed to fashion a crude bomb —one that would probably explode with the power of at least 100 tons of chemical high explosive. Once in possession of twenty to thirty pounds of weapon material, such as the plutonium produced in nuclear power reactors, a small group or even one person working alone could build such a bomb. Beyond the nuclear material and chemical explosive, the necessary materials could easily be bought from hardware stores and from commercial suppliers of scientific equipment for students.

In criminal hands, plutonium could be a danger not only as material for a bomb but also in a relatively simple radiation dispersal device. Because it is so extremely toxic, the amounts that could pose a threat to society are very small. A few ounces, or even a fraction of an ounce, of the stuff could be a deadly risk to everyone working in a large office building or factory, if it were effectively dispersed.[70]

The upshot of this is that small amounts of nuclear weapons material (plutonium, uranium highly enriched in uranium-235, and uranium-233) can be used to do nuclear violence—tens of pounds for a crude bomb, and ounce quantities of plutonium for a radiation weapon. Enormous quantities of these nuclear materials will be circulating throughout the fuel cycle for civilian nuclear power in the very near future. It is expected that by the mid 1980s, a hundred tons of plutonium will be reprocessed each year.[74]

International security and the growth of nuclear power: For almost a decade, the number of nuclear armed powers remained constant at five. Then in May 1974, India joined the "nuclear club." The spread of civilian nuclear power puts a potential for nuclear arms within the reach of many more countries. By 1980, at least 30 nonnuclear weapons states will have large nuclear power or research reactors. It is likely that two dozen more countries will have nuclear reactors by 1990.

The risk of diversion and proliferation of nuclear weapons is inherent in a commitment to civilian nuclear power. The best an international safeguard's system can do is to keep these risks as low as practical in the face of international political realities.

The present international safeguards system began in 1957 with the formation of the International Atomic Energy Agency (IAEA), and was reinforced in 1970 by the Nonproliferation Treaty. The greatest weakness of this system, which relies chiefly upon accounting procedures, is that its objective is to detect diversion rather than to prevent it.

Changing the system to make it more effective inevitably runs into complex problems of international politics. For example, it would be technically feasible to build large, secure nuclear parks under the aegis of the IAEA, to process and reprocess nuclear fuel for all the nonnuclear weapons states. The only nuclear material flowing out of the park would be fuels that could not readily be used for weapons; flowing back in would be material that is radioactively too hot to handle.

A neat system. But it would require an unprecedented degree of international cooperation. An international security force capable of preventing the unlawful diversion of nuclear material would require a still more revolutionary commitment to supranational authority on the part of jealous, contentious nation states.

The problem of nuclear theft: Nation states anxious to join the "nuclear club" are not the only source of concern in the matter of diverting nuclear materials to weapons use. A very real possibility exists that terrorists or criminal groups, or even a single fanatic, could steal nuclear materials to fabricate crude weapons. Without effective safeguards to prevent theft, the development of nuclear power will create substantial risks to the security of Americans and people everywhere.

The problem of nuclear theft in this country has only recently begun to receive serious attention. A few dramatic incidents, notably an Orlando, Florida, nuclear bomb threat in 1970, have prompted the AEC to pay more heed to safeguards against nuclear theft. In Orlando, an anonymous letter writer threatened to blow up the city with an H-Bomb unless he got $1 million and safe escort from the country. A convincing drawing of a device accompanied the threat, and neither the AEC nor the FBI could be sure that the threat was unreal. It turned out in fact to be a hoax, perpetrated by a 14-year-old boy who was an honor student in his high school science class. But such incidents prompted AEC action to supplement accounting methods (that may detect theft after it happens) with physical security measures (that seek to prevent it).

The study sponsored by the Energy Policy Project concluded that while the present safeguards[70] program does not provide a high level of physical security, a system can be developed that will keep the risks of theft from the nuclear power industry at very low levels. But a system of nuclear theft prevention must be worldwide to be really effective. Every country with nuclear power must develop an adequate system to prevent theft.

The basic components of an effective safeguards system should be:

- A system of physical barriers and other security measures designed to defeat a maximum credible threat of theft anywhere in the nuclear fuel cycle.
- A federal nuclear materials security force, which would have sole responsibility for protecting nuclear materials against theft.

The costs of a reasonably effective safeguards system need not unduly burden the economics of nuclear power. An effective national nuclear security force might cost as much as $70 million per year by 1980,[70] but this would be less than 1 percent of the cost of generating nuclear electric power in that year. The potential costs of ineffective safeguards, in terms of potential property damage and destruction of human life, are of course enormous.

Future options for nuclear fission

Present day light water reactors (LWRs) make use of about 1 percent of the potential nuclear energy stored in natural uranium. The high temperature gas cooled reactor (HTGR), now becoming commercial, requires about half as much uranium as the LWR per unit of electricity produced. But the potential of both these kinds of reactors as a long term energy source may be limited, if high growth in energy use continues.

One of the underlying reasons for the whole fission nuclear power program has been to provide a replacement for fossil fuels that will last for many centuries and not just a few decades. The breeder reactor is an energy technology that can do this, by ultimately using over half the potential energy stored in natural uranium. It does this by converting much of the otherwise unused portion of natural uranium into useful plutonium fuel. It is called a breeder reactor because it produces more useful fuel than it consumes. Because of the long term supply benefits offered by the breeder reactor, a major portion of U.S. energy R&D expenditures are being committed to its development.

Various breeder reactor designs are possible: the liquid metal fast breeder reactor (LMFBR), the gas cooled fast breeder reactor (GCFBR), the molten salt thermal breeder (MSBR), to name some of the more studied options. The most fully developed option is the LMFBR, which is now moving into the advanced stages of R&D, with the focus of the program being an LMFBR demonstration plant.

Many of the environmental problems of the LMFBR are similar to those of present day reactors. Problems of managing high level wastes are qualitatively similar, although spent fuel shipped from the breeder reactor is more radioactive and must be handled with greater care. Since more plutonium is produced in an LMFBR than in an LWR, the nuclear theft risks and potential plutonium health hazards are more intense with the LMFBR. The amount of plutonium discharged annually from an LMFBR is seven to eight times as much as a uranium fueled LWR discharges.[74] As LWRs begin to use recycled plutonium for fuel, the amount of plutonium they put out will be comparable to that discharged from a LMFBR.

Reactor safety problems in the LMFBR and LWR are qualitatively different.[77,78] In the case of the LWR, the accident of greatest concern is a pipe rupture followed by a loss of the high pressure coolant and subsequent core meltdown. The LMFBR's liquid sodium coolant is at low pressure so that a loss of coolant accident may not be as serious a problem. But a major concern is an explosive accident sequence that cannot occur in either the LWR or HTGR. A critical mass of plutonium conceivably could accidentally assemble, leading to an explosive release of nuclear energy. LMFBR safety research is directed at guarding against the potential for such serious accidents.

These and other safety and environmental questions must be resolved before the LMFBR becomes commercial. An important question is how much time is available to conduct the needed environmental and safety research on the breeder (and on existing nuclear plants as well). The answer is dependent on the energy growth pattern the country chooses.

Let us consider this question more closely. If the nation persists on the *Historical Growth* track with a heavy emphasis on nuclear energy, then annual nuclear electricity consumption by 2020 could reach ten times the total electrical energy consumption today.[79] If this nuclear electricity were generated solely by light water reactors, or even the more resource-conserving, high temperature gas cooled reactors, most U.S. uranium resources up to $50 per pound would be exhausted, according to present resource estimates.[80] (Today's price is less than $10 per pound.) Uranium at $50 per pound would increase the cost of electricity to the consumer by 10 percent, all other factors being equal.

Commercial operations require a substantial advance supply of uranium fuel in the form of reserves and potential additional resources. This suggests that under *Historical Growth* conditions the breeder reactor would have to be in commercial operation by the turn of the century to assure continuing low cost supply of nuclear electricity. Another economic incentive to employ the breeder reactor is that the plutonium that light water reactors produce is worth more as a fuel for breeders than for light water reactors.

Still another pressure to move to the breeder, under high growth, would be the significant land use problems. Present day reactors require 100 times as much land for mining uranium as breeder reactors.[48] If high nuclear growth persisted to the year 2020 with light water reactors alone, uranium mining could by then require about 300 square miles per year—more than twice the area surface mined for coal today—as compared with three square miles to fuel the breeder. A combination of economic pressure and the need to strip mine vast areas would offer a strong motive for proceeding rapidly toward a commercial breeder reactor. Under *Historical Growth*, a choice of holding off the breeder program until safety concerns are resolved might well be foreclosed.

The *Technical Fix* scenario presents a very different picture. Nuclear power requirements for light water reactors by the year 2000 or even 2020 would not overtax the uranium resource base. Breeders might well be commercial before then, but the nation could deploy them by choice, not through any misguided sense of necessity.

The importance of *Technical Fix* is that it would buy the time to be sure we could "test first and build later," before the breeder went commercial. It would also permit time for alternative breeders such as the molten salt reactor, to be developed, if the liquid metal fast breeder reactor program did not prove out. The R&D efforts should be diversified and pursued as rapidly as feasible. But under *Technical Fix*, the breeder could be held to stricter safety and environmental tests without endangering energy supply.

Nuclear overview

Despite a great deal of effort on the part of the AEC and the nuclear power industry, it is not possible to be sure that the public is being adequately protected against the risks of nuclear power. According to the critics, the safety features built into present systems may not perform as they are supposed to if called upon; and further, there are important accident possibilities that safety designs have not taken into account. The atomic energy community in government and industry contests these criticisms, but since an adequate experimental basis for nuclear safety systems is lacking, it is difficult to refute some of the claims. The high rate of "abnormal incidences" in safety-related areas at nuclear facilities is evidence that necessary levels of quality control are not being achieved.

The record is even more disturbing in the area of long term problems. An acceptable long term, high level waste storage scheme has not yet been found. The problem of decommissioning old facilities has not even been seriously thought about yet. There is now a rapid growth in nuclear facilities on several fronts, but they are not being designed for eventual decommissioning. This disturbing situation is reminiscent of our experience with coal surface mining in areas like Appalachia, where reclamation is now often nearly impossible because it was not planned as an integral part of the mining activity. The potential long term hazards of accumulating plutonium in commercial low level waste burial grounds arouses the same disquiet.

Nuclear violence is another problem the nation is not dealing with adequately—perhaps in part because people do not like to think about it. Solutions are inherently difficult. Especially disturbing is the prospect of widespread proliferation of nuclear weapons throughout the world through diversion of nuclear materials from power programs. It is difficult to see how to evade this prospect without a radical reshaping of international institutions, should world nuclear power development continue as planned. A hopeful development is that measures to cope with the nuclear theft problem in the United States are now being carefully considered. The vulnerability of nuclear systems to sabotage also needs close attention.

What is lacking in this area is a serious attempt to design nuclear facilities for inherent invulnerability to a broad class of threats of nuclear violence. Also lacking is consideration of a whole new dimension to the problem of guarding against nuclear violence—the institutional issues. In dealing with malevolent men, institutional controls will be at least as important as technological ones. It is important to know how far such controls can be pushed before the institutional changes themselves become unacceptable. A recent AEC report on nuclear theft released by Senator Ribicoff recommends that better intelligence on terrorist groups be maintained through infiltration and other operational and organizational means.[81] While such actions may be unavoidably necessary, recent experience tells us that covert surveillance activities under the guise of national security can be subverted toward political ends, in violation of individual rights.

These considerations point out the great number of problems that remain unsolved in the nuclear area. We must also consider these problems in the context of the benefits of nuclear power, which we mentioned at the outset of this discussion. And we must also keep in mind air pollution and other difficulties with fossil fuel alternatives. The potential benefits of nuclear power provide strong incentives to try to resolve the problems it poses. In the final analysis, the public must decide if the inherent risks are low enough to be acceptable, in light of both the benefits and the problems with alternatives.

We propose a four-part program to help inform the public, and lay the bases for these decisions:

- Public assessment of the entire nuclear power program (benefits and risks) as a basis for public hearings on the program's future.
- Formulation of an independent regulatory program to deal with the critical nuclear power social and environmental issues, coupled with a restructuring of the nuclear R&D effort to emphasize effective environmental controls on new supply technologies before—not after—they become commercial.
- Revision of the Price Anderson Act, which gives the nuclear industry financial protection in the event of catastrophic accidents, so that the industry has more incentive to achieve quality control in implementing environmental safeguards.
- Putting in effect an aggressive energy conservation program that will permit a more orderly, cautious development of nuclear power.

Assessment of the nuclear fission program: It is doubtful that very many citizens know either the potential role that nuclear energy can play in the energy future of the nation and the world, or what the environmental benefits, hazards, and social implications would be of this commitment. Broad public discussion of these issues is a prerequisite to the government's and industry's aggressive search for solutions to the nuclear problems laid out here. Without the pressures brought by public scrutiny, bureaucracies such as the AEC tend to postpone attention to serious, difficult, or long range problems in favor of more pressing day-to-day obligations.

Unfortunately, the AEC has failed to inspire the confidence of many citizens concerned about nuclear risks, in part because the AEC has been both the promoter of nuclear power and its regulator. The expected separation of promotional and regulatory operations into different agencies is an important first step toward restoration of public confidence. The new nuclear energy regulatory body could make a major contribution by undertaking a balanced and objective study of the impact of the entire nuclear energy program, as outlined here. This study could form the basis for serious public hearings on the nuclear power option.

Strengthening nuclear regulation and research: On the basis of our own research and work done by others,[f] we make recommendations for programs to reduce both particular risks and multiple problems in nuclear energy production.

The following programs are oriented toward particular risks:

• To deal with the reactor safety issue, the AEC should aggressively pursue the development of a better emergency core cooling system for light water reactors, as advocated by the Commission's Advisory Committee on Reactor Safeguards.[82] More fundamental approaches to reactor safety problem, such as underground siting of power plants, and development of a reactor inherently stable against a fuel meltdown, should also be researched.

• Experiments simulating spent fuel transport accidents should be carried out to realistically assess public risks from such accidents.

• Nuclear facilities should be designed so that they can be safely decommissioned. A system should be devised to include decommissioning costs as a cost of providing nuclear electric power. Decommissioning guidelines to these ends should be formulated and conformance to them should be a prerequisite to issuance of an operating license.

• Development of a long term waste disposal plan deserves high priority, for both high level wastes and low level actinide wastes.

• A major research effort is needed to assess the long term public health implications of a plutonium energy economy. Specifically, the study should identify weak points throughout the entire fuel cycle where plutonium might be released into the environment; the transport of plutonium should be better understood; and present plutonium health standards should be reevaluated.

• The vulnerability of the entire nuclear fuel cycle to nuclear violence, including sabotage and theft of nuclear materials, should be assessed, and stronger safeguards prescribed, including the establishment of a security force to guard nuclear shipments. The AEC should formulate a new program indicating the degree to which the risks would be

[f] See discussion by Alvin Weinberg in note 83, for an overview of what problems need to be solved.

guarded against, and subject them to public scrutiny —perhaps through a public rule making hearing.

Control efforts aimed at multiple problems could effectively supplement these specific measures. A basic approach toward nuclear environmental protection is to restructure nuclear systems so as to greatly alleviate or even eliminate some of the more difficult problems. The nuclear park idea, which involves locating nuclear power plants and fuel reprocessing and fabricating facilities at one site, is one such approach that deserves close attention. Some of the potential advantages of the nuclear park include:

- Elimination of the vulnerable transport links for spent fuel and nuclear materials suitable for weapons use—thereby greatly reducing risks of transportation accidents, nuclear theft and sabotage.
- Improved quality control, since the complex would have to be built over many years by a relatively stable and sophisticated work force that should become more expert as the work advances.
- Reduced public risks through the possibility of more remote siting.

Of course there are disadvantages of nuclear parks as well:

- Vulnerability of large blocks of electric power (serving perhaps tens of millions of people) to failures, such as terrorist action, earthquakes, accidents, and so on.
- Intense local thermal pollution arising from the high level of power generation.
- Increased costs of transmitting electricity, from the need for longer transmission lines.

On balance, the nuclear park advantages appear to outweigh the disadvantages. It may prove desirable to redirect nuclear power growth toward building such parks and thereby limit radioactive operations to a relatively few places, if the nation chooses to continue the nuclear option.

Revision of the Price Anderson Act: The difficulties the nuclear industry has experienced in achieving a high level of quality control in the design, construction and operation of nuclear facilities strongly suggests that it may need a greater incentive to apply the required high level of expertise and discipline to environmental control problems.

The production of nuclear energy is potentially a very hazardous activity. In most business activities, the potential liability, or the costs of insurance to protect against liability, are recognized as elements of costs by an enterprise considering whether to engage in a hazardous undertaking. If these costs are excessive, the market system automatically halts the hazardous activity. With an enterprise held strictly liable, the market system would be society's first line of defense against catastrophic accidents. In the case of the nuclear industry, this safeguard and its powerful monetary incentive for the industry has been replaced by a reliance on the effectiveness of AEC regulation, as provided for under the Price Anderson Act.

In 1957 Congress passed the Price Anderson Act to eliminate a major roadblock to the commercialization of nuclear power—the industry refusal to invest in nuclear power for fear of bankrupting liability in case of an accident that might cause billions of dollars of damage. The insurance industry at that time was unwilling to provide any more than $60 million in liability insurance. Under the Act, the AEC requires nuclear licensees to buy as much private insurance as they can get against damages that might arise from a nuclear accident. The AEC itself provides indemnity in addition, at cut-rate fees, to further cover the licensee.

To give an example of how this works, one recently licensed nuclear power plant pays an annual premium of $250,000 for the maximum insurance coverage of $95 million, and an annual fee to the AEC of $73,500 for $465 million indemnity coverage.

The Price Anderson Act provides that all liability is cut off at $560 million. If liability claims are more, the $560 million fund is apportioned pro rata among claimants. Thus, if an accident causes damages of a billion dollars, each person injured will recover only about 50 percent of his loss. Price Anderson therefore provides a substantial subsidy to the nuclear industry. The continuation of this subsidy to nuclear power, as this technology moves from the R&D stage to full commercialization, has been criticized by some experts.[84]

The future of the Price Anderson Act, due to expire in 1977, is again being debated by Congress.[85] In May 1974, the AEC issued a draft environmental impact statement on proposed legislation to amend the Act, laying out the merits of alternative options to its extension.[86] It is our view that the nuclear power industry is now sufficiently mature for a revision of the Price Anderson Act, so that the marketplace will reflect the potential social costs of nuclear power. If nuclear power is indeed as safe as the AEC and the industry claim, the risks are small enough to be borne by the enterprises involved. If the utilities are unwilling to build new plants on certain sites, or to buy reactors of certain designs, without the shield of Price Anderson, then those locations and those plants are too risky to be built.

A new insurance system to make the industry fully liable for damages should include:

- A shift of the liability from the taxpayer to the utilities and their suppliers involved in providing nuclear energy.
- A removal of the limit on liability.
- Extension of these provisions to all parts of the fuel cycle where serious accidents could occur, including transportation of spent fuel.

There are various schemes that could be formulated toward these ends.[86] Some system of shared liability among suppliers and utilities is probably essential, since any one utility may not be able to pay for the damages of a serious accident. With a revision of the Price Anderson Act along these lines, there would be public protection through both AEC regulatory activities and market incentives against nuclear mishaps.

The role of energy conservation in nuclear power development: While nuclear power supplied only 5 percent of U. S. electric power consumption in 1973, the expected growth in nuclear power is spectacular. The installed nuclear generating capacity of 25,000 megawatts at the end of 1973 is projected by the AEC to increase more than tenfold by 1985, and twentyfold by 1990.[87] This tremendous growth rate (averaging 25 percent a year to 1990) would make it nearly impossible to resolve and satisfy all the concerns about nuclear power we have discussed, in time to avoid catastrophe if these fears prove to be well founded.

If energy conservation should be pursued as a serious national objective, however, the country would get a breathing period for reassessing the entire nuclear program, without foreclosing any of the options regarding nuclear power development. A program of nuclear reassessment is compatible with both the high and the low nuclear options of the *Technical Fix* scenario. In the low nuclear option, installed nuclear generating capacity would be limited to what is operating or under construction today. But even the high nuclear option for 1985, under a policy of conservation, would permit a pause in the initiation of new power plant construction and a slow enough pace of development to make the reforms in a timely manner, or choose alternative sources of energy.

Environmental perspective

Many of the most conspicuous environmental problems resulting from our use of energy can be solved, if the nation is committed to doing so. But an aggressive program of environmental protection measures is not sufficient to assure full protection of the public health and welfare. Some pollution is unavoidable—either because we don't yet know how to control it, or because it is inherently uncontrollable. We can greatly reduce air pollution but we cannot do away with it entirely; even with the best reclamation efforts, it is impossible to restore surface mined land to its original condition; and even a supersecurity system cannot prevent nuclear theft by highly determined and clever thieves. Each supply option is usually associated with certain such "limiting" environmental factors.

The question, then, is what choices does a high energy civilization have to continue its benefits and still protect the planet and people that live on it. The *Technical Fix* scenario provides time for the country to reassess its future supply and make the best choices. The troublesome sources of supply need not be substantially enlarged before 1985, and even then we need not develop all of them. Specifically, the *Technical Fix* Environmental Protection supply option (outlined in Chapter 3) would have the following impact on 1985 supplies:

- Combined growth in coal and oil use to 1985 could be kept to about 1.3 percent a year, so that the small particle air pollution is not greatly worsened. The slower growth in fossil fuel usage would also ease somewhat the global atmospheric buildup of particles and CO_2.

- Western coal development can be kept to a low level. If western coal accounts for half of the twelve-year 1973–1985 growth in coal production, then western coal production in 1985 would be higher than the 1973 level by about 25 percent and would result in the surface mining of only about ten square miles per year.

- No new production on the Atlantic or Pacific offshore areas would be necessary.

- Nuclear power could be held at the level of plants now operating or under construction.

Not all these restrictions could be maintained beyond 1985 in the *Technical Fix* scenario. Substantial growth in at least two of three major new supply options—western coal, new Atlantic and Pacific offshore oil, and nuclear power—must be pursued if imports are to be restricted. If import restrictions were relaxed after 1985, the country would need to substantially develop only one of these troublesome new sources of domestic supply.

With the conservation action in the *Technical Fix* scenario, the country has the long term option of limiting imports and foregoing development of at least one major new energy source—if environmental or other concerns should so dictate. It could, of course, go ahead with all three troublesome sources on a reduced basis. But if the nation wished to do so, it could make a choice.

The choice as to which source of supply to forego is difficult and should reflect a broad public consensus based on informed public discussion. Ultimately, it comes down to a fundamental value judgment of which is the worst: the risks from nuclear power; or air pollution and the destruction of recreational areas and the fragile coastal environment from oil drilling; or air pollution and disruptive changes in the way of life in the Rocky Mountain region from coal and shale production? The scientific facts help, but they cannot by themselves give the nation the answer; they are inconclusive. There are serious unknown risks in any option, and values forever lost.[88] In a democracy, the choice should be made by the citizens.

Public Participation

We have suggested mechanisms for bringing about public participation in the context of particular problems in this chapter. What follows are suggestions for a general program to encourage public participation.

Full disclosure: The most basic ingredient for effective public participation is information. As James Madison wrote in 1822:

> Knowledge will forever govern ignorance. And a people who mean to be their own governors must arm themselves with the power knowledge gives. A popular government, without popular information or the means of acquiring it, is but the prologue to a farce or tragedy.[89]

The public needs to be informed of probable environmental risks and benefits of proposed activities. The

National Environmental Policy Act of 1969 (described in Appendix E) provides a basis for full disclosure through its requirements for environmental impact analyses on major federal actions. Several states have passed similar laws requiring the assessment of state actions that might have significant environmental impacts.

Unfortunately, all too often NEPA statements have been issued too late in the planning process to be useful in public discussions. Consideration of alternatives have tended to be rationalizations for actions already decided upon. To be effective, the NEPA statements must be issued well in advance of final decisions on proposed new activities. In addition, to broaden the scope of NEPA we propose that a NEPA–like statement be made periodically for energy policy as a whole, setting forth the basic options open to the nation, and highlighting essential environmental benefits and risks of alternative growth patterns and supply mixes.

Support for public interest groups: The requirement that agencies provide full disclosure of relevant information is not in itself adequate assurance that the public will be informed—or will even have sufficient opportunities to be informed. It can be expected that much of the data and analysis so disclosed will be inadequate, or even totally useless. Institutions that can provide independent technical assessments of NEPA statements or other disclosed information would be extremely useful. Such independent assessment, taken together with the official statement, could provide a basis for public discussion. If an independent group were to find grave fault with an official statement, the agency might be compelled to go back to the drawing board.

One serious problem is finding an independent, objective group. The "public interest" groups that have begun to emerge throughout the country, staffed at least in part by scientists and engineers, may be able to play this role. Many of these groups (for example the Environmental Defense Fund, the Natural Resources Defense Council, and the Union of Concerned Scientists) have a "pro environment" perspective and may therefore rightly be accused of representing only a particular sector of the public. Nevertheless, a technically competent assessment representing an environmental point of view is useful in helping provide a basis for a public discussion. Such groups could help provide information to the public on the environmental implications of alternative energy futures. We encourage private foundations to increase support for public interest activity in various areas that would be affected by energy policy proposals.

Public forums: It is desirable that information released to the public reach a substantial number of people to be affected by a proposed action. Televised public hearings would be one means of providing information to a wide audience. Such hearings should admit both technical testimony and adversary views. Hearings could be conducted at the national, regional, or local levels, depending on the issue.

The ballot box: Ideally, decisions made in the political process on energy and environment issues should reflect a public consensus that is based on a broad public understanding of the issues. The only opportunity citizens have to hold public officials accountable for these decisions is in the electoral process. Citizens can also use their voting power to initiate significant innovations relating to environmental protection through the initiative process. The California Coastal Conservation Act (enacted by California voters in 1972), requiring that coastal development proposals conform to regional and statewide coastal zone management plans, is a case in point. The public voice on energy and environment decision making is strongest in the voting booth.

Ultimately, the decisions should be made by the public—and its representatives. The major choices for energy growth are nuclear or fossil fuels. The dangers are quite different. The pollutants of fossil fuels, until and unless controlled, are a daily hazard to our health. The nuclear risks are those of a gambler: the odds are small, but the stakes are high. People should be able to decide for themselves which risks are preferable.

The Energy Policy Project is not trying to choose one side or the other and, indeed, we are not of one mind in this debate. It would be comfortable to say simply, "Let's wait and decide later." Unfortunately, later is now. We must begin now to engage in the public debate that will answer the question—"Which is safer?"

CHAPTER 9

Private enterprise and the public interest

Recent rises in prices and profits have sharpened concern over the economic power and political influence of the energy industry, particularly the oil industry. To some extent, this concern is part of a wider problem of social control over big business generally. But the energy industry has certain characteristics that distinguish it from other industries and that present the problem in a much more acute form. Energy is essential to the functioning of the economy. A breakdown in supplies must be avoided. A substantial part of our energy supplies comes from abroad, often from insecure sources. Distribution of these supplies is controlled by companies with worldwide, rather than domestic, interests and responsibilities.

Oil companies are among the mammoths in the corporate world. Of the ten largest industrial corporations in the United States four are oil companies. Seven of the fifteen largest multinational corporations are oil companies. Oil companies' incomes eclipse the wealth of states and nations. Exxon alone receives as much revenue as California, and the combined dollar sales of the top four U.S. oil companies were about $57 billion in 1973, surpassing the revenues of many national governments in the world.

This is a staggering amount of money and economic might. Money implies power; and the American public harbors a lasting suspicion that rich oil companies use their power to control the market, setting higher prices and profits than real competition would allow. This belief is not without official support. At this writing, the Federal Trade Commission has pending a complaint against the eight largest oil companies, alleging that they pursued a common cause of action to hinder competition at all levels of the petroleum industry.[1] The states of Connecticut and Florida have also filed federal antitrust suits against the large integrated oil companies.

But the verdict is by no means unanimous. The Treasury Department defends the oil industry against charges of anticompetitive behavior and says that companies have "merely been responding to Government laws and policies, and these laws and policies have been the real culprits."[2]

The Energy Policy Project's analysis suggests another view. Judging by the standard economic criteria, it appears that oil companies do not at present exercise decisive monopoly power over the market. But the industry does wield exceptional political power, enough to shape profoundly its own economic environment and to pursue very effectively its own goals, which are not always identical with the public interest.

In the following pages, we take the first question first: Are the energy companies, oil companies in particular, workably or effectively competitive? Then we explore the question of alliances between energy companies and government, ask what goals the industry has pursued through the political route, what success it has achieved, and what the public can do about it.

Economic theory holds that workable competition exists when no single company, or group in collusion, can through market control keep prices and profits artificially high. To decide whether an industry is effectively competitive, an industry's structure, its institutional arrangements, and its behavior must be carefully examined. A series of reports commissioned by the Energy Policy Project made an in-depth study of the various energy industries in terms of

Table 36–**Concentration ratios for the primary energy industry by major sectors, United States only**

	Uranium Mining and Milling		*Crude oil Production*		*Petroleum Refining*		*Gasoline Sales*		*Natural Gas Sales (Interstate)*		*Coal Production*		*Energy Production (Btu Basis)*	
	1967	*1971*	*1955*	*1970*	*1955*	*1972*	*1954*	*1972*	*1955*	*1971*	*1955*	*1972*	*1955*	*1970*
	(percent)		*(percent)*		*(percent)*		*(percent)*		*(percent)*		*(percent)*		*(percent)*	
Big 4	57.0	54.4	18.8	30.5	32.9	33.1	31.2	29.0	23.0	25.3	16.5	30.4	11.0	21.2
Big 8	78.7	78.5	31.1	50.1	57.5	59.0	54.0	51.6	35.0	42.8	24.0	40.4	19.7	35.0

Source: Thomas D. Duchesneau "Competition in the Energy Industry," draft report to the Energy Policy Project, 1974

these criteria.[3] One of these is the issue of seller concentration (see Table 36). Possibilities for collusion are greater when a few firms control a large share of an industry's activity (usually sales or production). The higher this level of concentration, the greater the possibility that sellers will be able to reach and maintain agreements on a collective course of action, rather than the independent decision making that characterizes effective competition.

It is impossible to designate a precise point below which an industry can be decisively judged competitive, or above which it can be judged uncompetitive, but a rule of thumb has been formulated which says that monopoly power begins to be felt when the four largest firms account for more than 50 percent of the industry's output, or when the eight largest account for more than 70 percent. By these standards, only two elements in the energy business—the uranium mining and milling industry, and the electric generating equipment industry are uncompetitive.

It is not surprising that concentration in uranium production, an infant industry developed under government aegis, is relatively high. Concentration in the electric equipment supply industry, especially the rapidly growing nuclear power supply industry, is still higher. In almost all stages of the industry, the four largest firms account for 100 percent of the market.[4] Two firms—G.E. and Westinghouse—in practice dominate the electric generating equipment business.

Institutional aspects of the electrical equipment industry suggest large possibilities for exercising market control. These include high vertical integration, interlocking directorates, and "tie-in" sales involving the obligatory purchase of a combination of services or equipment. Another factor that may weaken the force of competition is the nature of purchasers—public utilities whose rate fixing systems do not automatically promote cost cutting.

Concentration in the branches of energy production other than these two are well under the rule of thumb indications, particularly when all the separate branches are consolidated into a single "energy" industry. The four biggest fuel producers account for 21 percent of total energy production and the biggest eight for 35 percent. These ratios are below those obtaining in other major industries such as steel or automobiles. By such standards the energy supply industries are not overly concentrated.

But as Table 36 also indicates, concentration ratios have been increasing sharply in some branches of the industry—particularly in crude oil and coal production. The biggest eight oil producers increased their share of production from 30 to 50 percent of the total between 1955 and 1970; and the eight biggest coal producers from 24 to 40 percent. These rising ratios reflect the merger activity that took place in both the petroleum and the coal industry in the 1960s. Four massive mergers—Union Oil Company of California with Pure Oil, Sun Oil Company with Sunray DX, Atlantic with Richfield; and Phillips's partial acquisition of Tidewater (marketing facilities in the western states)—took place between major companies. In addition, petroleum companies also acquired $337 million of the assets of independent petroleum marketers and refiners. A series of mergers also resulted in greatly increased concentration in the coal industry, with the major oil companies playing a large role. By 1969, the oil industry accounted for almost one-quarter of total coal production, compared with a mere 2 percent in 1962.

Furthermore, the petroleum companies are expanding rapidly into all other branches of the energy field. Of the 14 largest petroleum companies (ranked by 1969 assets), seven (including the four biggest) had diversified into all other branches of the energy industry—gas, oil shale, coal, uranium, and tar sands.[4] The other seven companies produced oil, natural gas and at least one other form of energy. Table 37 identifies the big eight producers in each of the three fossil fuel sectors. Partly due to the recent entry of oil companies into coal production via the merger process, the eight largest fossil fuel producers consist of seven oil companies, plus Peabody Coal.[a] Exxon is the leading producer

[a] Peabody, the largest coal company, was acquired by Kennecott Copper in 1968, but at this writing, a Federal Trade Commission divestiture order has been upheld in the courts, and is scheduled to go in effect October 1974.

Table 37–**Big eight fossil fuel producers, 1970**

	Crude Oil	*Natural Gas*	*Coal*	*Energy Market*
1.	Standard Oil of N.J. (EXXON)	Standard Oil of N.J (EXXON)	Peabody Coal[a]	Standard Oil of N.J. (EXXON)
2.	Texaco	Gulf Oil	Consolidation Coal (Continental Oil)	Continental Oil
3.	Gulf Oil	Shell Oil	Island Creek Coal (Occidental Petroleum)	Texaco
4.	Shell	Standard Oil of Indiana (AMOCO)	Pittston Group	Gulf Oil
5.	Standard Oil Co. of California (CHEVRON)	Phillips Petro. Co.	U.S. Steel	Shell Standard Oil of Ind. (AMOCO)
6.	Standard Oil of Indiana (AMOCO)	Mobil Oil	Bethlehem	Peabody
7.	Atlantic Richfield	Texaco	Eastern Associated Coal	Atlantic Richfield
8.	Getty Oil	Atlantic Richfield	Ayrshire Collieries Corporation	

[a] Held by Kennecott Copper until divested by order of the Federal Trade Commission.

Source: Thomas D. Duchesneau, "Competition in the Energy Industry," a draft report to the Energy Policy Project, 1974.

of both crude oil and natural gas; it is also the leader in the energy market in total.

Among the 16 firms currently engaged in uranium mining and milling, three are oil companies (Kerr-McGee, Exxon, and Continental Oil Company). Kerr-McGee, though a relatively small integrated oil company, is the largest uranium producer. Six of the big eight majors—all except Texaco and Standard Oil Company—have acquired uranium deposits. These developments are reflected in the rise in concentration ratios for the consolidated energy industry. The share of the biggest four firms rose from 11 to 21 percent between 1955 and 1971, and of the biggest eight firms from 20 to 35 percent.

Ownership of reserves may be the crucial test of industry concentration for the future. Here public policy is handicapped by a paucity of hard facts, but the Federal Trade Commission data indicate that the eight largest oil companies own 64 percent of the oil and gas reserves.[5] Oil companies own over 50 percent of the uranium and tens of billions of tons of coal. Exxon is not among the top twenty coal producers, but its vast coal reserves assure it a leading position in coal production in the future.

The extension of oil companies into all branches of the energy industry has implications for interfuel competition. If an electric utility, for example, can choose among three kinds of fuels, the degree of concentration in the industry supplying each fuel may not be of much relevance. Even if one company has 100 percent control of one of the fuels, it would still have to face the competition of other fuel supplies in setting prices. If one company dominated supplies of all three sources of fuel, however, there would be no competitive safeguard and that company could set the prices it wished. The spread of oil and other companies into the production and reserve holdings of other forms of energy, becoming in effect energy companies, could in this way eventually diminish interfuel competition.

Concentration of sales or production is in any case only part of the story. It is at best an indirect indication of the possibility of anticompetitive behavior. Other factors can be of major importance. In the petroleum industry, for example, a series of institutional features may imply a much higher degree of corporate "togetherness" than is indicated by the market shares of leading producers. For instance, most of the pipelines carrying crude oil and oil products are owned by consortia of the major oil companies. Naturally enough, the major companies route the pipelines for their own convenience, which may put the oil supplies of independent producers at some disadvantage—even though the pipelines, as common carriers, are obliged to carry other people's oil.

Controversy abounds at present about whether the majors are using their dominant position in pipeline transport to squeeze out independent producers. Lack of data precludes arriving at a definite conclusion, though there seems little doubt that the majors' ownership position in the key transportation link of the crude oil business gives them a strategic position in organizing supplies.

Joint ventures are not confined to pipelines only; they play a large part in oil company operations throughout the world, most particularly in international crude oil production and distribution, but also in domestic oil exploration and development, bidding for federal leases, and refineries. The oil industry is the most joint venture prone industry in the entire economy. Of the 32 largest American oil companies, Exxon Corporation, for example, had 302 joint ventures with 27 other oil companies out of 31 possibilities. Mobil was found to have 300 joint ventures with 28 out of 31 possibilities.[6]

Major oil companies act in concert in other ways as well. Processing agreements, under which major oil companies agree to provide refining services to each other, are common; so are exchange agreements, by which firms swap crude and products, usually on the grounds of convenience to markets. Such agreements can offer considerable cost advantages. But they may have other less desirable effects, such as heightening regional concentration and reducing the amounts of crude products available for sale to other companies. The vertically integrated structure of the industry may place a severe limitation on the amount of crude oil and refined products offered for sale to outside markets, with the effects that nonintegrated firms may face severe obstacles in getting supplies. Interlocking directorships and the use of common financial agents and accounting services also promote considerable interdependence.[7] As already indicated, institutional aspects of the electrical equipment industry suggest possibilities for market control.

A further test of whether an industry is workably competitive is how easy it is for new firms to enter the business. If entry is easy, established companies—no matter how few and how large—would find it difficult to maintain artificially high profits. In crude oil and natural gas production, this question turns on whether it is possible to gain entry into production, more specifically at the point of access to new reserves of crude oil and natural gas.

The most productive unexplored oil and gas prospects in the United States are on the outer continental shelf, controlled by the federal government. Barriers to entry into this market consist of the capital requirement, in the form of a bonus bid, plus subsequent exploration and drilling cost. For many years these capital requirements were modest: the average bonus bid for tracts offered for lease between 1954 and 1974 amounted to approximately $5 million.[8]

But in recent years the cost has risen sharply. Joint bidding has enabled small companies to gain access to leases. But joint bidding ventures, by creating a community of interest among participants, especially the majors, also create the danger of anticompetitive behavior. The Interior Department is now taking steps to forbid joint bidding ventures by the largest companies. On balance the rising cost of bidding, exploration, and drilling tends to favor larger companies.

New entry into the refining stage appears to turn on access to crude supplies. During the 1960s there was no shortage of crude, and the operation of the quota system favored the independent refiners rather than the majors by awarding them a more than proportional share of the cheaper imported supplies.[9] Now, with crude in short supply, and imported prices higher than domestic, the majors, with their greater self-sufficiency in crude supplies, have the advantage. Access to supplies is also important to entering marketing. During the 1973–74 shortages of gasoline, the majors supplied their own stations first, which forced thousands of independents to leave the business.

The costs of new entry into coal mining (around $20 to 25 million for a mine in the three million ton class) are not large in relation to the financial resources of the petroleum and chemical companies who have recently been entering the industry. Nor are they prohibitive for the larger coal companies. Such costs are large in relation to the resources of small coal operators, many of whom have left the business in recent years.

As for uranium, excess capacity discourages entry at present; the excess stems from the fact that development of nuclear power has been much slower than expected. Successful entry into the electric generating business is also difficult, because of the large capital requirements and expertise needed.

A final criterion of the competitive behavior of the energy industry is its price and profit record, the presence of abnormally high prices, and consequently high profits generally being taken as a symptom of monopoly power. So far as prices are concerned, the energy industries have in one sense a very good record. Over the last twenty years, prices of all kinds of energy have either fallen, or risen less rapidly than the general price level. Beginning in 1973, of course, gasoline and fuel oil prices rose very sharply indeed. But until recently "real" prices of energy have constantly declined.

There is little doubt, however, that over the years crude oil prices were much higher than they would have been, had the industry not benefited from government price support programs such as market demand prorationing and import quotas. (The influence of the industry in securing such programs is investigated later in this chapter.)

Despite such protective measures, oil company profits during the 1960s and indeed up to 1972 were no higher than the average profit rate for all manufacturing industry (see Table 38). Since 1973 they have been rising and are now running ahead of the average. Profit rates by themselves are not an adequate indication of monopoly conditions within an industry. The cause of the recent rise in oil company profits was the OPEC imposed price rise for crude in the world market, followed by a sharp increase in domestic crude prices, granted by the Cost of Living Council. (The weighted average of crude oil prices rose from $3.40 a barrel at wellhead at the beginning of 1973 to $6.50 at the end.) These two actions, by OPEC and the Cost of Living Council, would have increased oil industry profits in the short term even if the industry had been a textbook

Table 38–**Profits as a percent of invested capital, twenty largest U.S. petroleum firms**

Firm (Ranked According to 1971 Sales)	*Average Profit Rate (1967–1972)*	*Profit Rate 1973*
1. Exxon	12.3	19.4
2. Mobil Oil	10.7	16.0
3. Texaco	13.8	17.6
4. Gulf Oil	10.4	14.8
5. Standard Oil of California (Chevron)	10.4	15.7
6. Standard Oil of Indiana (Amoco)	9.7	13.1
7. Shell Oil	10.6	11.1
8. Atlantic Richfield	8.4	7.8
9. Continental Oil	10.1	14.5
10. Tenneco	11.2	N.A.
11. Occidental Petroleum	14.2	7.2
12. Phillips Petroleum	8.4	12.4
13. Union Oil of California	9.2	10.1
14. Sun Oil	9.5	10.8
15. Ashland Oil	11.6	18.2
16. Standard Oil of Ohio (Sohio)	8.2	6.8
17. Getty Oil	8.5	9.1
18. Marathon Oil	11.4	15.9
19. Clark Oil	15.2	33.4
20. Commonwealth Oil Refining	11.5	20.5
Oil Industry Composite	10.8	15.1
Average: All Manufacturing (1967–1971)[a]	10.8	

[a] Note that Tenneco is a conglomerate with only about 15 percent of revenues derived from oil operations, *Economic Report of the President,* January 1973, p. 280.

Sources: *Fortune,* May issues and *Business Week,* March 9, 1974. Taken from Thomas D. Duchesneau, "Competition in the Energy Industry," a draft report to the Energy Policy Project, 1974.

model of competition. They were windfall profits. OPEC exercised monopoly power, and both international and domestic oil companies went along for the ride. Such windfall profit taking should be distinguished from monopoly profits, which are produced by industry through implicit or overt collusion.

This survey of the energy industries and some of their salient market characteristics yields ambiguous results. Compared with other major industries, concentration ratios are low, long term profit rates are generally average, and entry moderately free. On the other hand, concentration is rising sharply in crude oil and coal and for the industry as a whole—especially in reserve holdings, which foreshadow the future. Furthermore, the oil industry is typified by a series of institutional arrangements—joint ventures, swap agreements, joint services—that provide such a strongly interdependent framework that formal collusion may not be necessary for the various participants to understand each other.

It is important to note that those developments and institutions which have an anticompetitive flavor can be explained by economic and technical imperatives. It is not necessary to subscribe to a conspiracy theory to account for them. The extension of the oil companies into other fuels, for example, has been guided by a variety of motives: the need to diversify out of an industry with stagnating production; tax advantages applying to purchase of coal mines; the development of technical expertise in the case of shale and uranium mining; the desire to secure raw materials for coal gasification or liquefaction, in anticipation of future shortages of hydrocarbons.

But this is cold comfort to the ordinary citizen, who can appreciate that there are good technical reasons behind just about everything that happens, and yet remain suspicious about the political leverage and influence that the energy companies yield.

Economic and political power are, of course, closely linked. The greater the economic power, the greater the possibility that the economically powerful can arrange the legislative and administrative environment within which they operate very much to suit themselves.

How effectively economic power can be applied to the political process depends on both the nature of political decision making and the structure of the economic sector concerned.[10] It depends on how susceptible the political system is to economic influence, and how well industry is organized to take advantage of this susceptibility. Perhaps the outstanding characteristic of the policy process in American government is its highly fragmented nature. In Congress, business is channeled through specialized committees and legislative power delegated to committee chairmen. The same fragmentation of authority exists in the executive branch; administrative law is made and implemented in innumerable agencies.

Rather than a single, integrated policy process, what emerges instead is a series of largely independent "policy subsystems" linking portions of the bureaucracy, its related congressional committees, and organized clientele groups. The key to policy making power, therefore, is access not to the political system generally, but to the subsystems. The organized citizen can gain access to the large general system, but access to subsystems is much more difficult. Typically it is confined to those with expertise, resources, and influence they can use over extended periods of time. This means that the corporation is much better equipped to take advantage of what the system offers than is the individual citizen.

What features of an industry's structure are likely to affect its political effectiveness? To begin with, the size of the industry. Large industries will normally have more money, talent, and overall resources than small industries. The same applies to firms within the industry. The absolute size of available resources, hence firm size, is important because political involvement has certain large minimum costs attached to it. Small firms can rarely sustain such costs on their own; they generally have to operate through trade associations, with all the intraorganizational differences, lack of control, and consequent weakening of influence that goes with it.[11]

Community of interest is also an advantage. It helps to avoid the time consuming process of negotiating industry positions on political matters among numerous competing firms. A large industry, therefore, containing large firms with a high degree of community interest might be expected to have greater political influence than an industry of the same size but composed of more numerous small firms. On the other hand, the political effectiveness of the very largest firms may be diminished because of their extreme visibility.

In addition to these structural aspects, geographical dispersion can influence the level of political power, although this is a factor that can work both ways. Too much geographical concentration means that an industry is unlikely to muster broad based political support on the federal level. But unless the industry has the minimum of geographic concentration necessary to attract local attention, it may be ignored altogether. If the industry interest is strong enough to swing votes in a few states, it can prove powerful indeed—especially when the nation as a whole is unexcited about the issue.

The foregoing analysis suggests that the energy industry has unique advantages. First, the energy industry is clearly one of the largest. Oil refining alone is three times larger than iron and steel, and two times larger than automobile manufacturing, in assets. The industry is also large in terms of the number of people with a direct, personal stake in its prosperity, including 196,000 service station dealers as of 1967.[12] Individual firm size within the energy industry is also characteristically large. Of the ten largest U.S. corporations, four are major integrated oil firms.[13] But despite the large size of the firms, the majors can partially defend themselves from charges of bigness by pointing to the numerous smaller firms operating in the industry, and arguing that government policies that might hurt the majors could hurt the small producers even more. In many respects, the petroleum majors enjoy the best of both worlds.

The economic effect of the spread of the petroleum companies into coal, uranium, and so forth, may not be significant at this stage, but the political effect, by reducing conflicts within the industry, promises to be greater. The clashes over policy between the coal industry and atomic energy, between natural gas and electricity, and the like, are now muted. The industry trade associations tend to sing the same tune. The opportunity is there for the energy industry to exercise considerable pressure on the political process.

Geography has also favored the energy industry. No fewer than 32 states have oil and gas production. These states have enjoyed remarkable political influence in Washington as a result of the strategic positions held by their Congressional representatives. For example, of the nineteen members of the House of Representatives with the greatest seniority in the 93rd Congress, ten come from major crude oil producing states, and three from coal states.[14] The two major oil states—Texas and Louisiana—send to Congress the men who chair the House Appropriations Committee, the House Agriculture Committee, the House Armed Services Committee, the House Appropriations Committee's Subcommittee on Foreign Operations, and the Senate Finance Committee. Oklahoma, the fourth ranking oil producing state, is represented in Congress by the Speaker of the House. Recently, however, energy policy has encompassed such new areas as environment, price controls, leasing policy, and so forth. These matters fall within the jurisdiction of committees such as Interior and Commerce, whose key congressmen represent consumer rather than producer states.

The oil industry exercises its power by a variety of means. The coalition of industry interests and associations known as the oil lobby is to be found at every level of government, from the smallest county seat, up through state and federal government, and on through foreign governments.

The largest of the industry associations is the American Petroleum Institute (API). It includes representatives from all sections of the industry but is dominated by the majors. The API collects facts, conducts research, and presents conclusions directly to members of Congress. It also takes its case to the public. The organization spent more than $6 million on a three-year advertising program to tell America that "a country that runs on oil can't afford to run short." This sum does not include the advertising by individual companies such as Mobil, Shell, Exxon, and Atlantic Richfield, which spent many more millions advancing their views on price controls, environmental measures, and other issues of public concern.

Some of the industry's most effective associations with government are sponsored by the executive agencies themselves. The National Petroleum Council (NPC), appointed officially by the Secretary of the Interior, is composed entirely of industry representatives. It is charged with advising the Secretary on oil matters affecting the public interest. In 1972, the NPC set forth industry views in a massive report, *The U.S. Energy Outlook,* that was sponsored by the Interior Department. There is no comparable group to provide a public oriented counter balance to the input from the NPC.

The oil industry is also a major contributor to political campaigns. A recent study by Rep. Les Aspin (D-Wisconsin) showed that 413 directors, senior officials, and stockholders in 178 different oil companies contributed a total of $4,981,840 to President Nixon's 1972 reelection effort. Mr. Aspin estimated that in total the contributions by individuals directly interested in the oil industry came to at least $5.7 million. Officials from the ten companies that were the largest contributors gave a total of $2,668,424—70 percent or $1,883,172 of which were in secret contributions. Three oil companies—Gulf, Phillips, and Ashland—admitted illegally donating a total of $300,000 in corporate funds to the President's campaign. The contributions were later returned. Officials of one company alone (Gulf Oil) gave $1,172,500.

The coal industry is much less well endowed. Indeed, Carl E. Bagge, President of the National Coal Association, told the *New York Times* in 1973 that his organization has no advertising budget. "The industry doesn't have that kind of money," he said. But the incursion of the oil industry into the coal industry may mean the development of a united energy industry view in advertising, lobbying, and campaign spending.

In the past, the coal industry's political influence has been more effective at the state level. Until Congress passed the Federal Coal Mine Health and Safety Act in 1969, regulation for coal miners' safety was left largely to the states. Mine operators successfully resisted strict controls on the grounds that they were too expensive and would put coal out of business, and that mining was inherently dangerous anyway. The United Mine Workers, until the recent change in leadership, was no more safety conscious than the mine owners. The result was that mine safety in America, as measured in fatalities per million man hours, improved not at all from the early fifties till 1971, when the new federal standards began to have an effect.

Coal companies also used their influence on state governments to keep strip mine regulation at a rudimentary level. Even where states have passed enlightened laws, enforcement has been lax. "The general experience under previous laws has not been good. Unless stronger programs are instituted and carried out, more land will predictably be left damaged by surface mining," said the Council on Environmental Quality in its 1973 report on strip mining.[14]

A mass of favorable legislative and administrative programs attest to the effectiveness of the oil industry's long history of involvement in the political process. Market demand prorationing—the system of allocating production quotas between the producing wells of an oil field—is an important example. The policy had its origins in the chaotic conditions existing in the early years of the oil industry. At that time, enormous new oil fields in Texas and Louisiana had just been discovered. Panic took hold when the owners of each small piece of land lying above a single large oil pool tried to pump out the oil before the next man got it out first, like two people with straws in the same soda, sipping away for dear life.

The basic trouble was the peculiar "law of capture," which meant that the land owner who drilled first and fastest could drain the whole pool. The result was gross physical waste, as oil literally poured out onto the ground, and cutthroat competition, as prices fell to levels far below cost. As a Texas Executive of Humble (now Exxon) put it, "We had to let a president of Humble quit to become governor" to put an end to the mess.[15] Governor Ross Stirling, the Humble president in question, restored order by sending in Texas state troopers to shut down the wells; the state then established production quotas, or prorationing.

Prorationing thus originated to stop waste in the oil fields. But the same system soon came to be used very effectively, by Texas and Louisiana, to support crude oil prices. Production quotas were fixed exactly to match the demand, with all wells getting prorated share. The more productive wells were then restrained well below capacity to avoid any threat of price competition from a productive capacity that exceeded demand for many decades. At one point during the recession of 1958, for instance, allowable production days were down to eight per month, in order to hold the price of crude oil firm. The federal government also lent its essential cooperation to the system with the passage of the Connally Hot Oil Act of 1935, which forbids interstate sale of oil produced in violation of state quotas.

During 40 years of its operation, the system helped support the price of domestic crude oil. Combined with the oil import control system, it also insulated domestic oil prices from competition in the world market. The industry was thus able to exact higher prices from the public at large. That such a costly, government-created, price rigging system should have lasted for so long is certainly testimony to the political influence of the oil industry.

The history of the oil import quota also reflects the exercise of economic and political pressure. The growth of lower priced oil imports during the 1950s threatened to undercut domestic oil prices and sales by domestic producers. They demanded that government strictly limit imports. Initially, international oil companies opposed import controls, but as world crude prices continued to drop, and U.S. prices held steady, they learned to live amicably with the quotas. The value of their own domestic production and reserves, which were substantial, was protected by quotas. In addition, the quota system proved to be profitable for refiners throughout the industry. A certain quantity of imports was permitted, and was rationed to refiners roughly on the basis of previous import patterns, though small independent refiners won more than their proportionate shares. Import tickets, permitting a refiner to buy a barrel of foreign oil, became as good as cash. If U.S. crude sold for $1.25 a barrel more than imported oil, then the ticket was worth $1.25. Many independent refiners in the middle of the country far from ports swapped their tickets to the majors for guaranteed supplies of domestic crude.

In 1970, President Nixon's cabinet level Task Force on Oil Import Quotas estimated that consumers paid about $5 billion more in 1969 alone for oil products than they would have if trade had been unrestricted. The Task Force recommended that quotas be abolished. They suggested tariffs instead, if the domestic oil industry still needed protection, so that the U.S. Treasury and not private companies would reap the differential between U.S. and foreign oil prices.

The President did not accept the recommendations of his prestigious Task Force—a decision due, in no small part, to the influence of the oil industry, together with the indifference of the consuming public.

The political influence of the oil companies is also discernible in the tax treatment they receive, which is much more favorable than that accorded to industry generally, and more favorable than the tax treatment accorded to competing fuels.[16] The tax code provides a deduction for depletion of oil and gas deposits, either in the United States or abroad, calculated for each productive property as the larger of (1) a portion of capitalized cost of the property or (2) 22 percent of the gross value of production at wellhead, not to exceed 50 percent of the net income for a given tax period.

The first alternative is called "cost depletion" and the second, which is much more widely used, is called "percentage depletion" or, more popularly, the depletion allowance. Previous to 1969 the depletion allowance was 27.5 percent. Percentage depletion bears no relationship to the investment necessary to discover an oil or gas field. For any given productive oil and gas field, the sum of all percentage depletion deductions over the life of a field is likely to be many times the value of the original investment in the oil field.[17]

Other fossil fuels also receive depletion allowances: uranium at a statutory rate of 22 percent, shale oil at 15 percent, and coal at 10 percent. Effective depletion rates, taking into account minimum tax laws and the stage at which the allowance applies, enhance the relative advantage of oil and gas over uranium and particularly over coal.

In addition, the oil and gas industry benefits from the treatment of intangible drilling costs, which are deductible against income immediately rather than being charged off over some estimated useful life of the property. The intangible drilling costs include all drilling costs except for the cost of depreciable property used in the drilling, whether the hole is productive or is dry.

The special tax benefits of percentage depletion and the deduction of intangibles for oil and gas, plus the normal investment credit for tangible drilling costs, are equivalent to an investment credit to the oil and gas industry of about 50 percent. The cost of this special tax treatment of fossil fuels to the U.S. Treasury is estimated at about $3.6 billion a year, largely accounted for by oil and gas.

A subsidy of this size raises the questions, first, of why oil and gas should get special treatment, and second, has it been a successful policy?

Special tax treatment of the oil and gas industry is frequently based on the claim that the business is uncommonly risky, so that if special tax treatment were not given, the flow of investment into oil and gas operations would be inadequate. It is true that several hundred thousand dollars to several million dollars may be invested in a drilling enterprise only to find a dry hole; but the price of oil can reflect the risk of dry holes, which are a necessary cost of finding oil.

It is well known in advance that some wells will produce nothing. Oil producers know they must drill a large number of wells and that if they do a certain success ratio can be expected. Furthermore, high risk is not a unique characteristic of the oil and gas industry. Small businessmen, entertainers, and indeed many lines of business face far greater financial uncertainties. But even if risk is greater in the oil industry than in other industries, percentage depletion is an inappropriate policy because it helps the successful driller more than the one who may never get beyond the dry hole, and thus benefits the large firm more than the small independent producer.

More basically, the price of oil can and should reflect the risk and provide the appropriate financial incentives for exploration and development. The price can enable the industry to attract the necessary capital without tax exemptions that give high income investors a tax shelter and leave huge oil companies paying a lower percent of income taxes than a poor working man.

A major rationale is that the tax incentives are to spur domestic development as a national security measure to avoid reliance on foreign oil. Domestic sources are, of course, more secure than foreign sources, but ironically the depletion allowance also applies to overseas production, supplemented by even larger tax benefits than for domestic production. It is true that price alone will not automatically reflect the greater value of domestic oil to the nation. But as we have explained elsewhere in this book, stockpiles or a tariff on imported oil are a more effective (and less costly to the public) method of meeting the security need.

The arguments advanced in support of favorable tax treatment accorded to the oil and gas industry are therefore not compelling; in most cases, the purposes they are said to fulfill can be met by more direct policies that cost the public less money. The effect of current tax policies, aside from being discriminatory and expensive, cannot be shown to have been very successful. The production performance of the oil industry has been very indifferent in past years. Production declined since 1970, reflecting a prior fall in both oil finding rates, and number of wells drilled. The peaking out of domestic production cannot of course be blamed on the tax incentive. The point rather is that domestic production since 1970 has failed to meet the growth of domestic demand, despite the generous subsidies.

A further disadvantage of these tax benefits, from the conservation point of view, is that they tend to subsidize the price of oil and thus encourage greater consumption during an era when conservation should be our goal. Studies carried out for the Project suggest that crude oil prices were about 5 percent lower than they would have been without special tax treatment. Consumption was thereby encouraged and our resources of fossil fuels used up more rapidly than they would have been without such tax treatment. The subsidy for oil also distorts the competition between oil and other fuels, and thus increases dependence on oil.

On all grounds, then, it seems advisable to abolish percentage depletion and the expensing of intangibles both in this country and on the operations of American oil companies abroad, where the domestic security arguments cannot conceivably apply. This is an opportune time to take such a step. The recent sharp rise in crude oil prices means that the oil companies' cash flow is high enough to absorb the loss of these tax benefits without higher prices.

It is estimated that the value per barrel of crude of depletion and expensing is some 70 cents, whereas prices (old and new oil averaged) have risen by about $3 a barrel over last year from a level at which oil companies were still making profits comparable with the rest of manufacturing industry. The net increase will still be large enough to offer abundant economic incentive to further exploration. The withdrawal of the tax subsidies would have the additional advantage of capturing for the Treasury part of the very large transfer of purchasing power from energy consumers to energy producers which followed the sharp price rise.

Oil companies with foreign operations benefit from additional tax preferences. Like all U.S. business firms operating abroad, they are not taxed twice over on foreign income. The host country normally takes first cut at taxing income, and U.S. tax law allows a credit against the corresponding U.S. income tax liability for foreign income taxes already paid.

For the oil companies, the question at issue is whether their payments to host governments, which represent the major cost of operation, should continue to be counted as an income tax and thus a dollar for dollar offset against U.S. taxes. If they are considered to be a royalty, an excise tax, or payment for oil purchases, just another form of tax or royalty, then they are merely deductions as business expenses and worth only half as much in offsetting the payment of U.S. taxes.[b]

There are several reasons to consider that payments to host countries should be classed as business expenses rather than dollar for dollar deductions. Since payments are levied on each barrel, the "tax" in practice is more like an excise tax than an income tax. Also the governments in the OPEC countries are, in effect, the owners of oil in the ground, so that such payments resemble royalties rather than taxes. Foreign tax credits are not normally extended either to royalties or excise taxes.

---

[b] For example, assume a U.S. oil company earns $100 million abroad and pays a foreign tax of $60 million. If this tax were treated as a business expense, the U.S. tax would be levied on $100 million − $60 million = $40 million net income, which at the U.S. tax rate of 48 percent would be (48 percent times $40 million) = $19 million. But if the host country take is assumed to be an income tax, the same company would not pay any U.S. tax at all having already paid more ($60 million) income tax to the host country than the U.S. tax liability of 48 percent of $100 million. Instead, it ends up with credit of $60 million − $48 million = $12 million.

Furthermore, the tax paid by oil and gas companies in the OPEC countries (where most foreign oil and gas production occurs) is not a general tax, but applies only to oil and gas companies. In some of the OPEC countries, general business income taxes do not exist. In many respects the tax situation in the OPEC countries, especially in the Persian Gulf countries, looks like an accommodation by host governments to operating companies, at the expense of other taxpayers.

An important feature of these tax arrangements is that taxes so defined are much higher in host countries than they are in the United States, so that companies operating abroad typically have large excess tax credits that, under certain circumstances, can be used to offset taxes owed to the U.S. Treasury as the result of operations in other countries. Insofar as such tax provisions favor the vertically integrated rather than the nonintegrated company, they reinforce tendencies towards concentration in the oil industry both at home and internationally.

It might be thought that the expressed intention of OPEC to take over production facilities in the near future automatically solves the tax credit problem. But in the meantime—and the interim may last for several years—the oil companies stand to run up enormous tax credits following the rise in crude prices. For this reason, action should be taken immediately to include the major part of the payments to host countries as business expenses rather than dollar for dollar tax deductions. The retention of all these tax privileges in the face of constant criticism testifies to the political strength of the oil industry.

Another area where the industry's political influence appears to have had a crucial effect is in antitrust actions. There can be little doubt that such influence has been a major factor behind the failure of antitrust enforcement. This failure has been a constant through both Democratic and Republican administrations.

In the late 1930's, the government charged 22 major integrated oil companies and 379 of their subsidiaries with monopolizing crude oil production, transportation, and marketing. The Justice Department's suit asked for injunctions against anticompetitive practices and also demanded sweeping divestiture of transportation and marketing facilities. That action ended in a consent decree, which several of the government lawyers who worked on the case refused to sign.

International oil companies fought a Truman Administration antitrust suit in 1952, refusing to produce subpoenaed documents on the ground that they contained "national security" information that would benefit Communists. In 1953, the new Eisenhower administration dismissed the grand jury investigation, citing national security reasons. In 1972, the Antitrust Division prepared to investigate potential antitrust problems relating to the Trans Alaska Pipeline which will be largely controlled jointly by three companies: Exxon, British Petroleum (through its partially owned operator, Standard Oil of Ohio) and Arco. This investigation was vetoed by Attorney General John Mitchell.

The background to these events is complex and is open to many interpretations and there is no doubt an explanation that would justify each of these actions for many people. What concerns us is that they all lead to the same result—a failure to rigidly enforce the antitrust laws. The laws themselves, particularly Sections 1 and 2 of the Sherman Act and Sections 2, 3, and 7 of the Clayton Act (as amended), appear to be fundamentally sound. Moreover, the laws have even been reasonably successful in those industries where they were enforced. In the petroleum industry, they have not been rigidly enforced.

A survey of legislative and administrative regulations directly affecting the industry discloses that the only policy in which the oil companies have not been effective in imposing their wishes has been in the regulation of the field prices of natural gas. Even this was something of a fluke. Field prices of natural gas are regulated as the result of a 1954 Supreme Court decision which interpreted the 1938 Natural Gas Act to make it applicable to producer sales of gas in interstate trade. Congress, supported by the oil industry, has twice since passed legislation to amend the Act, but each bill met with a presidential veto—on the last occasion because of the attempted bribery of the late Senator Francis Case by a lobbyist.

Looking towards the future there would appear to be an even greater need for more rigid enforcement of the antitrust laws and other policies to protect the public interest. There is reason to believe that if events are allowed to take their course, the energy industries will become even more concentrated and more powerful.

In almost all branches of the fossil fuel industry, there appears to be a trend towards larger size firms and increased concentration ratios in production. Both contribute to economic and political strength. In coal, large producers are better equipped to fill the long term contracts that are favored by many utilities and are suited to the enormous needs of coal gasification plants. Reserve ownership of coal, which indicates concentration trends in future production, is much higher than the concentration based on existing production.

In oil, most considerations point to an increase in concentration. The Federal Trade Commission figures on reserve holdings show that the top four, eight, and twenty firms control 37, 64, and 94 percent of proven reserves, respectively—a much higher ratio than existing production ratios. Future lease sales and exploration are expected to come from the OCS, and the size of the sums required for exploration and drilling appear to give the advantage to large firms.

The future position of the independents in marketing depends partly on the success of independent refiners, from whom they receive the bulk of their supplies, and partly on the marketing practices of the majors (which in 1973–74 took quite a toll of the independents).[18] It is by no means certain that independents will be able to compete in the future on the scale of the past.

The future of competition in natural gas production is unpredictable. Larger firms have the advantage in developing offshore gas; much of it is found in connection with the search for offshore oil, which requires substantial funds. But much of the remaining gas is onshore and independents could grow in strength in gas production.

Prospects in the uranium mining and milling industry depend upon the rate at which nuclear generating plants are constructed and placed in operation. The structure of the uranium enrichment industry will turn primarily on the technology; the centrifuge process is smaller in scale than the gaseous diffusion process and is likely to bring more companies into the business.

A further factor is contributing to closer government-industry relationship and, therefore, to greater industry influences. The present price of Middle East oil is arbitrary, with no relation to its much lower costs of production, presumably reflecting a balance of political and economic needs of the OPEC member countries, plus testing what the market will bear.

We have concluded in another chapter that prices are likely to stay in the current range, or go higher, but so many factors are involved that the economic planner and businessman would be imprudent not to include the possibility of even a sharp price fall in his view of the future. In such circumstances, those contemplating investment in energy related fields may ask for some price or purchase guarantee from the government before they undertake investment. And it will be exceedingly difficult to deny it, especially if the nation continues on the *Historical Growth* pattern in energy development.

What can be done to assure that these vast industries are to be answerable to some sort of public control? We have noted that the power of the energy industries depends not only on their size and structure but also on the nature of the government decision making process. The first step, therefore, is to modify the political system so that their disproportionate influence is removed. Economic and political power must be uncoupled so that the economically powerful can no longer run the government to achieve the policies that suit them best regardless of whether those policies suit the rest of the people. These are fine words. How can they be implemented?

The oil industry makes enormous contributions to political campaigns. A crucial action, therefore, is a strong campaign finance law that would reduce the political power at present exercised by the big contributors. The seniority system in the Congress tends to stifle the wishes of the majority in a variety of issues because of the extraordinary power it confers on committee chairmen.

In the past, a disproportionate number of committee chairmanships and the positions of power have been held by congressmen from energy producing states. Diffusion of power in the Congress would reduce the influence of the energy industries.

If some uncoupling of political and economic power could be achieved, it would be easier to put in effect the measures needed to improve competitive forces and economic efficiency. Our survey of the structure of the industry indicated some of the areas of weakness in competition. Concentration of market power in the industry has risen, following the series of mergers of the 1960s. Further horizontal mergers between major oil companies (the kind which created Atlantic-Richfield, for example) should not be permitted. Also further mergers of coal companies with petroleum companies or those in other industries should not be permitted. Entry into the coal industry should be by establishing a new company and thus enlarging competition, not by merger.

On the other hand, we find no basis at present to

suggest the breakup of the majors into constituent parts. Rather, government policies which have contributed to bigness and vertical integration must be changed.

- Tax policies have encouraged integration by in effect providing the vertically integrated company with lower cost crude oil than is available to independent refiners; only conpanies that produce their own crude can take advantage of the tax benefits. The abolition of the depletion allowance and expensing intangibles previously discussed would eliminate a major financial advantage for the integrated company.

- The regulation of pipelines, including gathering lines, should be carefully investigated to see whether the majors are using their dominance to consolidate their own position. This is an area for stricter enforcement of existing laws.

- Present leasing policies favor the large companies. Royalty bidding on federal leases would be a possible means of encouraging entry. Joint bidding between majors should be forbidden.

- Though prorationing is not now operative, it is in ready reserve for use to control domestic production, just in case the world is ever faced with a surplus of oil again. Now is the time to remove the federal underpinnings to the system—the Connally Hot Oil Act.

- The question of interlocking directorates, which appears to be so important in the development of a common industry view, could be adequately dealt with by the implementation of the Clayton Act.

These measures are easy to list. But unless the political influence of the industry can be overcome, such a program has little chance of being implemented. For this reason, it is often suggested that in order to assume social control over the energy industry, government must take a more active role in energy supply, as indeed happens in many other countries.

Of course, such measures would encounter even stronger industry opposition. But a new element in the political equation is that with shortage and sudden sharp price rises, energy has directly touched the lives of people, leaving them angry and frustrated at failures of government policy, and making new policy departures politically feasible. The democratic system can really work when enough people become personally concerned. If citizens decide on a further degree of social control over the energy industry, a number of techniques involving differing degrees of government intervention may be considered.

The energy companies could be required to incorporate federally. State incorporation laws cover little more than internal relations between shareholders and officers. Many states, eager for corporate charter fees, have engaged in a competitive race to lure corporations, Delaware taking the lead with the least stringent requirements and the most corporate charters. If a state should wish to impose some tough standards or duties on the corporation it charters, companies can simply move to another state. Besides, a state with its limited geographical jurisdiction can hardly enforce obligations on worldwide companies. Says Ralph Nader,

> "Our states are no match for the resources and size of our great corporations; General Motors could *buy* Delaware . . . if DuPont were willing to sell it. . . . To control national or multinational power requires, at the least, national authority."[19]

Federal chartering would establish nationwide standards for major interstate corporations, governing not only relations between shareholders and officers, but between the corporation and the public too. The first thing federal chartering ought to require is fuller disclosure. Conglomerates present their financial figures lumped all together, with no separate accounting for their various diverse enterprises. Energy companies could be required to make public their reserves as well as their finances, thus filling in a big gap in the knowledge the nation needs to make energy policy. Federal chartering could make disclosure of stockholders a matter of course. It could include requirements for a publicly appointed member of the Board of Directors, to serve as a watchdog and voice of the public within the corporation.

A federal chartering agency for large interstate and multinational corporations would supplement, not replace, other regulatory bodies. No regulatory body is immune to capture by the industry it is meant to regulate. But an agency that deals with many kinds of companies is harder to co-opt. For example, compare the record of the broader agencies, such as the Environmental Protection Agency with agencies such as the Interstate Commerce Commission to see the point. Federal chartering could impose legal requirements to invest in developing more supply and set limitations on profit margins.

Oil and gas companies, unlike regulated public utilities, have no legal obligation to maintain adequate supply. A utility type of regulation has been suggested as a means of placing this responsibility on the oil and gas industry and at the same time controlling profits. A further step to assert public control over oil industry activities would be to establish a federal "yardstick" corporation for oil and gas development. The final option is nationalization of the industry. (These options have all been discussed in Chapter 2.)

No matter which route is taken to exercise public control of these large private industries, citizen participation is essential. In fact, without an aroused and interested public, the history of industry domination of national energy policy is apt to continue into the future, and none of the possibilities for reform that we have discussed will come to pass. Simply establishing a mechanism for greater public control is also likely to come to nothing if citizens assume that some agency in Washington will take charge of the matter and they can forget it.

Citizen interest and action must be connected to the decision making process on a continuous basis so that it becomes a lobby or a pressure group to be taken into account when decisions are made. This is not a utopian idea; everyone who reads this book can cite some personal experience in which determined citizen action changed the system. To take an example close to home, the difference between schools is often largely explained by the quality of continuous citizen participation in school and school board affairs. This is the sort of intervention that is needed if government is to stop being the ally of the energy industries and become instead the ally of the citizens who elect and support it.

CHAPTER 10

Reforming electric utility regulation

The electric utility industry has the distinction of being the biggest American industry, measured by total assets. It is also the most capital intensive, requiring more annual investment than any other industry, about 12 percent of the total each year. Since World War II, the electric utility industry has grown fast enough to meet a demand that doubled every ten years, while America's energy consumption as a whole was doubling in twenty, or more recently, fifteen years.

Electric utilities achieved the remarkable feat of keeping prices nearly steady for twenty years. Retail electricity prices went up only 12 percent from 1947 to 1967, while the general consumer price level advanced 50 percent. In some regions, householders' electricity bills actually went down.[a]

[a] As late as 1974, the bill for a typical all-electric house in Boston was $152 for the month of January, compared to $185 fourteen years earlier, in 1960.[1]

But the picture has changed in the past few years. In the mid 1960s, the long trend of improving efficiency in the generation of electricity came to a halt. Nationwide, rates began to rise, and after 1970 they rose faster than the cost of living. In New York, where residential rates are higher than in any other large city in the country, the average householder's monthly electric bill more than doubled in three years, from 1971 to 1974.[2] Outraged consumers who had long taken the good gray electric company for granted began to storm utility commission rate hearings. After a few years of sporadic brownouts, even the reliability of electrical service was no longer certain.

If consumers felt abused and overcharged, investors were thoroughly disenchanted. Industry earnings were on the decline, and some utilities began to experience great difficulty in selling their bonds, which once were deemed gilt-edged. The cost of everything that goes into building new capacity—money, labor, building materials—soared. No longer was it cheaper to produce electricity from new plants than from old.

The prices that utilities paid for fuels to run their turbines rose precipitously too. In 1973, fuel oil prices to large users increased 85 percent or more, depending on the region and the type of oil. Coal prices, largely covered by long term contracts, rose about 15 percent in most regions. (Some spot market coal prices doubled during the year.)[3]

Meanwhile, the environmental movement had stirred into vigorous life, and people became aware of the non-monetary costs of generating electricity. The electric power industry is the nation's largest fuel consumer by far, and one of its biggest land developers. By the same token, it is also a leading pollutor of the air and despoiler of the countryside. The industry, which once consulted no one but itself in planning new plants, now had to respond to environmentalists' concerns. And the smokestacks that once poured soot, ash, and sulfur oxides into the air had to be fitted with costly pollution controls.

If the past was serene, and the present is troubled, what about the future? It is promising. It is likely that whatever energy future the nation may choose, and however successful we may be at conserving evergy, an increasing share of energy growth will be in the form of electricity.

Electricity has the commanding advantages of flexibility. All sorts of energy sources, from abundant coal through nuclear power to geothermal, solar, wind, and recycled waste power, can rather readily be used to generate electricity. Electricity is also versatile in its end uses. It can cook food; heat, light and air-condition houses; turn on the television set; and run trains and subways as well. For transportation, its possible uses have scarcely been tapped. A further advantage of electricity is that it is clean at the point of use, even though its production may cause a great deal of pollution. If ways can be found to produce it with less pollution and less waste, its advantages will be even greater.

The public interest in electric utilities focuses on two points: the price companies charge, and the damage they may inflict on the environment. Less obvious, but also important in its ultimate effect on the consumer, is the organization of this vast industry. Utility regulation deals directly with these matters. This chapter is concerned with reforming utility regulation in an effort to better serve the interests of the public, and at the same time help the industry manage its serious problems.

Pricing

Electric utilities are "natural" monopolies. State and local regulatory commissions control their prices, setting rates to cover costs and allow for a fair profit, and protect the companies from competition. In return, utilities are legally responsible for "keeping the lights on "—that is, for anticipating and meeting demand. When the costs of operating old plants or opening new ones increase, utilities are obliged to maintain service, and regulators are obliged to raise rates to cover the higher costs.

So long as prices were low, the subject of a "rate design" for electricity was generally ignored. Regulatory agencies concentrated on the level of company earnings allowed and left them free to spread the price burden among their consumers, as they saw fit, as long as no unjust discrimination took place.

The most widely used rate schedule charges residential customers according to their use of electricity. The more they use, the lower the unit price. For example, the "declining block rate" system charges one price for the first 250 kilowatt hours, a lower price for the next block, and so on. And most utilities also offer lower rates for special uses such as electric home heating and hot water heating. The effect of such a rate structure is to promote the use of electricity.

In the past, these promotional rates reflected the trend of industry costs which were declining. Rates that promoted greater use gave the companies more income, which in turn enabled them to build large new plants that further reduced costs and rates. Recent studies conducted for the Project support the conclusion that this happy era has ended. The new cost trend is up. The maximum size of plants, about 1,000 megawatts, is no longer increasing because economies of scale are captured below that size. There may still be some economies of scale in distribution; it costs more to hook up many small users than a few big ones. But these economies are offset by rising power plant and fuel costs.

Under these conditions, promotional pricing is inappropriate. Present rate designs, which promote greater use through price reductions, eventually push prices higher for all customers. When enough customers take advantage of cheap promotional rates—for example, for electric heat—the resulting increase in demand forces costly expansion of capacity. Within a few years, this cost backlash can double the heating bill for an all-electric home. People who are thus victimized can become upset enough to turn utility rate hearings into near riots, as recently happened in New York's rich, respectable Westchester County.

Another fundamental problem confronting the utilities is uneven demand. Electricity demand fluctuates by season and even by hour during the day. There are times when everyone wants to use it at once, such as the early evening hours, or on very hot days when everyone wants to turn on the air conditioner. These are "peak load" times, which the electric system must be designed to accommodate, even if the equipment is not needed at other times. Without the extra capacity, brownouts (reduced power at peak times) will occur.

In most instances, today's prices for electricity make no distinction between loads that add to the peak and thus require more capacity to be built and those that are off-peak and do not. The industry's legal obligation to provide service includes building enough capacity to meet peak load, plus providing some designated reserves. Some small plants are built solely for use during peaks. If pricing provided incentives for consumers to use less electricity at peak hours and during peak months, these plants would not be needed. A pricing system that does not charge customers a fair and full cost for their peak period consumption, when it is worth more and when it costs more because extra plants have to be built simply to cover the peak, wastes money by causing overbuilding of capacity.

It clearly makes more sense to price electricity to encourage conservation—especially during peak periods, but also at off-peak times. A move away from declining block rates will help to eliminate the incentive for indifferent and wasteful consumption. The advantage of peak load pricing is that people who use capacity on peak days (air conditioning in summer or, in some areas, electric heat in winter) will actually pay for the extra capacity needed to serve them. With promotional rates, the peak load customer sometimes gets a cheaper rate because he is a large user—a case in point is air conditioning. High peak load rates may also encourage people to conserve energy by insisting upon more insulation, or better efficiency in air conditioners, or even solar homes. They might encourage industry to shift their use to off-peak periods or to design plants so that power can be interrupted during peak periods.

Peak load pricing would mitigate the strange "Catch 22" effect experienced by those customers who conserved electricity and were rewarded with higher electric bills. This happened because the reduction in use on an overbuilt system designed to meet high peaks and wasteful use resulted in higher costs per kilowatt hour to all customers. The rate system was designed to promote more use and when people started conserving, the system backfired on them. If rates were sensibly designed to discourage peak use and wasteful use, a consumer who saved electricity would save money as well.

There are, of course, difficulties in implementing a peak load system of pricing. It may require a time meter in every home or place of business so that the customer can be billed according to the coincidence of his use with the system's peak. There is also the suggestion that the pricing system may cause the peak hour to shift, resulting in unstable rate schedules.

But that "problem" does not appear to be a serious concern. It implies such a degree of success in reducing peaks that revising the rate schedule would be a trivial matter compared to the benefits. As to the metering problem, residential meters cost less than $60 and industrial metering costs are small compared to potential benefits. Several utilities are experimenting with time meters in the United States. In England, time meters are used for appliances that consume large quantities of electricity, such as hot water heaters. And it may be possible to approximate peak load pricing—at least for seasonal loads—for residences without a demand meter.

Research done for this Project suggests that drawbacks to the use of a peak load pricing system are not insuperable, even though it would, of course, take time and experience to be certain.[4] The important point about changing to a peak load system of pricing is that it will create incentives to use available capacity more efficiently.

The reforms we propose are fundamental. We would eliminate the sliding scale of rates, which reduce prices as consumption increases. We would base the price of electricity on a capacity charge for the power used on peak days plus a price for each kilowatt hour used which reflects its costs. Consumers who add to the peak load would pay much more than those who draw on capacity at slack times. Industrial customers who require no distribution facilities would, of course, not be required to pay the cost of an expensive distribution system to serve residential consumers. But no user would be encouraged to increase consumption through promotional rates.

This basic reform would eliminate a system of pricing that encourages greater consumption and would substitute a system designed to encourage more frugal use of electricity. Peak load pricing thus has the potential for holding down electricity prices. By reducing the number of new power plants required, it also can help protect the environment. This incentive system of pricing offers exciting potential for helping the electric industry thrive in an era of energy conservation.

Site selection and coordination among utilities

Environmental problems relating to plant siting and pollution are certainly not unique to the electric power industry. However, the large generating plants and transmission facilities are quite visible. In addition to being esthetically displeasing, they consume and pollute land, air, and water. (Chapter 5 discusses the specific environmental problems encountered with nuclear and fossil fueled power plants.)

These problems raise serious concerns about how decisions to locate and build these facilities should be made, and about who should have the final say. From the industry's standpoint, the present system fails to provide uniform rules and standards to guide the selection of sites. While the industry is still in charge of its expansion plans, it must obtain approvals from many government agencies at the local, state and national levels.

From the public's viewpoint, the worst result of multiple, uncoordinated, state-by-state licensing of plants is that there is no chance to select the best site in the region and require that the plant be located there. And there is no way to involve the public in such basic questions as whether a given plant is really needed or whether fundamental alternatives to plant or system development have even been considered. The responsibility for raising such questions and pursuing answers goes begging. As a result, power is needlessly expensive and the environment is inadequately protected.

Several states have passed laws centralizing the authority and procedures governing the licensing of power plants and transmission lines. State laws cannot, however, deal adequately with the regional or national ramifications. Yet most power plants are part of regional power pools, and decisions increasingly are—or should be—made within the regional context. There is an urgent need logically to extend the planning and decision making process to cover the broader geographical areas which now comprise reasonable planning units.

This, of course, was not always the case. Eighty years ago, the electric power industry was made up of unconnected individual companies serving separate urban areas. Over the years, to achieve economy and improve reliability, these companies hooked their systems together, first into larger companies serving several cities, and later into power pools covering broad areas of the country. The power pools are not single corporate structures, but are, in essence, loose confederations of individual utility companies. Although they have some shared interests, the companies are largely autonomous and retain control over expansion and siting decisions.

Yet, because the generating and transmission facilities are large and expensive, their benefits and problems spill across regions and sometimes affect the entire nation. Whether one part of a region rich in coal should serve, willingly or unwillingly, as the utility base for the densely populated coastal areas within the same region is at least a regional issue. And when the source of power supplies and the destination of the electricity lie in separate regions, the issue becomes one of national concern.

For example, is it in the nation's interest that the Southwest keep its energy resources and preserve its natural environment? Or is it preferable to exploit some or all of the Southwest's energy resources, and transport the energy out of the region to sustain the economic prosperity of citizens in the other regions far removed from the degradation and impacts of such development? It is clear that no single state utility commission can begin objectively to assess such broad issues.

While most power companies engage in some form of regional planning, many are still going it alone in developing construction plans. And the regional planning that does occur may not capture all the savings of a truly integrated regional generation and transmission system. The basic economic decisions are made by individual companies concerned with problems of local taxation, corporate rate base, and other parochial matters, rather than on a regionwide basis of maximum efficiency.

And it is doubtful that inter-regional power lines are being built to take advantage of diversity in loads among regions and cut down on the number of power plants required. Industry reserve margins are now 20 percent, as compared to a 15 percent target set in 1964 for a fully coordinated industry.[5] The 5 percent difference would save over $10 billion of capital investment, which means saving the equivalent of 20 huge power plants today—and more in the future.

Industrial organization and competition

The electric utility industry consists of about 300 privately owned companies that generate, transmit, and distribute about 75 percent of the electricity in the United States. In addition, there are some 150 small investor owned systems, over 3,000 governmental systems (mainly municipal), and over 900 rural electric companies.

Most of the public and cooperative entities are only distributors. A few government and cooperative entities generate and transmit electricity. These range in size from the largest generating company in the United States, the Tennessee Valley Authority, to small cooperatives. A few companies also sell natural gas and steam.

In most other industrialized nations, the generation and transmission of electricity is a separate function handled by one organization which wholesales power to local distributors. However, in the United States the great majority of consumers are served by unified "vertically integrated" companies that generate, transmit, and also deliver electricity to the ultimate customer. Superimposed on this system, and depending on the region, the utilities are also interconnected by a series of regional transmission grids that permit the flow of power among utilities and regions.

But bulk power supply—the generating plants and the power lines that connect them and deliver power to metropolitan centers—appears more and more to require some sort of regional management system to govern its expansion. While we expect bulk power supply to encounter healthy competition from on site generation in the future, central stations will also have to be built. A truly regional management system could better realize the economies of a completely integrated grid and provide "elbow room" for finding sites that fit land use plans.

Bulk power supply (generation and transmission) might be better handled by a dozen large regional companies, which could fully use the available technology and minimize their difficulties. Some financial and other organizational changes would also be needed to make such a system work. But the advantages of regional generation and transmission companies do not apply to the distribution of electricity in a metropolitan area. Distribution facilities are not characterized by substantial economies of scale, so consolidation of these facilities would not produce major savings. In addition, substituting a large distant utility for the local power distributor would run counter to consumer aspirations for more accountability and responsiveness.

Thus, while there is nothing to be gained and something to be lost in consolidation of distribution systems, it is desirable to examine alternative forms of organization that would permit regional management of bulk power supply under a single corporate entity. Existing utilities in a region, while retaining their distributor role, could establish a jointly owned single company for bulk power.

To ensure that these large and powerful private companies were responsive to consumer concerns, there should be public participation in their management. It would be appropriate to have a public watchdog and ombudsman within the highest management councils. One or two public directors, with full access to corporate files and board deliberations, would be in order. Publicly owned regional entities, such as the Tennessee Valley Authority, offer an alternative form of organization. But such a system would be a considerably more radical departure than the concept of privately owned regional bulk power systems.

Within such a revised framework, limited "competition" should be encouraged. This would assist the regulatory agencies and the public in their attempts to improve the industry's efficiency. On-site generation of electricity by industry (see Chapters 3 and 4) and total energy systems in some instances can provide more efficient power at competitive costs and should not be denied to the public in the name of protecting a "natural" monopoly. Direct competition between power companies could occur if the transmission grids were considered "common carriers." Innovative distribution systems could obtain power from alternative suppliers. Careful study would be required before each innovation was implemented. The ultimate goal is building an efficient overall regional generation and transmission system.

Another aspect of competition in the electric industry requires reform. The era of promotion is over and should give way to conservation. As nonprofit public agencies, publicly owned power agencies such as TVA and Bonneville Power and the thousands of small municipal systems are exempt from federal income taxes. While these agencies and systems make payments in lieu of state and local taxes, the price they charge for electricity should also include a fair share of the cost of maintaining the federal government. Low priced power that doesn't reflect society's total costs no longer serves the public interest. An excise tax should be imposed on electricity sales by publicly owned agencies. The tax rate should be the same percentage of revenues that a private power company would pay in income taxes if it were serving those customers. This excise tax would help promote energy conservation and spread the federal tax burden fairly to all consumers.

Regulation

The electric utility industry is subject to several layers of federal and state regulation. The Federal Power Commission (FPC) regulates the price for interstate wholesale sales of electricity, which are a small part of all electricity sales. Most sales are at retail or are made direct to industry. These are regulated by state utility commissions. The FPC can, under certain conditions, order interconnections between the systems of two power companies. For companies subject to its jurisdiction, the FPC supervises the utility accounting and may investigate matters of general importance to the industry. The U.S. Atomic Energy Commission licenses the construction and operation of nuclear power plants. The Securties and Exchange Commission approves issuances of securities for utilities not subject to FPC jurisdiction. The Rural Electrification Administration (REA) provides financial assistance to rural electric cooperatives and the Department of the Interior controls the budgets for the Bonneville Power Authority and a host of other regional marketing agencies that sell power generated by government projects, mainly hydroelectric facilities.

At the state level, retail electricity prices are normally subject to approval by a state or local regulatory authority. A few state agencies have authority to license sites for generating plants or transmission facilities. The Environmental Protection Agency, state and local environmental agencies, and local zoning boards all now play a more vigorous role in land use decisions, largely in response to local citizen pressures. This crazy-quilt of uncoordinated government actions does not add up to a coherent program to assure either reasonable prices or environmental protection.

The key question is how to make regulations effective, in light of the problems of inflation, the environment and the need to move toward greater conservation of energy.

The current system of utility regulation that relies primarily on the states may no longer be appropriate for an industry that commonly has firms with operations extending into several states. Plant siting issues should initially be examined by a public agency on a regional basis. However, these issues are normally examined only after sites in a state are selected by the utility. State regulatory agencies are relatively powerless to control such regional activities because their jurisdiction is strictly bounded by state laws. And one state is less likely to innovate by requiring pricing to encourage efficiency and conservation if neighboring states still permit rates that promote greater consumption. State regulatory agencies in most instances are undermanned. Almost half the agencies do not license new construction. They lack the manpower or authority to scrutinize a regional power pool, and in rate cases are largely confined to a postaudit of expenses already incurred.

For these reasons, a study commissioned by the Project recommends the establishment of a regional regulatory agency in place of existing state systems.[6] Because utilities are increasingly operating their generation and transmission facilities on a regional basis, it is sensible to extend the scope of regulation over the same geographical and political boundaries. Such a regional agency, with state government participation but under a federal charter, could marshal the necessary staff and other resources to become a knowledgeable and effective regulator. It could insure that expansion plans were truly regional in scope and were taking full advantage of the economies in integrated planning and operation. Most important, it could provide regional projections of needed additional capacity, and provide a forum for definitive answers about the need for more capacity and feasible alternatives.

Since utilities are increasingly operating their generation and transmission facilities on a regional basis, having the scope of regulation over the same geographical and political boundaries makes sense. Obviously, regional regulatory agencies, like regional generation and transmission companies, have limitations. They are farther removed from local control. To be effective, they would require enthusiastic support from the states. Regional regulatory commissions are obviously more appropriate in some areas than others. New England, for example, is a region of several small states and it has taken some tentative steps in this direction. In contrast, California is in some ways a "region" by itself. But even California depends increasingly on power generated far from its boundaries.

In all cases, regulatory agencies should be able to respond to the need for conservation and environmental protection, and the state regulatory commissions usually cannot. Regional regulatory bodies and regional generation and transmission companies seem better equipped to meet the challenges of utilities management in the coming decades.

Of course, these problems do not have automatic solutions and waving a magic reorganizational wand will not perform a miracle. However, we believe our basic suggestions for reorganization would clearly move the industry in the right direction. Other reforms are worth consideration.

A more simplified licensing process, fewer separate proceedings, and utility expansion shaped to fit regional and national land use plans would improve efficiency. At the same time, licensing procedures should be designed to include a broader range of social considerations than is usually the case today. Government agencies should be aware of the utilities' long range construction plans, and the public should have a role in agency hearings. Several specific and detailed recommendations are made in a study for the Project by the Public Interest Economics Center.[7]

Another utility regulation issue deserving the most serious attention is the relation of the large regulated power companies to on-site production of electricity. The utilities' promotional rates have discouraged the widespread application of this potentially efficient source of electricity. New York's huge but ailing electric utility, Consolidated Edison, provides an illustration in its dealings with the World Trade Center, the twin skyscrapers in lower Manhattan which are famous for using enough energy to supply the needs of the city of Schenectady. Con Ed gave substantial rate reductions to the World Trade Center, and to another very large development, Co-op City in the Bronx, to get their business. Both projects had seriously considered generating their own electricity on site.[8]

The rate design changes recommended earlier would abandon promotional rates and make on-site generation more competitive. Some of the most promising opportunities for saving energy involve building small electric power plants next to industries, within shopping centers, and in apartment complexes. By generating electricity on site, as industry did 30 years ago, many companies would get the use of the electricity and the process steam, and still might have electricity left over to sell to the central power systems. Similarly, total energy units can efficiently generate electricity for some large apartment houses and shopping centers, possibly from trash, by using the waste heat from the electric generating plant to supply heat and cooling requirements.

At present, these sources cannot be assured of a hookup to the central system. Without such standby power at reasonable prices, the market for the more efficient integrated and total energy systems is limited to those places where the risk of total breakdown of service could be tolerated, where the price of electricity is so high that the small integrated energy system is competitive without hooking up to the main system. Since major industries building on-site generation capability may well have surplus power to sell, they will need interchange and backup arrangements on reasonable terms.

The regulatory agencies should make certain that central utility systems act reasonably and cooperatively in supplying backup power and interchange agreements to all types of on-site power installations. Naturally a utility worries about losing business, but it is in the best interest of the nation, and the utilities as well, to achieve maximum efficiency in the use of scarce resources. Regulators should ensure that the utilities cooperate in these opportunities for conservation.

Although the recommendations in this chapter may be somewhat modest, the path to the "best" public policy to govern the electric utility industry is not entirely clear. We believe the regulatory reorganization and electricity pricing reforms proposed in this report can prove effective. But much will depend on the kind of energy future the public wants. As with all reforms, little can be achieved unless able men can be called upon to manage and regulate this major sector of the energy industry.

Vertically integrated electric firms can fail to serve the public, especially when regulation is weak. By severing the industry into its two functional parts—generation and transmission on the one hand, and distribution on the other,—we see potential for improved service. Regional entities concentrating on building a fully integrated generation and transmission system could realize the large opportunities for savings, help keep the price of electricity as low as possible, and, with public representation in management, be more responsive to environmental concerns. A transmission network with common carrier obligations could foreclose redundant, duplicative lines and provide economical transfers of power between regions. Finally, if distribution companies were solely concerned with the ultimate customer, management would devote more attention to the quality of service and consumer needs.

It is not a simple task to improve the operating efficiency and public responsiveness of an industry as large as the electric utility industry. But we believe that our proposals for regulatory and pricing reform will assist the transition to a more functional industry structure, able to deliver more service at lower total cost.

CHAPTER 11
Federal energy resources: protecting the public trust

Several turns in the course of history have made the federal government the dominant energy resource holder in the United States. The energy future of the nation was surely far from the mind of President Thomas Jefferson when he negotiated the Louisiana Purchase in 1802; yet with those vast lands, combined with later cessions from Mexico and Texas, came most of the nation's coal, oil shale and uranium. Nearly a century and a half later, in 1945, President Harry Truman unilaterally proclaimed jurisdiction over the underwater resources of the U.S. continental shelf; he realized that this contiguous offshore region probably contained some of the world's richest oil and gas deposits.

Over the years, many of these lands and resources have passed into private ownership, but the bulk are still under federal control. They are the property of the people of the United States, and their proper management becomes perhaps the most important of the many energy policy responsibilities of the federal government.

The U.S. Department of the Interior has had principal charge of these resources for more than a century. Its execution of this responsibility has been marked by controversy. The mixing of oil and politics in the 1920's culminated in the Teapot Dome scandal, when the Secretary of the Interior was convicted for taking a bribe to allow a major oil company to drill a Naval Petroleum Reserve. This was the Department's lowest ebb. But even in recent years, the disposal of public resources with an inadequate return to the Treasury has been a constant issue. And the disastrous oil blowout on a federal oil and gas lease in the Santa Barbara Channel in 1969 cast doubt on the Department's ability to protect the environment while executing its resource responsibilities. Recent fuel shortages have called into question whether resource exploitation is taking place at a rate commensurate with national needs.

Enormous pressures are now being exerted to open these resources to much more rapid exploitation. National priorities will be affected because the pace at which federal lands are opened can play a key role in determining the overall rate of energy supply growth, the mix of fuels, and the degree to which the nation must depend on imports. It will largely determine whether environment values are respected, what regional values take priority over others, whether the people receive a fair return, and how much of the resource will be left for future generations.

The resources

The federal resource base is extensive, containing altogether over 50 percent of the fossil fuel energy resources in the United States. This resource domain includes all the public lands,[a] the offshore area known as the Outer Continental Shelf (OCS),[b] and the energy resources under-

[a] Those lands acquired by the United States through cession or conquest which have not been disposed of by the government. Included are national forests, national parks and other reserves, in addition to vast areas not designated for any specific use. Some, such as national parks and the Naval Petroleum Reserves administered by the Navy Department, are not open to mineral development. 633 million acres are subject to development under the mining laws. Almost all these lands are in the western states.

[b] Those offshore lands beyond state jurisdiction (usually beyond three miles) out to the limit of United States jurisdiction as defined by international law.

lying some private lands.[1] It is impossible to be sure just how much the public owns, since the government's knowledge of the resources it controls is, in general, inadequate for good resource management. Table 39, based on the best public data available, shows the percentage of federal ownership of various key energy resources, and the 1972 rates of production.

Table 39–**Federal resource ownership and 1972 production**

	Reserves	*Percent of Domestic Total*	*Resources*	*Percent of Domestic Total*	*Production*	*Percent of Domestic Total*
Oil[a] (Billions of barrels)						
OCS	8.7–10.7	11	58–116	30	0.41	10
Onshore	2.8– 3.3	4	15– 30	8	0.22	5
Total	11.5–14.0	15	73–146	37	.63	16
Gas[a] (Trillion cubic feet)						
OCS	57.8–76.8	15	355–710	36	3.04	16
Onshore	24.2–31.2	6	75–150	8	1.06	6
Total	82.0–108.0	21	430–860	43	4.1	22
Coal[b] (All categories, billion tons)	186.9	48	Not available		10 million tons	2
Oil shale[c] (Billions of barrels oil)						
25 gallon-per-ton shale (10-plus thickness)			480	81	(No commercial production)	
15–25 gallons-per-ton shale (15-plus thickness)			900	78		
Geothermal	No breakdown available—approximately 50 percent of domestic total				(No commercial production on federal lands)	
Uranium	No breakdown available—approximately 50 percent of domestic total				Not available	

[a] USGS Press Release, February 1974. Reserves include measured reserves, indicated reserves and inferred reserves. (See definitions, Appendix D). Percentages computed using mid range of estimates.

[b] USGS. Circular 650; Federal reserve estimates derived from "Draft Environmental Impact Statement for the Proposed Coal Leasing Program."

[c] Derived from U.S. Department of the Interior Environmental Impact Statement, Prototype Oil Shale Leasing Program. Estimates for Colorado, Utah, and Wyoming.

Currently, the most sought-after domestic oil and gas deposits are found on the OCS. Since the OCS Lands Act was passed in 1953, about ten million acres have been leased for oil and gas exploitation. About one-third of all remaining domestic oil and gas resources are thought to be in the OCS. The OCS probably contains a still larger share of the most promising areas for near term development. But these estimates are speculative. Much more exploration is necessary to ascertain the true extent of the OCS resource base. In 1972, these lands produced about 10 percent of all domestic oil and 16 percent of all domestic gas; the share of production of both from the OCS is growing annually.[2]

By comparison, onshore federally controlled oil and gas resources are presently insignificant. Under the provisions of the Mineral Leasing Act of 1920 and earlier laws, these resources have been exploited for more than 70 years; in 1972 they contributed only about 5 percent of total domestic oil and gas production. There are, however, vast natural gas resources in "tight formations"[c] in the Rocky Mountains on the federal domain. The Naval Petroleum Reserves managed by the Department of Defense, particularly Reserve Number Four in Alaska, are potentially significant sources of future supply.

About one-half of domestic coal reserves are under federal control in the west. More important, about 85 percent of the strippable low sulfur deposits are in the public domain; these are now in increasing demand. Vast amounts of coal have passed into private control under the Mineral Leasing Act, but relatively little is currently being developed.

About 80 percent of the high grade oil shale is controlled by the federal government. There is so far no commercial shale oil production, but if production of shale oil becomes commercially feasible, federal policies will directly control the extent of its development. Five leases out of a planned total of six in a prototype oil shale program were offered in early 1974, four of which were sold. The program is designed to determine whether oil shale can be developed in an economically and environmentally acceptable manner, and whether a full scale oil shale industry is feasible.

About half the nation's geothermal resources are on

[c] These are deep deposits where the gas is trapped so as not to be commercially exploitable by conventional fracturing techniques. Nuclear devices and hydraulic fracturing have been used experimentally to attempt to free this gas.

public lands, and a leasing program was inaugurated in early 1974. The general commercial and environmental acceptability of geothermal development is also uncertain. An estimated 50 percent of the domestic uranium supply is likewise in the public domain. Most of this is not subject to the leasing laws, but may be "claimed" by any prospector under the general mining laws.

There are other lands and resources which are not directly under the control of the Interior Department but which will be strongly influenced by its policies. The Interior Department is in a trustee relationship with the Indian Tribes and their lands, many of which are rich in resources. Traditionally, the Department has been a powerful influence on tribal councils in their decisions about resources. A number of tribes, with Departmental approval, have leased extensive coal reserves (approximately five billion tons) to coal companies, without any government review of impacts or desirability. Also, many private lands are commingled with federal lands and cannot be economically exploited for their mineral wealth without concurrent sale of federal resources.

In the semi-arid western states where coal and oil shale is being extracted, water is a critical factor. Water in abundance will be required for reclamation of strip mined lands, generation of electric power from coal, and production of synthetic fuels. It is highly questionable whether adequate water exists to support massive development of federally controlled fossil fuel resources in the west, without the extensive reordering of regional water use priorities.[3] It might even require federal funding for huge and costly interbasin water diversion projects.[4]

Most important, extensive production of the vast federal resource base will set the pace for the nature and quality of life in some regions of the country. Economies will shift from agrarian to industrial; large population influxes will follow; and all the human and environmental problems associated with industrialization and a mining boom can be expected.[5]

In managing the national treasure of publicly owned energy resources, the government is responsible first and foremost to all the nation's citizens. The public, nationwide, has an interest in receiving the benefit from these energy resources at a reasonable price; in its position as the dominant resource owner in the nation, the government can powerfully influence price. The general public has another interest in safeguarding the human environment to the fullest extent possible. Third, the public has an interest in lightening the tax burden by maximizing the net revenue from sale of public resources. There is a further important public interest in extending the capability of the nonrenewable resource base as fully as possible, both for this and future generations.

The regional public—those who live in the vicinity of publicly owned resources—have an additional set of interests, focused on minimizing the adverse effects of development and on channeling benefits to those who live and work in the region.

Similarly, state and local governments have obligations to protect the health, safety, and welfare of their citizens. To exercise these responsibilities properly, they must have a role in establishing federal policy and in seeing that the federal government, as landowner, pays the cost of providing necessary services. Finally, those who will use the resource base for economic purposes have an interest in knowing the ground rules under which they will be allowed to use it.

These obligations make the government an anomaly in the marketplace. The government must respond to much more than normal price signals in selling public resources. It cannot, and should not, act as a monopolistic private owner, seeking simply to maximize profits. On the other hand, its obligations go beyond simply responding to energy industry sector demands for the resources.

The Interior Department has three stated policy objectives that are to be reconciled in its leasing program: orderly and timely resource development, protection of the environment, and receipt of fair market value. These stated policy goals for the federal domain are often in conflict. Decision making should be a process of making intelligent choices among and between them. It requires good information and good planning, neither of which currently exists in the federal government's resource programs.

The policy fulcrum of the resource programs is the rate at which the resources are sold out of public ownership for private development. Leasing the federal domain to developers faster than necessary makes it difficult to plan for environmental protection, to ascertain the value of the resources, and to promote competition. It tends to reduce the price paid to the government and to encourage private speculation in these resources at the public expense. Leasing too slowly, on the other hand, could lead to scarcity of these resources for the consuming public and increased prices.

Historically, the Interior Department has sold mineral rights almost entirely in response to demand from the private sector, with minimal government planning. There have been some recent changes in this approach in leasing OCS oil and gas resources, and in the new oil shale program. But the dominant philosophy, with historical roots going back to laissez-faire policies of the 1800s, is that the government acts as interim holder of the land and resources, a conduit for passage into private ownership. Consequently, the pace and manner of development is determined largely by grossly inadequate government planning and strong development pressure from the private sector. Corporate objectives are much narrower than the multiple objectives the government should assert, and the public interest suffers in the process.

A recent decision to increase by tenfold the acreage leased annually on the OCS (from a 1973 leasing rate of about one million acres per year to ten million acres per year in 1975) is illustrative. Following an overall policy goal of domestic energy self-sufficiency, the apparent guiding philosophy behind this decision was to release as much of the resource as could be sold, with little concern for the revenue impact of flooding the market—and with no assurance about when the oil and gas would be produced, or what price the consumer would eventually pay for it. No overall environmental impact assessment preceded the decision, announced in a Presidential Message on January 23, 1974.

The remainder of this chapter will explore the principal policy problems of federal energy resource management. Market forces will be an important element in determining the rate of federal resource exploitation. But the system must also accommodate the important nonmarket objectives we have identified.

A major finding is that the existing Interior Department managment mechanism is not capable of addressing and resolving the crucial policy issues inherent in expanded exploitation of the federally controlled energy resource base. It lacks adequate resource and environmental data, a sound preleasing planning system, stringent postleasing regulation, and in general, a consolidated structure to make sound public policy and carry it out.

The present exploitation system

The laws that form the basis for the exploitation of the public resource base have their roots in several different eras, and reflect differing and often competing policies.[6] Onshore oil and gas, coal, and oil shale are governed by the Mineral Leasing Act of 1920, the product of an era of great oil surplus and incredible waste in production. It sought, with some success, to check the complete giveaway of public resource lands and to promote good conservation practices. The Outer Continental Shelf Lands Act of 1953 reflects the post–World War II energy growth ethic and states that the urgent need for exploration and development of OCS oil and gas should be the major policy determinant. The environmental awakening of the late 1960s is reflected in the National Environmental Policy Act, which requires a number of actions to ensure environmental protection in the resource programs.

A principal thrust of the laws is to ensure that exploration and development take place in a timely manner and that the public, through a system of competitive bidding, receives a fair return from the sale of its energy resources. Conservation of the mineral resources, other adjacent natural resources, and the general environment are themes that permeate the laws. Because of weak and, oftentimes, unenthusiastic administration, these legal objectives generally have not been realized.

The Interior Department decision making structure for energy resources itself has made it difficult to effectively administer the laws. Policy for the resource programs is not established in any coherent fashion. In theory, two Assistant Secretaries of the Interior have had policy responsibility: the Assistant Secretary for Land and Water Resources who is to provide guidance on the rate and manner of leasing, and the relationship of energy minerals to other resources and the environment; and the Assistant Secretary for Energy and Mineral Resources, who has responsibility for overall energy policy within the Department, as well as for overseeing the regulatory aspects of the leasing programs.

In actual practice, this formal policy structure is more often bypassed than used, especially when an issue becomes politically sensitive.[7] For nearly a decade, oil shale policy was handled by a series of special Secretarial level working groups and task forces; a program was finally implemented in 1974. The recent decision to increase by tenfold the amount of acreage leased on the OCS was also made at the Secretarial level, with no apparent consultation of agencies responsible for implementing the program. This has the obvious effect of discouraging initiative and a sense of responsibility in the designated program agencies.

The bifurcation of line agency authority between the Bureau of Land Management (BLM) and the Geological Survey—each reporting to different Assistant Secretaries—makes the problems in program management worse. There is an artificial division of responsibility, with BLM making leasing decisions, and the Conservation Division of the Geological Survey assigned to evaluate resources before leasing and to regulate operations afterwards. In practice, of course, these functions overlap, but each agency closely guards its prerogatives, and open exchange of information is often lacking.

In addition, the philosophy of the two agencies is quite different. BLM sees its function as land and resource management under multiple use principles. Its programs are dominated by resource managers and economists. In contrast, the Conservation Division is staffed primarily with petroleum engineers, who are by training oriented toward production of oil and gas. This in itself is an anomaly within the Geological Survey, whose orientation otherwise is scientific and not regulatory.

The unsurprising result is that often there has been no agreement on critical aspects of program management. For example, the agencies take fundamentally different approaches to appraising the value of the resources before they are leased. Geological Survey's estimates have been extremely conservative, often turning out to be a small percentage of the actual worth; they are based primarily on engineering criteria and poor geologic information. Such appraisals make it difficult to know if the government is getting a fair return on its sale of the resource; but because the Geological Survery has authority for resource evaluation, its appraisals carry the day. BLM would opt for applying standard economic assessments, but it cannot implement these because it lacks access to the basic geologic and engineering data necessary to such evaluations, and there is no policy direction to implement that approach.

In examining critical policy points, it must be realized that the Department has no "resource program" as such. What it does have is five separate programs, one for each resource (OCS oil and gas, onshore oil and gas, coal, oil shale, and geothermal resources). These are administered with little integration—there are even three separate divisions within BLM which handle these programs. There is no method to assess the effects of one program upon the other or the total impact of all of them on the national energy supply picture.

In order to look at overall energy resources policy, the rest of this section will examine, in general, major elements of the exploration, prelease planning, leasing, and lease development that occur in each of these programs.

Exploration

Exploratory activities vary according to resources, with great differences in legal and administrative approach.[d] But for all resources there is relatively little government effort, compared with industry.

Usually, industry must acquire an exploration permit before it can enter the federal domain. The permit systems are quite different, depending on where or what resource is being exploited. Outer Continental Shelf exploration is carried on under an informal set of orders issued in the 1950s, despite the fact that the OCS Lands Act calls on the Secretary of the Interior to issue formal geological and geophysical exploratory regulations. Permits are issued on an ad hoc basis, often to large industry joint ventures. But the permits confer no ownership rights.[8]

Permission to explore for coal in "unknown" areas (where the government has inadequate data to assess the existence of commercial coal deposits) is granted with "prospecting permits." If the permittee discovers "workable, commercial" coal, he is entitled to a "preference right," noncompetitive lease for the lands included in his permit. He receives this free, subject to a nominal rental and a royalty on the coal subsequently produced. Of course, the coal is not hard to find; so these permits confer very valuable rights free of charge. Furthermore, according to Interior Department officials, there is a great amount of unauthorized and, hence, illegal coal exploration carried on without permits. This gives industry a better assessment of the value of coal deposits in a particular region. The Department has insufficient personnel to police this practice.

---

[d] Exploration can be of two types. Geophysical exploration uses sound waves to ascertain the possible existence of resource-bearing structures, and is used primarily for oil and gas operations. Geological exploration covers a variety of activities, from actual drilling into a supposed resource-bearing structure, to bottom sampling and testing of surface outcroppings.

Onshore oil and gas present an even more haphazard picture. Oil and gas on many of the public lands are sold either on a first-come, first-served basis or by lottery. Leases sold by this system require only minimal fees and almost entirely omit incentives for development. Exploration of the lease takes place entirely at the will of the lessee.

This haphazard exploratory system gives rise to two major policy problems. First, there is no assessment of the environmental impact of preleasing exploratory operations; only minimal efforts are made to control the adverse effects. Exploratory drilling, in particular, can harm the environment, as can the indiscriminate use of exploration vehicles on arid or fragile western lands.

Most important, based on the Interior Department's own rules, data acquired from exploration for publicly owned resources remain an industry trade secret, unavailable to the government agency responsible for managing the resources, or to the public. There are no apparent legal obstacles to requiring that this data be submitted to the government for its use. The information need not be made public in any way that would directly affect the proprietary interest or competitive position of the company involved. But the government could learn a great deal more about what it owns before it sells it—or gives it away.

The result of the situation is that the government knows considerably less about its resources, including both their quality and quantity, than the industries which are seeking development rights. This lack of data puts the government at a severe disadvantage in judging whether it is getting fair market value for its resources; and it hampers the effort to establish reasonable leasing rates.

Independent government efforts to broaden understanding of its resources have been highly limited. Too low budgets and too little manpower limit the effort, a fact in part explainable by the low priority the Department assigns to resource evaluation. In recent years, some funds have been found for evaluating OCS potential; but the Geological Survey is restricted to buying some raw data from broad area seismic surveys, usually conducted by industry joint ventures. It has not had the funds to purchase or conduct more detailed surveys, and it has only a limited capability to evaluate the data it does acquire.

Big fluctuations in recent Geological Survey estimates of the oil resource potential of the Atlantic OCS—which changed from 114 billion barrels in 1972 to "8 to 16" billion barrels in 1974—illustrate the magnitude of the government's ignorance about the resources in undeveloped regions. There is no government sponsored program for detailed seismic work, bottom sampling, or shallow core drilling to better understand either the amount or the nature of the resources in the OCS.

The government's efforts to improve the knowledge of its coal holdings are similarly limited. Of course, the government knows quite a bit about where its coal deposits are, but it does not know much about the quantity or quality of the coal. Only since 1972 has there been a sampling and drilling program, and this has been focused primarily on the coal-rich Powder River Basin area of Wyoming and Montana. Other federal coal lands have been virtually ignored. For planning purposes, the government has had to rely heavily on estimates submitted voluntarily by industry, an uneven and potentially misleading system at best. A case in point: as the Montana Bureau of Land Management activated its multiple use planning system for the Decker-Birney unit (a coal field containing some of the most sought after coal in the country) it was forced to get most of its data from those industries willing to submit them voluntarily, rather than from the Geological Survey.

The coal picture is further clouded by very complicated land and resource ownership patterns. Since the turn of the century, the government has retained the mineral rights to vast amounts of land where the ownership of the surface passed into private hands under the homestead laws. But it neglected to make maps or even keep detailed records of these retained mineral rights. For example, in the Powder River Basin the government controls as much as 85 percent of the coal but only 20 percent of the surface. As a result, a major record searching and mapping effort will be required before the government fully understands how much coal it controls and where it is located. This search will have to be conducted even as the government is under enormous pressure from industry to lease the coal—the extent and quality of which is likely to be better understood by industry than by the government.

Addressing national needs

Rates of development must be established according to an overall estimate of national energy needs; trade-offs among and within potential producing regions must be made; and decisions must then be made on the sale and regulation of specific tracts. However, the present Interior Department planning system is not designed to help make such decisions.

Presently, only the OCS leasing program has even a primitive mechanism to project the rate of leasing and relate it to national energy needs. As we mentioned earlier, that mechanism is by-passed more than it is used. In developing a five-year lease schedule in 1971 and a subsequent schedule in 1973, the Interior Department did attempt to assess national and regional demand for oil and gas, and production from existing OCS leases. Non-OCS supplies were then evaluated, and new OCS sales were proposed as a means of closing the demand gap between domestic oil and gas production and domestic demand—an impossible goal as it happens. No such basic analysis was undertaken in connection with the 1974 Presidential decision to increase by tenfold the amount of acreage leased by 1975!

The key factor in the analysis is the projected rate of demand growth. The demand figures now being used for resource programs are based on Bureau of Mines projections developed in 1972.[9] These, in turn, are based on historical energy growth patterns, and project continued energy demand growth at an average annual rate of 3.6 percent a year (or 4.2 percent in one recent program announcement) through the year 2000. This projection fails to take into account the possible dampening of demand from higher energy prices and the effects of conservation measures, which are documented in this report.

Unhappily, the only analysis of energy alternatives in any leasing program takes place, as a result of a 1971 court order, in the context of environmental impact statements.[10] Even for this analysis little demand flexibility is assumed, and environmental trade-offs are juggled under high growth projections.

Lower growth rates are thus not even considered. But what is more frustrating is that it would make little difference if they were. While Interior officials are obligated under the National Environmental Policy Act of 1969 (NEPA) to consider a broad range of energy conservation options, the decision making process within the Department pays no attention to them. In the absence of a national energy conservation policy, dominant factions within the Department are oriented exclusively toward development at historical demand growth rates.

Actually, the Interior Department's planning program for leasing is largely reactive. By and large, Interior responds to industry's expressions of interest in a geographic region, or a more narrowly defined area such as the actual leasing tract. Only rarely does leasing of a specific area take place upon the motion of the Interior Department.

The process can take place in a number of ways. Policy guidance to the Secretary of the Interior often comes directly from industry organizations. For example, in 1972 the National Petroleum Council, a permanent oil and gas industry advisory body to the Secretary,[e] recommended increases in the amount of acreage leased on the OCS from a projected level of less than one million acres per year to 1.6 million acres per year by 1980, and to 2.3 million acres per year by 1985.[11] Shortly thereafter, the Interior Department announced that the rate of leasing would be increased threefold, a rate which would comport roughly with the NPC recommendation. The subsequent 1974 decision to lease ten million acres per year is far in excess of the NPC recommendations, which envisioned a total of 21 million new acres leased by 1985—compared with the total of approximately eight million acres leased from 1954 to 1972.

Interest can also be expressed through the formal administrative process of the leasing programs themselves. Under recent reforms, industry and general public comments are invited for new offshore regions. Then the oil and gas industry is given an opportunity to nominate the 5,000-acre tracts they would like to see put up for lease, and the Bureau of Land Management uses the nominations as the principal criteria to decide which tracts to offer at the next lease sale. The number of tracts nominated by the government on its own is relatively insignificant. It is interesting to compare the tracts nominated by industry with those nominated by the Department. Industry nominations are far more reliable indicators of the best tracts. While leasing those tracts of interest to industry is important in insuring competitive interest, heavy reliance on industry interest precludes making important environmental tradeoff decisions among tracts.

By comparison to OCS leasing, the other ongoing resources programs are chaotic. The coal leasing program

---

[e] The establishment of an equivalent advisory body of consumer-citizens has never been seriously considered by the Department.

has functioned completely in response to private requests with virtually no government involvement in establishing the rate of leasing. Prospecting permits, which by Departmental interpretation entitle the holder to preference right leases upon discovery of commercial quantities of coal, were for years granted routinely to all eligible applicants. "Competitive" lease sales for land containing known coal deposits were held in direct response to requests by private parties with little prospect for much competition among bidders. Vast amounts of coal at an average price of less than 1/100th of a penny per ton have been leased under this system.[12] So little of this coal was actually being developed that the Bureau of Land Management in 1971 imposed a near moratorium on further coal leasing while it reevaluated the program.

Onshore oil and gas leasing is conducted with no planning. More than 15 percent of the total public lands are covered by noncompetitive oil and gas leases, which are dealt out automatically to whoever asks first. Where this is more than one applicant, leases are awarded by lottery. The leases carry a ten-year term and no production requirements. The government receives only 12 ½ percent royalty if there is production; it gets nothing for the leases. Competitive leasing is highly limited.

Reliance on direct and indirect industry demands to determine the rate of leasing has been a matter of some necessity for the Interior Department, because in most regions and for most resources the Interior Department simply does not have the necessary data to establish a national leasing policy. Thus, planning for development of publicly owned resources takes place primarily in the private sector. Corporate planning and decision making generally will reflect the objectives of the individual company, leaving out important national goals. The environment and the public treasury have suffered the most from this approach.

Values and regional choices

Hard choices about development among producing regions and within particular regions are critical to a national leasing policy. The cumulative effects of proposed coal extraction in the Northern Great Plains region illustrate the problem of fragmented planning. Most of the coal is federally owned; much of the remainder cannot be developed without some federal action, and yet the federal government does not know what to expect and has no overall plan to shape a pattern of development that best serves the public interest.

But planning at the state level is very limited. The states of Montana, Wyoming and North Dakota are dotted with proposed projects on the drawing boards of dozens of companies: electric utilities, integrated energy companies, coal operators, gas pipeline companies. Some plans would export coal from the region, some would burn it there for electricity, others would convert it to gas or liquids. Still others would do nothing with it for some time, but hold it to speculate on its increasing value. In total, these proposals would exert a tremendous cumulative impact on the water resources of the region, resources which are already hard-pressed by existing demands. Further, the air pollution potential is tremendous, far exceeding that from the high polluting Four Corners complex in the Southwest.

No mechanism exists at the federal level to comprehensively examine regional problems such as those in the Northern Great Plains; there is no way to assess whether the region can absorb the planned level of development without suffering serious and irreversible damage, or whether in fact the planned energy growth is even necessary or desirable. The energy planning that does take place is highly fragmented. It is usually done by local offices of federal agencies, or by state agencies which have limited ability to enforce their decisions on federally owned resource areas.

For example, under its comprehensive siting law, Montana requires all energy companies to submit plans for new facilities ten years in advance of construction. Montana is well ahead of most other states in this regard, but the system stops at its borders. It is grossly under-funded, and it has yet to sustain a major "test" of its practical ability to withstand development pressures.

The Interior Department has never yet implemented a comprehensive land use planning system for areas of potential resource exploitation. The Bureau of Land Management does have a rudimentary, multiphase land-use planning system which, if properly funded, could be used to balance the many resources' values on some production areas. But it is confined to limited area "units" within the states.

NEPA could and should provide a vehicle for sound preleasing environmental planning for federal resource programs. NEPA requires that, prior to any major federal action, the responsible federal agency prepare an environmental impact statement that assesses the environmental and socio-economic effects of the action, and consider reasonable alternatives to the proposal. It could be used to judge whether it makes sense to develop a particular region; and, if so, what stipulations should be imposed.

But application of NEPA to the resource programs has been uneven at best. In most cases, the NEPA evaluation takes place after the major policy decisions have been made, so that it can have no bearing on the fundamental decision of whether to go ahead with development; and the broad regional concerns addressed in the impact statement are largely ignored in the development programs that follow.

The OCS leasing program is a partial exception to the rule. An impact statement must be produced for each lease sale. Since 1970, this process has grown progressively more detailed, in part to comply with court decisions. Unfortunately, the environmental impact statements do not serve as planning tools in the Interior Department, although on occasion a few tracts are eliminated from a sale. Generally speaking, there is little relationship between the environmental assessments and decision making.

The trouble is that the analysis takes place after most of the key decisions—whether to lease at all, and which areas or specific tracts to offer—have been made. The Department's perfunctory production of "boilerplate" impact statements, which change little from sale to sale, demonstrates a lukewarm commitment to environmental protection. Such statements are convenient and do limit the administrative burden, but they miss an important chance. The NEPA analysis could be used to come to grips with basic choices in resource programs.

A still more pointed example of neglect is the fact that by early 1974, three years after passage of NEPA, not a single impact statement had been completed for the coal or onshore oil and gas leasing programs. NEPA has been treated as an obstacle to be overcome by producing legally acceptable environmental impact statements. Environmental planning and management is not yet a reality in the Department's resource programs.

Leasing

"Leasing" is something of a misnomer. It actually means selling minerals from public ownership into private ownership. Leasing is divided into competitive or noncompetitive systems. The law prescribes the general features of these leasing systems, but allows for a considerable amount of administrative flexibility. In practice, there has been virtually no change or experimentation with a leasing procedure once it is instituted.

Competitive procedures are used for oil and gas lease sales in the OCS, for some coal, for very limited amounts of onshore oil and gas, and for oil shale and some geothermal steam. Cash bonus bidding has been the invariable rule for lease sales. Under this system, bidders for a sale offer a single cash payment (either orally or by sealed bids) for the right to develop the resources within a defined tract. The government also collects a preestablished royalty, based on a percentage of the value of the resource produced. For example, the royalty on OCS oil and gas is 16 ⅔ percent of the value at the wellhead.

The leasing laws allow other types of competitive leasing. One is royalty bidding where rights are sold to the party offering the highest royalty shares on production; bidders in this system must also pay a preestablished, fixed cash bonus. The fact that almost all the federal oil and gas resources onshore, and most of the coal and uranium resources, are leased under noncompetitive systems means that, for all practical purposes, these resources are not sold—they are simply given away.

Under a competitive system two prerequisites help to assure a fair market value return. The first is a sizable number of bidders. In past resources sales, the size of the high bid has usually been directly related to the number of bidders for the tract. This has been true for all resources where competitive bidding has been used.

The second is to reject bids if they are too low. When few bidders are present and competitive interest is cool, the government has the power either to reject the bids for lack of competition, or accept only those bids that are higher than the appraised value of the resource being sold. But since the agency which is responsible for the government's appraisal is poorly informed and does not have an economic evaluation program to make long-range value predictions, bids that exceed its appraisal could still be a giveaway. In these circumstances, an adequate number of interested bidders is vital if the government is to receive fair market value. In several cases we have studied, the Geological Survey's appraisals did not relate in a statistically significant way to the high bids in competitive sales.

The government's best performance in assuring fair market value has been in the leasing of oil and gas on the OCS, generally in blocks of about 5,000 acres.[13] It has proceeded for 20 years under a cash bonus bidding system. In 1971, reasonable leasing targets were established under a five-year lease schedule. The competitive environment improved markedly, with the average number of bidders per tract increasing since 1970. In addition, the average high bid in recent sales has been about $3,000 per acre—double the average during the previous decade. Bonuses in recent sales have gone as high as $40,000 per acre with many bids of more than $10,000 per acre. But the pace of leasing, as mentioned earlier, is accelerating so rapidly that the market for leases in the coming years is apt to be flooded. Revenues to the government will probably drop as competition decreases; this has already occurred in an early sale under the accelerated program. If the pace of leasing is not trimmed down to meet reasonably defined needs, the government faces a reversion to a noncompetitive environment in selling its oil and gas.

Because the bonus payments represent very high "front end" costs for the industry, the cash bonus system has drawn two important criticisms. One is that the bonuses tie up excessive amounts of capital, precluding its use for exploration and development of the tracts after leasing. It is argued that this results in a slower rate of development of the leases, and consequent delays in producing oil and gas. In a recent report the American Petroleum Institute estimated that the $2.16 billion in bonuses paid during 1972 could have drilled 3,391 test wells on the OCS—over three times the 1974 total, which was 993.

The high bonuses may also work against some smaller companies who wish to enter the leasing competition. While a substantial number of newcomers have entered the bidding in recent years, they have come in primarily as joint ventures, and it appears that much of their bidding has been centered on low value tracts. The higher bids, particularly in previously undeveloped and riskier areas, have been dominated by the larger companies, often in joint ventures as well. Oil and gas production is concentrated in fewer hands in the federal offshore than on land.

These criticisms need in-depth analysis before the bonus bidding system is eliminated or overhauled. It is questionable whether many smaller firms have the capital to operate independently on the OCS, even with the elimination of bonuses. Financial requirements for rigs, pollution liability insurance, and other equipment probably bar the entry of many smaller companies, except in large joint ventures, and these they can already enter. Availability of rigs and skilled manpower is at present a much bigger obstacle than lack of cash to increased exploration and development.

There is a valid economic criticism that bonus bidding places all the risk of failure on the lessee. If there is no oil and gas, he loses all his bonus. Our judgment is that deferring payment of part of the cash bonus will answer many of the objections in developed provinces where risks for the industry are not too great. In new provinces, royalty bidding, or bidding based on a share of profits, would spread the risks between industry and government.

Coal leasing has been similarly devoid of competition. Again, lack of resource data is a hindrance, since coal can be leased competitively only if the government knows that coal exists in commercial quantities and is "workable." Otherwise, any person may apply for a prospecting permit, and upon discovery of commercial quantities of workable coal has an automatic right to a lease. Of the 530 coal leases issued as of 1973, exactly half were competitive and half were noncompetitive. Of the 265 "competitive" leases, 193, or three-quarters, had only one bidder, often the adjacent owner or the surface landholder. The market was, in effect, noncompetitive, even though the Department went through the motions of a competitive lease offering.

Leasing for onshore oil and gas is almost entirely noncompetitive. If oil or gas is located on a "known geologic structure" of a producing oil or gas field—that is, one where the government knows there is oil or gas which is capable of commercial production—then leasing must be competitive. But inadequate resource information, coupled with a very narrow technical definition of the term "known geologic structure," has greatly limited the areas where the government requires competitive leasing. Of the 104,218 onshore oil and gas leases in effect in mid 1973, only 5 percent had been issued competitively.

The first leases issued under the new oil shale and geothermal leasing programs reveal some interest. Future competition is uncertain. The first oil shale lease sale in the prototype program brought in eight bidders and high bid of $210 million dollars. Up to 40 percent of the bonus may be forgiven and never paid if development costs exceed the amount of the bonus; and it is likely that only $135 million will be collected by the Treasury. Competition decreased significantly in the next three oil shale sales on more marginal tracts. The fifth sale brought no bidders.

Despite the strong competitive interest in offerings on the OCS in the recent past, it is clear that many public energy resources have been leased for private exploitation with little or no competition, with inadequate evaluation of the resources being sold, and, inevitably, without anything approaching a fair return to the public treasury. And the rapid pace of leasing now contemplated raises the real possibility that competition will be weaker, not stronger, in the future.

Lease development—ensuring production

Once the resource development rights are sold, the public, as consumer, has an interest in the development of those resources for its use at an early date and at a fair price. Assuming that energy prices are on the rise, if the lessee is allowed to sit on his lease without developing it, it will be costly to the public in two ways. The lessee will have "purchased" the development rights (if, in fact, he paid anything) with a lower payment to the treasury than at future prices; and by not producing his lease until some future time, he can sell the resource back to the public at a higher price than if he develops it immediately. The legal system is designed to encourage early production of the leased resources and thus discourage such private speculations. But Interior Department policies are having the opposite effect, contrary to the intent of the laws.

By not requiring the lessee to either develop his lease or surrender it, the lessee, not the government, is allowed to speculate with the public's resources. The nature of this speculation is quite simple. The resources are often leased at no cost or low cost relative to their actual present value. The initial lease terms are quite long and extensions are liberally granted. Holding costs, in the form of rentals and minimum royalties, are extremely low. Requirements for speed and diligence in developing the leased tracts are either lacking or not enforced. And leases can be transferred freely from owner to owner with a higher royalty to them than they pay to the government. The existence of such overriding royalties is in itself a good indication that leases have generally been sold at less than their market value.

The coal leasing program presents a clear picture of

private speculation at the public expense. In past decades, but particularly during the 1960s', vast amounts of federal coal passed freely into private ownership under situations of little or no competition and extremely low payments. About 15 billion tons of reserves are presently under lease; and an additional seven billion tons are subject to prospecting permits with applications pending for preference right leases. Since the Department has no authority to deny such applications, some 22 billion tons[f] of federal reserve coal are presently committed to the private sector. Only ten million tons of this were produced in 1972, an incredibly low 2 percent of the coal produced in the United States.

This type of private holding without development is possible because coal leases are issued in perpetuity, although the terms and conditions are to be reviewed every 20 years. The law states that "leases shall be for indeterminate periods upon conditions of diligent development and continued operation of the mine." This requirement can be abrogated only by a finding by the Secretary that it is in the public interest to do so. Contrary to the clear legal intent, the Department has allowed lessees to pay an advance annual "royalty" and has never enforced diligency requirements. The "royalty" is really simply an annual rental which, for most leases, is one dollar per acre.[14] The situation is similar for onshore oil and gas.

So far, failure to produce has not been a clearly identifiable problem with OCS oil and gas leases. High cash bonus bids tend to motivate the lessee to get back his bonus payment through early production.[15] This is, of course, less true when bonuses are low, as when there is little competition for leases. Also, the five-year primary lease term required by law, although loosely enforced, is intended to force development.

Nonproducing leases are likely to be a greater problem in the proposed, massive OCS leasing program. If anything approaching ten million acres is leased in 1975, it will be impossible to develop all the leases in the five-year primary lease term. This will inevitably result in enormous pressure to extend the leases until such time as industry can

---

[f] This estimate is probably quite conservative and stems from lack of information of the workability of coal in the lease tracts, which can only be determined upon mining. The actual volume of coal under lease is much greater; the issue is whether it can be mined economically. The greatly increased prices being paid for coal now and in the future should revise this figure upwards by a substantial amount.

proceed with development. Historically, the Interior Department has never denied extensions.

The problem, then, is quite sweeping: extensive leasing, combined with insufficient competition, little or no understanding of the value of the resources being leased, and no assurance of development. The public treasury does not receive a fair return for its resources nor does the public receive the energy from the resources under lease. The lease holders can sit on, or trade, their inexpensively acquired leases, waiting for higher energy prices before producing the resources.

Lease development—environmental protection

Environmental protection should be a subject for consideration at each stage of the resource development process; it requires both planning and regulation. Environmental planning under the National Environmental Policy Act (NEPA) has already been discussed. This planning ought to guide the decision which is primary from an environmental perspective—whether to proceed at all. If the decision is made to go ahead, specific lease terms and regulatory actions to protect the environment should be clearly outlined in the NEPA environmental impact statement.

Once again, the Department is making its best showing in the OCS program. Its origin was a matter of necessity. Immediately after the Santa Barbara blowout, because of the public clamor and the ensuing political pressure, the Department brought to a halt all further OCS leasing activity and set about the business of revising its environmental protection system for offshore operations. New regulations established strict liability for the cleanup of oil spills and required a number of new safety measures. The inspection force was increased from seven in 1969 to 43 in 1974. Helicopters were rented to carry on safety inspections.

Nevertheless, recent technical studies by varied groups have found weaknesses in the new OCS regulatory scheme.[16] The Geological Survey has made a good faith effort to respond to the studies and improve its offshore regulatory activities. Operating orders have been reviewed regularly and updated. But manpower limitations are a critical restraint. For example, there is not enough staff to make a detailed environmental assessment of each leased tract to determine specific environmental protection measures. Oil spill cleanup capability, although improved by cooperative industry efforts, is still very limited. A study on oil spills made for the Project concludes that oil spill technology is still in a primitive stage.[17]

By comparison, an environmental protection program hardly exists for other federal resource areas. The federal government is far behind some states, particularly in the areas of strip mined land rehabilitation and energy facility siting. It has shown no leadership in controlling the adverse effects of resource development; it has often abdicated its responsibility to the states. In fact the better state systems—Montana's reclamation controls, for example—are put into effect on federal lands. It is important to understand that the application of stricter state laws is primarily politically motivated. Little or no attempt is made to apply the best state practices to federal lands outside the states where there is compelling local pressure to do so. This hybrid regulatory system is a poor substitute for strong federal leadership on federal lands.

Broad regional evaluations need to be supplemented with environmental impact analyses that focus on specific sites. The National Academy of Sciences' study on rehabilitation after surface mining concluded that the ability to reclaim mined lands is critically site specific.[18] The implication of this finding for federal energy resource development is that detailed land use and environmental assessments are necessary before leasing, not only to determine whether the land should be leased, but also to serve as a basis for postlease planning, especially for measures to ensure subsequent rehabilitation.

The Interior Department issued stiff regulations for coal lands in 1969 which envision a system of site specific controls. Thus far, they have had little effect.[19] The greatest shortcoming is that the regulations are applied only to leases issued after 1969; this includes only about 10 percent of leases currently in private hands. There are nearly 500 leases to which they do not apply. Yet there are broad environmental protection provisions in the older leases that would allow the Department to include those leases under these regulations, if it chose to do so. In fact, retroactive application is probably mandated under the NEPA. But the Department has not ordered it. The result is that environmental regulation of the old leases is left primarily to the states—a sad commentary on the strength of environmental protection in the federal domain.

Difficulties have arisen even where federal regulations are in effect. The mining plans for both mining operations and land rehabilitation are inadequate. They are usually brief documents prepared by the operators themselves, subject to only minimal standards. Lessees frequently alter them as operations progress. The Geological Survey's field staff is grossly inadequate to evaluate the plans and, afterwards, to enforce the regulations. For example, the District Engineer with responsibility for Northern Wyoming, Montana, and North Dakota—the largest area of coal under the federal domain, and an area where strip mining is rapidly expanding—has only three part time men for this task.

A fundamental problem is that the Geological Survey is not, either by philosophy or composition, a regulatory agency. It is an organization largely composed of scientists with little inclination or skill as regulators. The Conservation Division, which has regulatory responsibility, draws heavily on the regulated industries for its own personnel. Relationships in the field tend to be hand in glove with the industry, partly because the government's limited personnel force it to take this approach, and partly because this is the most expedient way to operate. Furthermore, the Geological Survey is caught in a net of conflicting functions. Its promotional functions include working with industry to acquire data, encouraging production, and evaluating the resources before they are sold; then it must regulate the industry to protect the environment and human safety.

Alternatives for the public's resources

Under the existing federal resource policies, the country has no way of determining an optimum leasing program. High historical energy growth rates are the only foundation for developing a program, but there is no attempt to come up with the best mix of sources. Alternatives, such as they are, are buried in standardized environmental impact statements.

The present policy thrust is toward leasing all federally controlled resources simultaneously, at an extremely rapid rate without any regard for how fast industry will develop the leases. Ten million acres per year of OCS lands; a proposed new coal leasing program to supply large mine mouth power plants now (and later for development of synthetic fuels from coal); the movement toward a full scale oil shale industry in the 1980s; a major program of geothermal leasing—all point in this direction. These programs, aimed at the bulk of the remaining domestic fossil fuel resources, are harbingers of America's energy future.

Alternatives do exist

The scenarios we have developed in Chapters 2, 3 and 4 indicate that there is great flexibility in choosing the nation's energy future. There is a broad range of alternatives on the demand side, all the way from continuation of historical rates of growth to zero energy growth. Equally important, a wide range of options exist to supply the various demand levels.

These alternatives, even under historical growth assumptions, lead to two broad propositions. There is much flexibility in determining how much of the federal resource base should be exploited within a particular time frame. And resource policy can be selective in determining regional priorities for resource exploitation: offshore oil and western coal serve as general illustrations.

Offshore Oil

Federal offshore oil now constitutes 11 percent of total domestic resources. Regionally, on the basis of the scanty available data, the breakdown is as shown in Table 40.

The supply options for the scenarios present a range of alternatives for domestic oil production. Oil is a relatively scarce domestic resource, and there is no question that offshore oil will become increasingly important. But the

Table 40–**Federal OCS oil**

Region	*Reserves* *(Billions of barrels)*	*Resources*
Atlantic	0	8–16
Gulf of Mexico	5.5–6.5	18–36
Pacific	3.3–4.2	4–8
Alaska	0	28–56

Source: U.S. Geological Survey Press Release, February 1974.

rate of development of federal oil could vary considerably, depending on the particular supply case. A decision to exploit the Atlantic or Pacific offshore or the Gulf of Alaska must be weighed against taking measures to conserve the energy, and also against the level of oil imports, the use of coal or nuclear power as opposed to oil for electric power generation, and the development of a synthetic liquids industry based on oil shale and coal. The pace of leasing must also be trimmed down to the industry's ability to expand production.

Under the highest historical growth supply case we have projected, domestic oil requirements would be 5.76 billion barrels per year by 1985. If, optimistically, the OCS is to supply 20 percent of these requirements, the total OCS lands under lease should be about 21 million acres. This could be satisfied by a leasing rate of 2.5 million acres per year between now and 1985 (one-fourth the rate projected by present policy). With more intensive energy conservation, or with higher petroleum imports (as the supply scenarios set forth) the leasing rate could be held at about 1.5 million acres per year.[g]

A leasing rate differential of nearly one million acres a year could be crucial to decisions on opening the Gulf of Alaska or Pacific offshore to exploitation. Under a conservation oriented energy policy, or a decision to permit higher imports, the current nine to eleven billion barrels of reserves in the Gulf of Mexico and California offshore regions would provide adequate supply—at least until 1985. Reserve figures will increase substantially as oil prices, including the recent dramatic increases, are included in the calculations, and new discoveries are made in these regions. New leasing could be limited to the Gulf of Mexico for the next decade, under the supply cases that demand less domestic petroleum.

This would allow time for a more thoroughly planned approach to the Atlantic, Pacific and Alaska regions, including detailed exploratory operations before issuing development leases; for environmental and technical studies to deal with hazardous operating conditions; and for comprehensive regional planning to mitigate the onshore impacts.

---

[g] The National Petroleum Council estimated that to meet a high domestic oil case of 5.13 billion barrels per year by 1985, a leasing rate of 1.5 million acres a year average would be necessary between 1972 and 1985. These projections comport roughly with those used herein.

Coal

Federally controlled western coal presents an even greater range of options. Coal is vastly more abundant than oil and gas as a national resource (see Appendix D). With federal coal reserves estimated at 187 billion tons (at 1972 prices), about 50 percent of the national total, there inevitably will be some shift toward federal coal; exactly how much will depend on a number of factors that policy makers must consider.

One consideration is the pace of development of coal synthetics technology. Most important are choices among potential coal producing regions. For example, vast amounts of Appalachian and Illinois Basin coal (with recoverable reserves estimated at 125 billion tons) could serve many of the same markets as western federal coal. Much of this eastern and midwestern coal appears less economically and environmentally attractive at the present time. However, these objections could be overcome with improved deep mining and coal burning technologies. There may be no need for a massive shift to western coal.

The range of coal requirements varies widely under the potential supply scenarios we have projected. For 1985, coal requirements might vary from 1.12 billion tons per year (about double the 1972 rate) to 640 million tons per year (only about 30 percent higher than 1972 rates of consumption). But for purposes of resource policy planning, a simple look at 1985 requirements will not suffice to determine the rate of leasing.

Coal requirements in the more distant future provide a more meaningful indicator for the leasing program in the coming years. The highest coal case for the year 2000 is 2.12 billion tons per year; and the lowest case is 680 million tons per year.

Even taking the extremely high assumption that 50 percent (about one billion tons) of coal in 2000 is to come from western federal sources, there is no present need for a major new leasing program. Under our highest *Historical Growth* case, about one billion tons of coal per year would come from federal lands. We have already observed that about 22 billion tons of coal, which would be economical to produce under 1973 conditions, are already committed to private industry. Additional coal underlying the already leased areas could well become economic reserves.

Between now and 1985, a leasing program that would assure operators enough coal for contracts lasting beyond the year 2000 can be very limited indeed. It is unlikely that more than a few billion tons of additional coal need be committed prior to 1985, even to meet our high demand case that requires the most coal. Under the *Technical Fix* and *ZEG* cases cited above, essentially no additional federal coal leasing is necessary.

Under all cases, a "go slow" approach is warranted by the need to assess new coal technologies, potential regional impacts, and economic dislocations, and to make sure that production from existing federal leases is undertaken with fullest protection of the environment and maximum benefit to the region. In any case, industry ought to disclose specific production schedules before more leases are granted.

Facing the future

The foregoing sections have delineated policy problems that make it impossible to realize the stated objectives of the federal energy resource programs. Understanding of the resources is grossly inadequate. Leasing generally has taken place without regard to national need. With the exception of OCS oil and gas (and perhaps the recent oil shale leases) almost all these resources have been virtually given away, with no assurance that the resource will be developed. One result has been extensive speculation at public expense. In addition to this, environmental understanding, planning, and regulation vary from limited to nonexistent.

Broadly, there are three alternative policy approaches the government can take in managing the public energy resources. First, development rights can be granted to the private sector for laissez-faire development with minimal government intervention. This is the traditional approach; it has predominated for coal, onshore oil and gas, and uranium. Second, development rights can be sold competitively, subject to comprehensive government planning and regulation. Parts of this approach have been used in leasing Outer Continental Shelf oil and gas, and for the new oil shale and geothermal leasing programs. Finally, the government itself can assume the exploration and development role through the vehicle of a federal energy corporation.

The approach of massive leasing of all resources, without any appointed times for development, would mean that production would be dictated solely by market mechanisms—i.e., when the price is right.

In our judgment, this approach would make it

difficult or impossible to realize the nation's chief resource policy objectives. Fair market value from the sale of the resources would be out of reach—competition would be too diluted. Even with the present level of leasing, both the government's knowledge of its resources and its ability to judge their long term value are highly limited. Given higher levels of leasing, this weakness could be disastrous. Private speculation would be encouraged at the expense of the public treasury. Similarly, this approach would require leasing in areas where environmental data and analysis are virtually nil, and the social desirability of development highly speculative. We believe a much more gradual approach to leasing is warranted.

We feel that the program objectives can best be met, and the public interest best protected, by greatly improved planning based on a significantly expanded data base, a firm commitment to improved regulation, and some reorganization of functions.

Decisions would follow a three-stage process: first, developing an overall program that ties the rate of leasing federal resources to national energy policy objectives; next, regional resource assessment projects directed at the questions of if, where, and under what conditions resource exploitation should take place; and last, planning for specific sites to set leasing conditions and criteria.

A federal energy resource plan

It is our conclusion that alternatives based on energy conservation can and should be built into the federal resource programs. It would be helpful to have congressional guidance, including delineation of energy conservation policy goals, as suggested in the conclusions to this report. If these take into account the full potential for energy conservation—including the effects of higher energy prices—the basis will have been established for orienting the resource programs to national needs.

Once we gain some understanding of the demand side of the equation, we can examine alternative supply options, and weigh their consequences against other national objectives. In the context of these supply options, federal resource planners should assess the regional demand in those markets to be served by each federally controlled resource. The appropriate rate of leasing will then start to take shape.

An important planning tool would be to establish leasing targets for each resource in a five-year time frame and lay out tentative development objectives for fifteen-year periods. The plan should be updated annually, as national and regional forecasting become more sophisticated, and as new data are received from those parts of the system that deal with regions and specific sites. This multiphase plan could give industry a reasonable expectation of future federal actions for its planning purposes, and lead to improved public understanding and discussion of resource related issues.

To improve understanding of the resource base, two actions are urgent. The Geological Survey must expand its capacity to conduct its own exploratory operations in the public domain. It should augment this wherever necessary by contracting with states and private operators, but it can no longer leave the exploration process up to the developers alone. New regulations, to be issued immediately, should require private developers to submit all raw exploratory data to the government; all permits for exploratory operations would contain this condition. This would provide an excellent cross-check on the validity of the data acquired in the government's own program.

The federal energy resource plan would also rely on inputs from the private sector. Submission of industry planning projections would be mandatory, to fit in with the five-year and fifteen-year horizons. The plan would be developed in close coordination with other federal agencies and in accordance with the NEPA. Continued public participation would take place through a permanently constituted and funded Federal Resources Development Advisory Panel, with broad public membership, which would have its own staff, hold periodic public hearings, and submit an annual report to the Secretary of the Interior. We envision this body as a citizen advisory council, a counterpart to long-established industry advisory groups, such as the Interior Department's National Petroleum Council.

The five-year leasing schedule developed in 1970 demonstrates that this kind of planning is very much in the realm of possibility. Another fledgling system, known as the "Energy Minerals Allocation Recommendation System," is on the drawing boards in the Interior Department. It attempts to assess regional coal demand in developing a leasing program. Neither of these systems is, by any means, perfect. But they tell us that national policy planners need not start from scratch.

Assessing regional impacts

Regional planning for the resource programs is both a tool to assess regional impacts and an input into developing a national plan. It is virtually nonexistent today. As a result, national decision making usually ignores regional impacts. Consequently, it is impossible to channel the course of resource exploitation in the most acceptable manner, or even to decide whether or not to develop a particular area.

The relationship between coal, air and water is critical. In some arid localities, it is impossible to reclaim strip mined land because of water scarcity. Large scale coal development may well place demands on water resources that exceed the capacity of the region. A large influx of coal-burning plants may reduce the air quality of a region to unacceptable levels. Offshore development of oil and gas resources will require extensive onshore facilities such as pipelines, refineries, storage areas, and support facilities, the construction of which should take place only in the context of total regional coastal zone planning and management. This capability simply does not now exist.

An important step toward assessing regional impacts was taken in the multiagency Northern Great Plains Resources Program, which was designed to gather all existing data related to coal development in this important region. But it suffers from two principal defects: it is limited in duration and scope; and it is confined primarily to data gathering, not resource planning. A few more planning systems exist in rudimentary form. The BLM and Forest Service do multiple land use unit planning, and innovative regional planning is beginning in the oil shale country of western Colorado.

To cope with decisions that must be made within and among regions, the Interior Department, with assistance from the states, should establish and fund regional assessment projects. These should be ongoing, not limited in duration, with sufficient funding from resource revenues to have full time professional staffs and to pay for studies when necessary. State governments and other federal resource management agencies such as the Forest Service, the Bureau of Indian Affairs, and the Environmental Protection Agency should actively participate in the projects.

Legislation may be needed to authorize the creation of project managers for each region. These managers would be Presidential or Secretarial appointees, subject to

Senate confirmation. This would serve to upgrade the importance of the activity and clearly fix responsibility for regional planning within the Interior Department. (The Bonneville Power Administration has a similar function for federal electric planning in the Northwest.) Industry would participate through submission of development proposals, with an opportunity to support their plans in public hearings.

In developing recommendations, the project would have primary responsibility within the Interior Department for NEPA, and would assess all environmental, social, and economic impacts, not simply limited targets such as individual lease sales. The projects would also consider federal revenue sharing to help state and local governments.

Development: how and by whom?

The sale and development of individual leasing tracts present some of the hardest issues. Does the present competitive system hinder effective development because of high bonuses or size of leases? Is there in practice an optimal level of competition and a fair return to the treasury? What environmental controls should apply to which specific sites?

One reform is obvious. All noncompetitive leasing should be terminated. The Interior Department itself has recognized this, and has recommended that the Mineral Leasing Act be amended to this effect. Congress has failed to act, but the same result could be achieved immediately by administrative action without new law.

In areas where the resources are relatively well understood, it is our conclusion that competitive leasing should continue to take place by cash bonus bidding. Rather than a single bonus payment, as is now the case, payments should be spread over the first five years of the lease. The lease schedule should not flood the market. All leases should carry a primary five-year lease term, with no extensions granted where nothing is being produced. This will enhance competition by focusing bidding on the more promising leases that can be explored and brought to production within the five-year period.

Where the resources are poorly understood and there is a higher element of risk for both industry and government, we suggest a limited experiment with royalty bidding and profit share bidding to measure the effects of these approaches on competition and return to the treasury.

Bidding on a royalty basis, with a fixed (but low) initial bonus to deter irresponsible or purely speculative bidders, is permitted under present law. This approach would minimize the risk by deferring all substantial payments to the government until the time of actual production. This system has some administrative drawbacks. To ensure marginal production, once a lease becomes less profitable, it would be necessary to develop a system to reduce royalties on a sliding scale, based on thorough engineering-economic evaluations of each operation.

Under profit share bidding, prospective lessees bid a percentage share of their profits from each lease. Administration of this system is primarily a bookkeeping problem, and should be more manageable than royalty bidding, once a system of accounts is developed. Profit share bidding would require a change in the law for most resources.

The size of the leasing blocks for OCS lands also should be reconsidered. The present system of leasing a host of 5,000-acre blocks works reasonably well for the extension of well understood onshore geological trends into offshore areas such as the Gulf of Mexico. But for unexplored and poorly understood areas, larger acreage offerings would encourage exploration and early production. This approach could simplify government planning and reduce production costs, thus ultimately lowering prices to consumers.

The most critical need for better regulation is a leadership commitment. We feel this can best be accomplished by splitting off regulatory from promotional functions, placing responsibility for regulating environmental, safety, and conservation aspects of federal resource production in the Environmental Protection Agency. As a final reform, the policy and management functions for the resource programs should be consolidated into a single Federal Resources Administration, eliminating the present bifurcation of responsibility and debilitating jurisdictional conflicts.

Meeting the goals we have described for protecting the public trust in federal energy resources will require revitalized administration, which can best be achieved in a new, mission oriented agency. That agency should function as part of the Department of the Interior, or as part of an overall Department of Energy and Natural Resources, if that needed reform is brought to fruition by Congress. Coordination with other surface resources agencies should take place primarily in the field, through the regional resource assessment projects.

The new approach we have suggested is a departure from past policies of unplanned development, but it is hardly a radical change in the system. There is ample precedent for all the recommendations. The key is a new priority emphasis on public energy resources, a vital cog in the nation's energy future.

CHAPTER 12

Energy research and development

We no longer rely on individuals cloistered in their laboratories, pursuing their individual intellectual interests, to discover new scientific and technological knowledge. Rather, research and development (R&D) is the organization of scientific and technological talent by economic and political forces. The complexity and cost of developing new technologies, particularly for the last stages of their introduction into the economy, requires such organization. Even so, R&D does not always succeed—our efforts to tame nature and bring her to the marketplace on a leash sometimes fail.

What is the purpose of R&D? Its aim is not intellectual inquiry as an end in itself, but rather the widespread introduction of new machines into society. Therefore, questions of economics, efficiency, environmental protection, and other social goals must continually guide R&D efforts.

Funds should be invested in R&D to help society solve problems.

From this perspective, energy R&D has been sadly out of step with society's needs, not only in terms of the paucity of funds—which is being remedied—but in the allocation of those funds. Only one source of energy supply—atomic power—has been pursued with substantial federal funding, largely as a result of the initial federal development of the bomb. Private enterprise has invested in extensive product improvement as well as in oil and gas and central station electric technology.

But the overall record of government and industry alike is one of neglect of new sources of energy. The nation has neglected coal conversion and solar energy, and smaller scale energy technologies. R&D in energy-conserving technologies has been minimal. R&D has been starved for funds needed to solve the environmental and health problems of existing sources.

This misdirection has taken place because of the absence of a coherent national energy policy. Within the resulting policy vacuum, energy R&D efforts are dictated by narrow economic interests in the private sector, by established vested interests in government (of which the Atomic Energy Commission has been the outstanding example), or by a confluence of these narrow corporate and governmental interests.

In 1973, the federal government spent $642 million on energy supply R&D.[1] A full 74 percent of this expenditure ($480 million) was on nuclear energy, and most of this was on a single technology—the Liquid Metal Fast Breeder Reactor (LMFBR). (See Table 41.) The total expenditure in fiscal 1973 on solar energy, geothermal energy, energy from organic wastes, wind energy, and so on, amounted to less than $10 million. R&D for energy conservation was funded at about $20 million. The expenditures for systems research on health and ecological effects (about $20 million) declined between 1972 and 1974, and the environmental research on industrial energy use had the benefit of a few hundred thousand dollars of funding.[2]

R&D in the energy industries presents much the same picture (Table 42). Much of what is called research amounts to solving technical problems in the operation of existing systems. Unverified industry figures indicate that there was an expenditure of about $1.4 billion in 1973, but this sum included large expenditures in chemical research by the oil and gas industry. About $100 million was spent

Table 41–**Federal energy R&D funding—fiscal year 1973**
(millions of dollars)

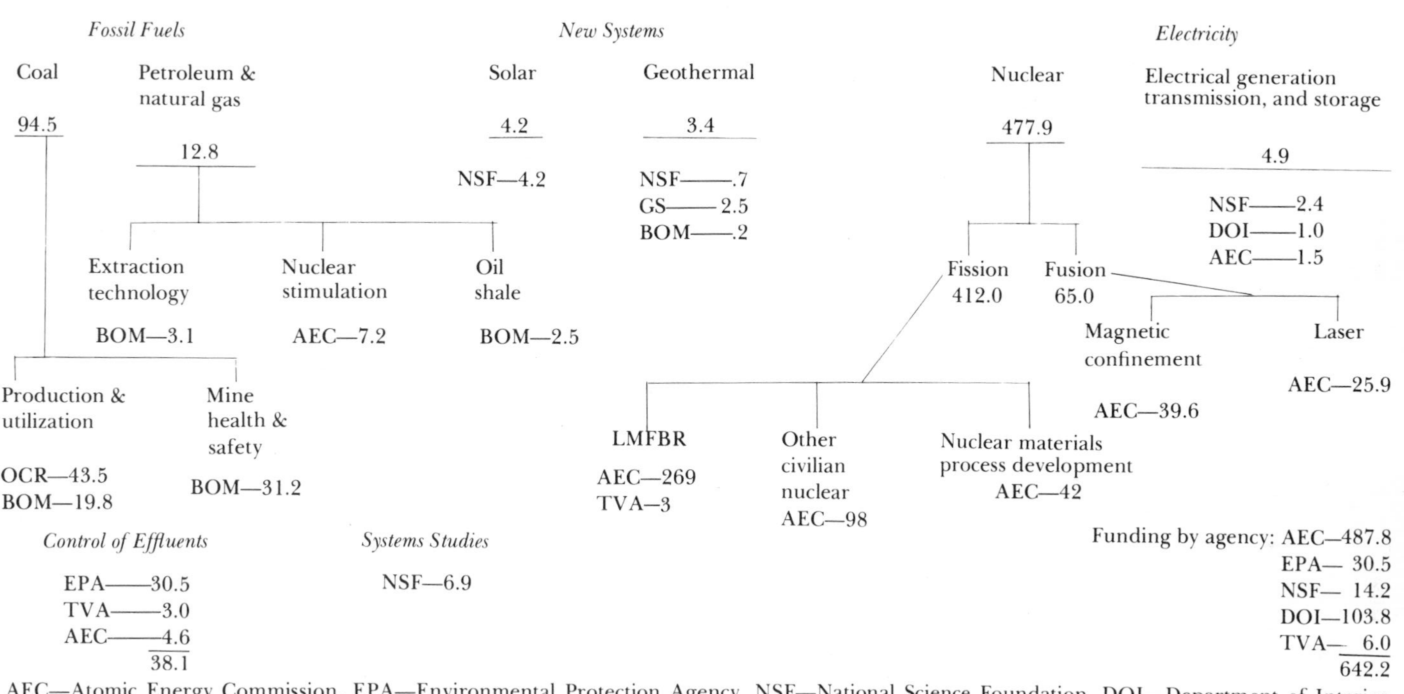

AEC—Atomic Energy Commission, EPA—Environmental Protection Agency, NSF—National Science Foundation, DOI—Department of Interior, TVA—Tennessee Valley Authority, BOM—Bureau of Mines, OCR—Office of Coal Research

Source: Herbert Holloman, et al., "Energy R&D Policy Proposals," a draft report to the Energy Policy Project, August 1974.

Table 42–**Estimates of R&D funding of the private sector, by industry group**

	Millions of Dollars
Oil	660 [a]
Oil equipment	50 [b]
Natural gas	100 [b]
Gas equipment	25 [b]
Gas transmission	50 [c]
Coal	6 [b]
Electric utilities	150 [c]
Electric utility equipment suppliers	350 [d]
	1,391

[a] American Petroleum Institute (API) estimate. This figure includes R&D on synthetic fuels, oil shale, tar sands, coal gasification and liquefaction, and expenditures by the petroleum industry on chemicals.

[b] Oil Information Center, University of Oklahoma.

[c] Federal Power Commission.

[d] Informed Industry source. Includes expenditure on nuclear power.

Source: Holloman et al., "Energy R&D Policy Proposals."

on environmental R&D—mostly coal and thermal pollution. The efforts were almost all directed towards supply. Research on energy use was a minuscule amount of the electric utilities' R&D expenditure of $150 million.[1] Amounts of reported private energy R&D expenditures should be interpreted with caution. "Oil and gas" research, as noted, includes large expenditures on chemical research. Moreover, R&D accounting procedures are not uniform among the energy companies and often not even within a single company.

While expenditures alone do not signify progress, they do indicate priorities. Without R&D expenditures, the development and introduction of new energy technology on a large scale cannot be accomplished in our energy intensive society. Thus, while solar house heating systems[3] and methods to produce methane gas from organic wastes[4] have been explored and partially developed by dedicated individuals and small groups, these promising technologies remain in their infancy for want of funds. Meanwhile consumers suffer high home heating oil prices, and industries such as the fertilizer producers run short of natural gas.

To remedy the current misallocation of R&D resources, we must define a set of R&D goals within the framework of a national energy policy. These goals lie in five areas.

- Assuring energy supply at reasonable prices;
- Using energy efficiently;
- Compatibility of energy systems with a clean and safe environment;

- Foreign policy concerns, such as safeguarding against supply disruptions and devising energy systems that benefit the nations without abundant energy resources;
- Diversity in energy supply technologies.

The first two goals are dependent on whether conservation or rapid growth in demand becomes our national policy. Present supply oriented R&D programs (public and private) are much like supply oriented capital investments: they both tend to be self-fulfilling prophecies by promoting high rates of growth in energy use. Shifting the federal R&D investments to energy saving technology can be just as effective in shaping the future. The last three goals are not dependent on the precise level of energy use.

Environmental protection is not now an integral part of energy production, processing or use, but it should be. In the past, environmental considerations have come as an afterthought. As a result, unsatisfactory and often expensive cleanup technologies have been installed in existing power plants and automobiles—originally designed with no thought to preventing pollution. The current controversies over the efficacy and cost of auto emission controls and sulfur dioxide removal from stack gases underline the failure of the patchwork, post hoc approach to environmental problems. Environmental protection must be a high priority goal of the R&D process from the very beginning.

In the short term, the aggressive pursuit of R&D on technologies for cleaning up existing energy systems is the price we must pay for our previous neglect. In the longer term, environmental considerations must become an integral part of the development of new energy technologies such as the breeder reactor, fusion power, coal gasification, and the rest. Not only should there be preliminary evaluation of a technology's environmental impact—as required by the National Environmental Policy Act—but ongoing evaluations throughout the R&D process through to the technology reaching the marketplace. This means both continual environmental evaluations by personnel who are part of the R&D team, and regular public disclosures of the results. Hence, energy technologies should come to the marketplace "clean"—or be stopped if they cannot be made "clean." New technologies should not be considered commercial until they can satisfy health, safety, and environmental standards, the setting of which should be part of the R&D process.

World events have shown the vulnerability of oil supplies in most oil importing nations. R&D can help safeguard against supply disruption by developing alternate energy sources.

Developing domestic oil supply technologies—particularly coal liquefaction[a] and shale oil[6,7,8]—may, in conjunction with stockpiling, be a reasonable solution. But this does not mean that the federal government must fund this particular R&D effort. The investments for developing coal liquefaction to date have come primarily from private corporations,[2] and ample incentive for private efforts in this field continues to exist. The major oil companies with their large cash flows are financially capable of developing coal liquefaction.

Shale oil development is another area where the economic incentive would appear to be favorable for development by private industry. The recent bids on federal government leases of shale oil tracts in Colorado indicate that the oil industry is interested in pursuing commercial shale oil technology. With shale oil, the uncertainties lie in the environmental concerns associated with that development.

The R&D program of the federal government with respect to shale oil and coal liquids should therefore focus on ensuring that adequate environmental protection is built into the systems and that the pace of development is compatible with maintaining continuity of supply in the event of oil import disruption. The government can also investigate what appear to be more satisfactory (from an environmental point of view) alternatives, such as *in situ* recovery of shale, and can conduct studies to determine the environmental effects of technologies developed by the private sector.

Many other nations share our concern for security of supply and are embarked on R&D efforts for new sources of energy. In fact, all industrialized nations[9] as well as many underdeveloped countries have energy R&D programs. At the very least, a regular and thorough program of information exchange, including exchanges of personnel, could reduce much of the waste and duplication now prevalent in R&D programs. Such exchanges of information have spurred progress in programs such as controlled fusion by helping to eliminate fruitless lines of research and opening up new ideas for inquiry.

---

[a]Coal liquefaction is the chemical conversion of coal to oil. Current estimates of the price of synthetic crude oil suitable for boilers (but not for refineries) are in the $5–$10/bbl range, in 1972 dollars.[5]

Bilateral and multilateral projects between and among industrialized countries could substantially reduce the funds required to develop new energy technologies by eliminating duplication. For example, automation of underground coal mining and mine safety have been major features of coal research in Great Britain, and the United States could certainly learn much from the English. (See Chapter 8.) Solar energy, a field that all nations have neglected and from which most could benefit, offers a splendid opportunity to initiate collaborative efforts.

The poor nations should also be considered, not only for humanitarian reasons but also because they are among our important customers for new technology. The poor and rich nations could cooperate in energy R&D to develop uses—particularly in agriculture—for indigenous fuels, to develop labor intensive technologies, and to develop technologies that require little maintenance. All this could go at least a little distance toward mitigating the need for long term food aid and capital subsidies.

Diversity in energy sources—the last of the goals listed earlier—is the key to greater flexibility in our energy system. The nation must have a wider range of options to enable us to deal with unforeseen problems in the future. We suffer from our preoccupation with the atom and neglect of every other option. If nuclear energy proves to be unsafe, we are in trouble. More emphasis on decentralized sources would increase individual options as to energy sources and technologies. R&D programs by both government and industry have so far been almost exclusively concentrated on technologies such as coal gasification or nuclear power that have significant economies of scale. Technologies that are not very sensitive to economies of scale, such as wind power, use of solid and organic wastes for energy, and harnessing solar rooftop energy, have not had significant R&D funding.

R&D diversification could have a positive effect on the balance of payments. For example, developing energy sources such as solar and wind could also create export markets, particularly in those nations without large fossil fuel resources. An indication of the potential is that in 1973 nuclear power technology exports amounted to $700 million.[10]

Advocating diversity in energy R&D is easy enough, but our resources are limited and the rub is deciding how to allocate them intelligently among the myriad ideas that the fertile minds of scientists and engineers conceive. The allocation of R&D money must be decided on the basis of how much energy we want and how we decide to use it—that is, within the framework of the scenarios for differing growth rates in the demand for energy.

The supply of large amounts of energy that is usable in an environmentally sound manner is an essential part of all the scenarios discussed in this book. From a worldwide perspective, the need for additional supply in decades hence is enormous; without new, cleaner sources of energy, most people on earth will never enjoy any of the material benefits that most Americans consider necessities. R&D on energy supply and conversion technologies is, therefore, a vital part of whatever scenario the nation chooses. Without it, the tendency toward crises in our energy supply and in environmental problems will surely be aggravated.

The problems of supply in the United States in the next quarter-century and beyond, will of course be considerably more pressing in the *Historical Growth* scenario than in the others. A policy decision in its favor must, therefore, be accompanied by greater R&D emphasis on the short and medium term supply technologies and their environmental problems. Although the current rapidly expanding energy R&D budget reflects an emphasis on supply (implying a policy favoring historical growth), the mix of the R&D budget does not adequately reflect the potential of a number of promising sources to contribute significantly to energy supply in the next twenty years.

The LMFBR will not make a substantial dent in the nation's energy supply before 1995 or 2000. And there are no allowances in this schedule for unforeseen delays and technical and environmental problems.[11] The dominance of atomic energy research is demonstrated by the fact that the budget proposed for the breeder program is more that 25 percent of the entire expanded energy R&D program.[12] Controlled fusion, which is not yet a scientifically proved concept, will certainly not make a large contribution before the end of this century—yet its share of the proposed budget is more than 14 percent. Thus, if this allocation of funds is adopted, about 40 percent of the R&D expenditures will be on technologies that will not bear fruit in the near and medium term.

The budget for R&D to conserve energy is slim, which is consistent with a high or historical growth policy. But even if that is the policy option the government favors, its R&D budget is misdirected. The R&D budget in the *Historical Growth* scenario should be redirected to place more emphasis on short and medium term supply options

—particularly those that will tend to mitigate serious pressures on the environment in the face of rapid supply development.

We do not question the importance of pursuing the R&D in technologies such as the breeder and fusion, which can bear significant fruit only in the next century. Certainly their potential benefits to mankind are so great, if environmental technical safety and economic hurdles are overcome, that continuing sufficient funding for these programs to complete them can be justified. Our point rather is that it is wrong-headed to concentrate only on options that will take decades and ignore options that are closer at hand and necessary to meet needs in this century.

The remedy, in our judgment, is to substantially enlarge the federal R&D budget for certain solar technologies, energy from organic and urban wastes, and geothermal energy, which are the most promising. These efforts should be upgraded and vigorously pursued in the *Historical Growth* scenario or any other scenario we choose, because they can help in the next five, ten or fifteen years.

Solar energy is the world's most abundant renewable energy resource. The solar energy that falls on the Saudi Arabian desert each year is about equal to the world's entire proved reserves of coal, oil, and gas.[13] There are a great many methods to convert sunshine to usable fuels, but they are in varied stages of development.

The simplest direct technology is the use of flat plate collectors to heat water or air, which is then used to heat homes and commercial buildings. This technology is on the verge of commercial implementation. Preliminary indications are that solar heating and cooling of buildings (including the heating of water) is competitive with oil and electricity in many parts of the country. However, we do not have enough field experience with pilot and demonstration projects to confirm this conclusion, and some technical problems still remain, particularly with solar cooling.

Because of the lack of institutional support in the past, there is very little commercial activity in the mass production of most of the components required for solar heating and cooling systems. Although support for this field has already increased somewhat, a rapidly accelerating program could have a significant effect on energy supply by the end of the century. Particular attention needs to be paid to the development of systems that can be installed on existing houses and toward institutional mechanisms designed to encourage builders to adopt this technology.[14]

Other solar energy technologies—direct conversion

to electricity, the production of high temperature heat to generate electricity, the conversion of sea thermal gradients to electricity,[15] the direct production of hydrogen from photosynthesis (photolysis), and solar energy from satellites in space—are further off. Photolysis and the solar station in the sky are still just ideas, but solar cells, ideas for central station solar power plants, and sea thermal energy to produce electricity are on the threshold of rapid technical advancement.[15,16,17,18]

Solar energy R&D funding bears little relationship to this promising array of opportunities. While there has been some acceleration of funding, the program does not reflect the medium term supply potential of sunshine. For the next several years, the government solar energy R&D program should provide the funding to develop the requisite infrastructure in industry, the national labs, and the universities to create a base from which rapid development of the technology can take place. It was just such a base program of upwards of $100 million a year for atomic power that made it a reality.

This seed money is vital to develop industry capability for producing the equipment for using solar energy. A good example is solar flat plate collector technology. Basic research, pilot plants, feasibility studies, and the training of personnel—particularly students in the universities—must be initiated rapidly if many of the promising solar energy technologies are to bear fruit in the next decade or two. As with any R&D venture, these actions do not guarantee success, but success will be difficult and slow without them. Of course, the necessity of solar R&D applies not only to the supply oriented *Historical Growth* scenario, but to the conservation options as well because of the many inherent environmental and foreign policy advantages of solar energy. As with any energy source, we must carefully assess the environmental effects of particular solar energy technologies.

The conversion of organic wastes to fuel is yet another example of a technology that is a promising short run (ten years) supply option but that has lacked adequate institutional support and does not have an industrial base. There is evidence—from some laboratory work here and small scale plants built around the world—that the technology could be commercial in the near future.[19] Yet the lack of experience with pilot plants that approach the size of commercial operation leaves a big question mark about costs. With a modest investment in pilot plants and systems design, this renewable resource could probably be well on the road to supplying several quadrillion Btu's of energy by the end of this century—as much as all hydroelectric power provides today. Research on using more diverse organic materials and raising plants for fuel or feedstocks should be initiated at once.

Geothermal energy is another neglected source. Production of electricity from dry geothermal steam is already commercial, but only a few dry steam fields have been discovered. The larger geothermal resources are hot water, hot rock, and geopressure (high pressure water) energy.[20]

The use of hot water for the production of electricity carries with it the serious problem of corrosion of materials; hot water, however, could become a significant regional energy resource in the western United States and should receive much stronger public and private funding. The use of hot rocks—a potentially huge source of energy—is more speculative and is considerably farther from commercialization; but with adequate R&D, the technology might well become commercially feasible in a matter of years, not decades. A serious evaluation of geothermal resources and the associated environmental problems ought to be undertaken immediately so that an orderly program of their commercial development can be initiated by the end of this decade.

Coal is a major energy resource in all the scenarios; but in *Historical Growth,* coal use and the substitution of coal for other resources are particularly important and should receive considerable attention. Pressing safety and environmental problems accompany coal production, processing, and use, and their resolution should receive a large chunk of the federal R&D expenditures.

R&D for the conversion of coal to fuel gas (both low Btu and high Btu) has been funded by the Office of Coal Research in the Interior Department, the Environmental Protection Agency[2] and by private industry. Low Btu gas with combined cycle power generation[b] can supply clean fuel for producing electric power at higher efficiency than current plants do. Except for application to some Appalachian coals, low Btu gasification is now a commercial technology.[21] The high Btu coal gasification program is also approaching the stage where industry can and should take over the great bulk of funding. They should both continue to receive some government support so that environmental

---

[b] In a combined cycle system, both a gas turbine and a steam turbine are used to produce electricty with high efficiency. The hot exhaust gases from the gas turbine are used to raise steam to drive the steam turbine.

safeguards such as small particle removal and controls for other air pollutants at the conversion plants can be incorporated.

In the next two decades, coal will play a crucial role in our energy supply. It is vital therefore not only to put in effect rigorous measures to protect health and safety of miners in existing mines, but also to pursue R&D for new, fundamentally safer mine technologies and for reclamation of strip-mined land. It is also essential to install pollution control devices to reduce the health hazards of burning coal, while rigorously pressing R&D to deal with air pollution problems that are still unsolved. (See Chapter 8.)

The present R&D effort is primarily aimed at supplying historical growth. We shall discuss R&D for energy conservation later on. But to achieve diverse and sufficient supplies of clean energy, we believe the following revision in the sense of direction in the ongoing national effort are needed for all scenarios:

- A much stronger effort in solar energy, organic wastes, and geothermal;
- Reduced federal support for coal and nuclear technologies that are near commercial feasibility and appear economical;
- Greater stress on the environmental, health and safety problems of coal and nuclear energy and environmental research and evaluation in general;
- More international exchange of information and cooperative projects;
- A significant new effort in basic research in the hard sciences and the social sciences directed toward energy systems;
- Creation of an Office of Long Shots.

The last item needs special mention. The Office of Long Shots (OLS) is designed to remedy one major defect in our energy R&D: there is presently no mechanism for institutional support of new ideas in their infancy. Ideas that do not square with current thinking, or that are dismissed by government and industry bureaucrats as naïve or unworkable at best, languish for years before they are supported by experiments and evaluation. At worst, they are never pursued. Two examples will suffice.

The first is geothermal energy which, until a few years ago, was usually dismissed as a small source of energy, of regional importance at best. A major energy study published in 1960[22] did not even mention geothermal energy.

It is now clear that geothermal resources (including hot rock) are large and fairly widely distributed.[20] Secondly, ten years went by before the concept of solar sea thermal power was recognized as legitimate, and it took the dogged determination of a few individuals to obtain funding for a feasibility study. Neither option may prove out but we have lost a decade in finding that out.

To correct this deficiency in energy R&D, we propose the establishment of a federal Office of Long Shots (OLS), say, within the National Science Foundation, separate from the main governmental R&D agency but with sufficient funding to go beyond mere paper studies. Research would feature new ideas such as the conversion of the energy in ocean waves to electricity,[23] or flywheel powered automobiles.[24] Ideas with economic promise should be transferred to the main R&D program after a preliminary environmental evaluation.

The study of the use of energy and other resources is intimately associated with the energy conservation scenarios, particularly the *ZEG* scenario. Neither energy R&D nor energy consumption nor energy conservation are ends in themselves. We spend energy and raw materials to buy goods and services that we want. However, there is no intrinsic reason why a comfortable and safe automobile (or a similarly flexible transportation device) could not weigh a few hundred pounds or get a hundred miles to the gallon. If economic growth is to accompany zero energy growth into the far future, then basic research of the relationship of the use of resources and energy to the economy must be begun as part of the energy R&D program.

Basic research on the energy needed to produce various industrial materials and manufactured goods would yield insights into the processes by which efficiency and raw materials use can be improved. The theoretical amounts of energy needed to produce most goods are much smaller than present technical sophistication permit, even after the improvements we have discussed in chapters 3 and 4 are incorporated. There is, therefore, ample room for new ideas for more efficient use of energy and other resources in industry.

Supply R&D discussed for the *Historical Growth* scenario is also imperative for the *Technical Fix* and *ZEG* scenarios with one important difference: since the use of energy is not as great, the energy R&D program will not be under as severe pressure as in the *Historical Growth* scenario. This does not relieve us from doing supply oriented R&D;

it merely gives us some more time to develop clean sources of energy.

To sustain the *Technical Fix* option in the long run, R&D on energy conservation is essential. Failing this, rapid energy growth will resume, once the conservation measures that are feasible now have been implemented. Technical energy conservation measures are also an integral part of *Zero Energy Growth.*

On the basis of the Thermo Electron study[25] and Energy Policy Project research, the following areas of R&D in energy conservation appear to be of major significance.

- *Heat transfer technology:* Basic research on heat exchange materials, heat pipes, heat transfer properties of various heat exchanger configurations, heat transfer fluids, and low temperature heat exchangers would be applicable not only to industry but also to improving the efficiency of electricity generation and to new supply technologies such as solar sea thermal power.

- *Low temperature heat applications:* Enormous quantities of low temperature heat are discharged by power plants, homes, and so on. Today, the use of low temperature energy is expensive and the applications are limited. Heat transfer technology R&D can help make low grade energy use cheaper. We also need research on new insulating materials and the cheap transport of low grade energy so that it can be economically used for applications such as space heating that do not intrinsically require high temperature heat.

- *Energy savings in space conditioning:* R&D on the application of the basic heat transfer technology research to heating and cooling homes with advanced heat pumps could yield substantial energy savings. It is theoretically possible to heat and cool homes and office buildings with just a fraction (on the order of 10 percent) of the fuel we now use.

- *Industrial steam production technology:* The application of solar flat plate collectors and low temperature energy storage to industrial steam production could result in significant fuel savings. (Steam production consumes more energy than all passenger cars.) These collectors would also be usable in the residential and commercial sectors.

- *Integrated power generation:* Development of technology for integrating decentralized electricity generation with electricity distribution systems.

- *Improvements in efficiency of power genera-*

tion: Improving the efficiency of electric generation by the use of fuel cells, topping cycles, bottoming cycles,[c] improving the load factor of generating plants, and a miscellany of such items are important to reduce the energy requirements for producing electricity and for building power plants.

- *New manufacturing processes:* Cooperative R&D efforts by the energy intensive industries to develop and demonstrate fundamentally new manufacturing processes that will save substantial quantities of energy. Federal leadership and seed money would be needed.
- *Transportation technology:* Improving the efficiency, safety, and environmental characteristics of automobiles and airplanes would significantly reduce fuel consumption. As mentioned previously, with a goal oriented, well funded R&D effort, the cars of tomorrow could achieve a fuel economy of 50 to 100 miles per gallon. We should work toward increasing the flexibility of mass transportation through technologies for door-to-door service to make it much more attractive without large energy penalties. The R&D could address the use of energy sources such as methanol, hydrogen, or electricity as a substitute for oil. This has the potential of reducing the consumption of oil.
- *Conservation strategies:* Research on the institutional problems of introducing conservation technologies in the marketplace is cheap but important.

The major shift in direction in an R&D program to support *Technical Fix* lies in the expanded commitment and funding for these conservation programs. Conservation oriented R&D must be considered a major program objective, with federal funding comparable to the funding for nuclear energy or coal over the next decade and beyond. But the federal funding need only be a fraction of the national effort. Federal leadership can elicit large industrial investments in R&D for saving energy.

Perhaps even to a greater extent than with R&D on energy supply, private R&D funding of energy-conservation technologies for specific end uses should be encouraged. The government should concentrate more on the development of basic technologies, on those specific technologies that are not close to commercialization, and on the technologies that do not have adequate representation of vested interests in the economy. The role of government

[c] With topping cycles and bottoming cycles the temperature limits for heat engines can be extended to make fuller use of the energy value of the fuels in producing electricity.

R&D is to create diverse and diffuse vested interests in the private sector for those energy supply, conversion, and consumption technologies that satisfy the general objectives of clean, reasonably priced, and efficiently used energy. R&D, whether it be for energy supplies or energy conservation, should not be the sole province of government.

Broadly speaking, the government R&D effort must fully support basic research directed at acquiring scientific knowledge, and development of basic, widely applicable technologies such as research on the heat transfer properties of materials, or flywheels for energy storage. This is not to exclude private investment in such efforts, but apart from isolated cases (Bell Laboratories, for example), private industry is not likely to invest heavily in areas that only have long term profit making possibilities. This stage of R&D is relatively inexpensive and we can afford to investigate even remotely promising avenues of research, provided they have a clear bearing on our energy R&D goals.

R&D becomes more expensive as technological development proceeds to the pilot plant stage, but understanding of the economics of commercial development improves. If the industrial infrastructure exists, the federal effort should be able to elicit financial support from private industry in the pilot and demonstration plant stage.

The energy technologies that we develop must not only meet stringent standards of health, safety and cleanliness, but also the acid test of the marketplace. The government should, therefore, phase out its financial commitment in the demonstration phase where private industrial vested interests are established in the economy.

A thorough assessment of the costs and benefits, and the health, safety, and environmental problems associated with the technology should be undertaken by an independent technology assessment office before any funding for a demonstration project is authorized. Such assessments should be followed by full public debate under congressional auspicies on the merits of proceeding with a demonstration plant. Further evaluation should be undertaken, if necessary. If the decision to proceed is made (by Congress), and the requisite industrial infrastructure exists, government funding for demonstration plants should be limited to 25 percent of the initial estimated cost. The remaining 75 percent and any cost overruns should be borne by the participating industry or industries. This procedure is not now a part of our energy R&D efforts. It should be implemented immediately.

The federal R&D effort must be sensitive to the need to transfer technology from the federal government to the private sector and to state and local governments so that the technology can be put promptly into commercial use. This is particularly true for technologies applicable to environmental controls and to the use of urban and agricultual wastes. However, where the technology appears economically feasible and environmentally desirable, but the industrial infrastructure to build and operate the plant and equipment does not exist, a greater share of federal R&D money for demonstration plants is justified. Such is the case today with solar home heating systems and with urban and organic waste use. In the latter case, further justification exists since it is likely that the purchasers of the technology will be state and local governments, which may not be able or willing to finance the new technologies because of limited resources and other more pressing obligations.

When the government decides to support a particular area of R&D, it contracts with its own laboratories, universities, private R&D firms, and industrial concerns to do the job on an ad hoc basis. This is a flexible approach since the government can select the most appropriate organization for a particular project. The ease of transfer to the private sector depends primarily on the contractor. Laboratories whose main business is R&D are naturally reluctant to part with their funding by declaring a technology commerical. Such organizations are, therefore, best suited to such basic R&D as proving scientific feasibility of controlled fusion or studying the properties of a new heat transfer fluid. When a technology approaches demonstration plant status, it should be taken from the professional R&D organizations and placed in the hands of potential commercial users.

Contracts to firms that manufacture energy equipment (boilers, for example) may help transfer a specific application of an energy technology to the marketplace without delay. However, contracts to one of several firms can adversely affect competition. Federal R&D programs must give a high priority to enhancing competition and be quite sensitive to the danger of funding the monopolists of tomorrow.

Such government contracts also sometimes suffer from management indifference, because to the company the award is practically "free money" and they do not put their best people on the job.

One way to assure management interest is to require

private firms to share the costs and risks. Management would consider market forces, its need for profits, and how the technology can be marketed when it commits significant internal funds to the effort. The most common approach to joint funding in the United States involves the government's sharing the cost of a new pilot or demonstration facility. The industry pays at least what a conventional substitute would be worth, and the government picks up some or all of the difference. One variation used abroad provides for repayment of the government investment in the event of success. The government advance could be treated either as a loan or as a share in the venture. The specific arrangements for a jointly funded project should depend on the stage of technological development, the costs involved, and the relation of the cooperating firm or firms to the marketplace, as discussed above.

All the previous mechanisms have focused on the R&D phase of innovation rather than its results. Through its purchasing power, government can also encourage innovation by purchasing automobiles, buildings, and other energy intensive equipment that incorporate new, more efficient technology. The Defense Production Act of 1950, up for reconsideration in 1974, enables the government to offer premium prices and other economic incentives to encourage production of strategic commodities. Thus, if the development of synthetic oil was important enough to the nation that we wanted to be sure of its development, the government could take bids and award contracts for specific quantities and then resell it in the market. In this way, market guarantees could be used to promote private development of new technology rather than pushing it through a federal R&D effort.

In addition to purchasing commodities or subsidizing a particular plant, the government might provide specific tax benefits such as depreciation of the facility, which would involve individual contractual arrangements for specific objectives rather than broadly based tax benefits such as tax credits for R&D.

Once the technology has been effectively taken up by industry, no more federal money should be appropriated for the development of the technology except as necessary for evaluating progress and for the implications of that technology for the overall government R&D effort.

We cannot leave the resolution of health, safety, and environmental problems to the private sector, because investments for such social concerns cannot be justified by profit making companies. They are the essence of governmental responsibility. Standards and taxes have a role in mitigating these problems when the requisite technologies exist for solving them. But if the technologies don't exist, taxes could result in simply passing on the costs to the consumer—along with the ill health and pollution. Without the technology to implement them, standards become the object of political strife and controversy; there is continued pressure for delay in their implementation in order to save jobs, protect the economy, achieve self-sufficiency in energy, and so on. While these reasons may have some merit, they do not achieve safe mines or clean air.

Making existing energy technologies—coal burning power plants, automobiles—compatible with health and safety should become a much larger part of the government's energy R&D. And in developing new technology, safety and environmental concerns must be a major objective of the design. These concerns are inseparable from the basic R&D effort. Nevertheless, experience with the AEC has shown that the "inventing" agency tends to cover up the problems with its brainchild.

An independent technology assessment office is needed to evaluate the efforts of the energy R&D agency from an environmental and safety perspective, and to calculate the costs and benefits of technology. This monitoring function must be continuous, it must be independent, and it must be public. This office would have responsibility for ensuring that technologies (new and existing) are compatible with health and safety. It ought therefore to be a part of an agency such as the Environmental Protection Agency, which should have and exercise the environmental responsibilities of the nation's R&D effort, both public and private. As a guard against politicization of federal agencies, all R&D proceedings and facilities, both environmental and developmental, must be public.

CHAPTER 13
Conclusions and recommendations

As a result of a two-year study, the Energy Policy Project has reached a number of conclusions on major issues that go to the heart of the debate over national energy policy. We hope these conclusions will be useful to citizens as they make the choices that will add up to an energy policy for the nation.

The major finding from our work is that it is desirable, technically feasible, and economical to reduce the rate of energy growth in the years ahead, at least to the levels of a long term average of about 2 percent annually, as set forth in our *Technical Fix* scenario. Such a conservation oriented energy policy provides benefits in every major area of concern—avoiding shortages, protecting the environment, avoiding problems with other nations, and keeping real social costs as low as possible.

The future rate of growth in the GNP is not tied to

energy growth rates. Our research shows that with the implementation of the actions to conserve energy in the years ahead, GNP could grow at essentially historical rates, while energy consumption grows at just under 2 percent.[a] And employment opportunities would also be essentially the same as under a continuation of the historical pattern of increased energy consumption. Investments would shift from more power plants to energy saving technologies. In fact, the *Technical Fix* scenario would result in a net saving in capital investment requirements of some $300 billion over the next 25 years. The capital requirements for the energy industry, which would otherwise absorb 30 percent of total capital, would be reduced to about 20 percent, or near the current percentage.

The Project also finds that it appears feasible, after 1985, to sustain growth in the economy without further increases in the annual consumption of energy. Such a *Zero Energy Growth* scenario can be implemented if needed for reasons of resource scarcity or environmental degradation, or it may occur as a result of policies that reflect changing attitudes and goals.

The great bulk of the savings over historical growth in energy consumption can be achieved by "technical fixes" in three key areas:

- Construction and operation of buildings to reduce energy needed for heating and cooling;
- Better mileage for automobiles;
- Greater energy efficiency in industrial plants through investments in new technology and self-generation of electricity to use waste heat instead of more fuel to make process steam.

We have given a great deal of thought to the key question of how to make these opportunities for saving energy, which are technically feasible and economical, become a reality. We believe that market forces can and should be encouraged to help balance energy supply and demand, but there are a number of specific areas where market forces are ineffective.

---

[a] The term "consumption", as ordinarily employed in energy matters, includes every Btu burned during the period in question, wastefully and usefully alike. With greater efficiency in production and conversion of energy (as described in the *Technical Fix* scenario), the Btu's actually available to ultimate consumers would increase more rapidly than 2 percent a year. Furthermore, increased efficiency in using this energy would mean that the beneficial services energy performs would be essentially the same as with *Historical Growth*.

Environmental degradation and foreign policy concerns are not automatically reflected in the market price of energy. Electric power and natural gas utilities are natural monopolies and, therefore, governmental price controls are necessary to protect the consumer. And the market is also weak in areas such as investments in new, energy-efficient buildings, where the investor has a stake in reducing initial costs and often passes up economical investments that would save energy. Nor is it clear that major savings from improved gas mileage for automobiles can be achieved on the most rapid timetable without supplementing market forces with specific performance requirements or tax incentives.

Furthermore, we do not write on a clean slate. There is a whole host of ongoing governmental controls and interventions in the marketplace that need to be removed or reformed, if the nation is to balance its energy budget in a satisfactory manner. Tax subsidies to the petroleum industry, promotional rate making policies by the utility commissions, leasing policies of the federal government, one-sided investments in R&D, and counterproductive restrictions on the railroads by the Interstate Commerce Commission are all examples of where reform is needed. Government policies and regulatory actions already play a major role in both the availability of energy supplies and the quantities that are consumed. Legislative and executive actions are needed no matter what energy policy the nation adopts.

We have examined three widely different patterns of future growth in demand. Each scenario will require government actions; the degree of governmental interaction with the marketplace is no greater in the *Technical Fix* scenario than in *Historical Growth*. The governmental actions differ for the different scenarios, but they each require certain specific items of legislation, administrative action, industry initiative, and citizen action to "make them happen" in a manner that serves the public interest.

The package of problems that is called the energy crisis constitutes one of the most formidable challenges facing this nation. Meeting future energy requirements without shortages, unnecessary environmental degradation, or adverse impacts on foreign policy will not be easy. Producing the energy required for even the lower energy growth options will be a tremendous task for industry. We recommend a specific set of actions because if we simply drift, the nation will inevitably suffer a series of energy related crises in the years ahead.

Energy conservation actions

To achieve all the energy conservation goals of the *Technical Fix* scenario requires a broad spectrum of policy actions. However, the most substantial energy savings could be achieved through the pursuit of four goals:

- Setting prices to eliminate promotional discounts and reflect the full costs of producing energy—especially important in achieving industrial energy conservation goals;
- Adopting national policies to assure the manufacture and purchase of more efficient automobiles;
- Developing incentives for increased energy efficiency in space conditioning of buildings;
- Initiating government programs to spur technological innovation in energy conservation.

To achieve these goals we recommend the following specific policies.

- *Changes in energy pricing*

Redesign the rates for electricity to eliminate promotional discounts and to reflect peak load costs.

Eliminate subsidies to energy producers such as the depletion allowance and expensing of intangibles, unwarranted use of the foreign tax credit, and cut-rate government accident insurance for nuclear power.

Enact pollution taxes supplementary to regulatory actions to reflect environmental costs of fuels extraction and energy operations.

Reflect in the price of oil the costs of stockpiling oil to guard against emergencies.

- *Incentives for more efficient space conditioning*

Establish a federal loan program so that easy credit is available to householders and small businessmen to make economical energy saving investments in existing buildings.

Revise FHA standards for mortgages to specify minimum levels of heat loss and gain for buildings and minimum efficiency of space conditioning systems based on life cycle economics.

Initiate federal, state and local government programs to provide credit to builders and owners to finance energy saving technology, to upgrade state and local building codes, and to provide technical assistance to builders.

• *Government action on automobile performance*

Enact minimum fuel economy performance standards for automobiles, supplemented by taxes and tax credits, to encourage the manufacture and purchase of more efficient cars (so as to achieve an average fuel economy of at least 20 miles per gallon by 1985).

• *Government programs to spur technological innovation*

Shift a sizeable share of federal R&D funds to development of energy conservation technology and research on problems of implementing it.

Direct government purchasing toward energy conserving equipment—efficient cars, tighter buildings, efficient space conditioning systems such as heat pumps, recycled materials—to provide a market for the most advanced energy saving technologies that are feasible on the basis of life cycle economies.

It is important that Congress debate and enact legislation which declares that energy conservation is a matter of highest national priority and which establishes energy conservation goals for the nation.

The goals should provide generally that:

• There should be significant reductions in the average national rate of growth in energy consumption as compared to historical trends. A target for the long term growth rate should be set at 2 percent per year and reviewed annually by the Congress;

• All possible measures should be taken to encourage the most efficient production and use of energy;

• Each sector of the economy should achieve the lowest possible energy requirements subject to economic efficiency and the state of technology.

All federal government energy-related programs should be coordinated, so that in their cumulative effect they fall within the national goal.

Congress should establish a new Energy Policy Council within the Executive Office of the President, with responsibility for developing and coordinating national energy policy. It would be responsible for translating the national conservation goals into guidelines that would be useful as a reference point to both government and private planners.

The guidelines should be broken down by major sectors and by geographic regions; and should be developed with input from governments in the regions, industry, and public hearings. An essential element of such a program is to institute a uniform system of accounting for energy in our economy so that we know better where and how energy is used, and in what sectors of the economy it will be needed in the future. We also need hard facts on the energy required to produce all the various energy sources so that we know how much net energy the economy gains from various supply options.

The conservation guidelines would provide government planners and decision makers a yardstick against which to measure their programs and regulatory actions. At the federal government level, they should be mandatory for program planning. Thus, for example, a federal coal leasing program would be based on the guidelines for coal consumption for the geographic regions to be served. State and local governments could also make use of these criteria in exercising their energy-related responsibilities. Thus, state utility rate making commissions (or, preferably, a regional agency) could assess projected capital expansion plans of utilities, on the basis of projected needs for regional electric growth under a policy of energy conservation. They would have a basis for deciding how many power plants are really needed.

The Energy Policy Council would also perform a comprehensive energy monitoring function. It would continually evaluate growth trends by sector and region, and would modify the conservation guidelines as necessary. It would report annually to Congress and the public on the nation's energy situation; and would recommend any needed legislative reform. It would be assisted by a Citizens' Advisory Board.

Supply actions

Our work indicates that with the achievement of the energy conservation opportunities in the *Technical Fix* scenario, energy supply will need to be approximately 28 percent larger in 1985 than in 1973. Achieving this increase in supply over the next decade will require a strong effort by the energy producing industry. But unlike *Historical Growth,* the energy savings in this scenario will make unnecessary additional developments which threaten serious environmental damage, or increased oil imports which pose foreign policy concerns.

The lower rate of growth in the supply requirements from now through 1985 could be filled without massive new commitments to energy supply systems that are the source of major controversy: large scale development of western coal and shale where land cannot be reclaimed; imported oil; nuclear power; and presently undeveloped offshore provinces such as the Atlantic and Pacific coasts and the Gulf of Alaska. During the next decade, new supplies would come from the following sources:

- New discoveries of oil and natural gas in the lower 48 states and Alaska onshore, and offshore in the Gulf of Mexico;
- Coal from deep mines and areas where surface mining reclamation is feasible;
- Electric power plants that are already in some stage of construction;
- Secondary and tertiary recovery of oil and gas from existing wells.

The development of these supplies will not prevent shortages in the very short term because of the lead times involved in any new developments. But if no effort is made to improve efficiency in energy consumption, any near term shortages would be worse.

In the period after 1985, significant development of substantial additional supplies from controversial sources will be required even to support 2 percent growth. But the lower growth rate compared with the *Historical Growth* scenario permits much greater selectivity. The nation will be able to pick and choose, avoiding the most undesirable sources that would still be needed under historical growth.

In this same post-1985 period, some supplies can be expected from unconventional sources, including solar energy, geothermal energy, and solid and organic wastes. However, total energy requirements even at the lower rate of growth will be so large as to require continued expansion of conventional supplies. We must either make major commitments to at least two of the four troublesome energy sources noted earlier—oil imports, nuclear power, the Rocky Mountain coal and shale, and drilling in the Gulf of Alaska and off the East and West coasts—or we must go ahead with all four on a more moderate scale. In addition, coal production will be required approximately to double from current levels by 2000. If pollution technology to control small particles, especially sulfates, can be available by then, increased coal production in the latter part of the century could come from midwestern areas, where reclamation can be readily accomplished.

Our judgment is that the oil and gas resource base in this country is far from exhausted and can supply over half the U.S. energy supply in the *Technical Fix* scenario for the remainder of the century. Limitations on oil and gas availability are likely to stem from a combination of environmental, social, and political constraints on rates of development rather than from a physical limit on the quantities in the ground that could in theory be available. In the long run, when oil and gas prices rise relative to more abundant energy sources, oil and gas may have even greater value as chemical feedstocks and protein sources—uses that may be expected to take an increasing fraction of available supplies.

The nation is gradually moving toward a predominantly electric energy economy. This places a premium on reforms in the pricing, regulation, and institutional arrangements of the electric power industry to assure maximum efficiency and environmental protection in its production, and maximum efficiency in its use. Electric power has steadily become a larger and larger share of our energy supply, and now over 25 percent of all fuel is converted to electricity. This trend has been due to the unique flexibility of electricity for use in industry and in the home, as well as to a long period of falling prices relative to other fuels. The trend to electric power is apt to continue even under the lower growth rates of the *Technical Fix* or *Zero Energy Growth* scenarios.

One important conclusion from our work is that the expansion program of the electric power industry now underway is substantially greater than needed to supply the electricity that the *Technical Fix* scenario requires. Demand for electric power in this scenario would grow faster than 2 percent per year overall growth rate; but it would still amount to only about half the 7 percent which is the electric power industry's historical growth rate. Power plants now on order for completion by 1980 could satisfy the demand for electricity until 1985 under such an energy conservation policy. This would mean that a pause of several years in new power plant starts is possible for the nation as a whole. During this period, technical progress could diminish concerns about the safety of nuclear power and about air pollution from burning coal or oil in power plants.

In our view, at the present time there are ample incentives in the existing price of crude oil and coal for industry to produce the required quantities of fuel. Indeed, the incentives are excessive, in view of the existing tax subsidies, and taking into account that prices of new production do not reflect free and open competition, but are the same high prices that were fixed by a cartel of oil producing nations. However, a few specific governmental actions are required to stimulate the necessary growth in environmentally preferred energy sources without underwriting excessive profits to industry:

- Adoption of a combined policy covering oil and natural gas pricing and federal income tax payments, with the purpose of eliminating special tax advantages, yet providing high enough prices to attract sufficient capital for the development of these resources;
- Establishment of oil stockpiles to provide at least a 90-day backup to imports as a safeguard. The federal government should adjust the size of the stockpile in the future as an integral part of its energy policy. A tariff on imported oil should be levied to finance the stockpile program;
- Redirecting research and development of new sources of energy to enlarge the effort on near and medium term opportunities, such as organic wastes and geothermal energy, and toward solving environmental protection and safety problems with existing sources.

Zero Energy Growth

The policies for the *Technical Fix* scenario and *Zero Energy Growth* are virtually identical for the next five years. However, in the years that follow, additional energy conservation measures and small, but important, shifts in the pattern of economic growth would be required to move toward a stable level of energy consumption with a healthy economy. Our work suggests that such a policy direction would be desirable to meet certain social goals that would improve the quality of life, at least in the view of a growing number of citizens. We also believe that, with a transition over a ten- to fifteen-year period, it could well be technically and economically feasible to achieve stability in energy consumption while continuing healthy economic growth. Furthermore, it is altogether possible that one or more environmental concerns or resource constraints will force us to such a policy, whether we like it or not.

It is therefore our recommendation that the new Energy Policy Council, as an undertaking of the highest priority, make an intensive, continuing study of the desirability, feasibility, and necessity of moving to zero energy growth. The Council should publish annual reports on their studies to the Congress and the American people; these could serve as the basis of widespread public discussion and congressional hearings, so that timely decisions could be made if the nation decides that zero energy growth should, or must, be our national policy.

We stress that the quantity of energy required by the United States when it reaches a stable level as presented in the *Zero Energy Growth* scenario is by no means sacrosanct. The quantity we use, based on near term technology, is meant to illustrate the point that we can level off energy consumption and continue with an economy in which consumer well-being continues to improve. We have not examined in detail what might be accomplished in the way of additional efficiencies (what might be called a "super technical fix"), but our research suggests that a satisfactory economy after the turn of the century might be possible with appreciably less energy than shown in our *Zero Energy Growth* scenario.

Specific findings

Social equity

• The more money people have, the more energy they use. But the poor spend almost 15 percent of their household income on energy, while the high consumption of fuel by the rich typically accounts for only 4 percent of their incomes. Any major energy price increases will thus cause hardship to poor families, since their energy use levels do not include a margin of extra amenities easily done without.

• Government contingency planning is needed to help lower-income families cope with shortages and sharp price increases. Emergency policies might include a system of "energy stamps" (similar to food stamps), as well as special grants or fuel allocations to low income persons who demonstrate potential hardship as a result of shortages or price increases for energy.

• Social equity concerns also require a strong effort to save *all* consumers both energy and money, particularly for the most essential uses. This effort should include labelling appliances to show their energy efficiency, making credit available to low income homeowners for energy saving home improvements, and setting performance standards for automobiles.

- The social equity problems of our nation go far beyond energy, and cannot be solved through energy policy. Since energy is essential and comprises a large—and growing—amount of poor people's budgets, we conclude that the social equity implications of high energy prices should be resolved by a national commitment to income redistribution measures, such as a guaranteed minimum income or a negative income tax.

Energy, employment, and economic growth

- U.S. manufacturing has realized significant energy savings in the past, in a period of stable or declining relative prices of energy. The past rate of improvement was 1.6 percent, and a pre-embargo study indicates a 2 percent rate is probable out to 1980. Given much higher prices, fear of future shortages, and explicit government actions, even greater energy savings are likely.
- Energy saving is economically attractive today. Energy conservation measures should pass the test of economic efficiency as well as thermal efficiency. Our studies indicate that the conservation measures in the *Technical Fix* scenario (see Chapter 3) meet this test.
- It is reasonable to expect that energy conservation in the most energy intensive manufacturing industries will have little, if any, adverse effect on employment.
- Government policy measures to stimulate capital investments by business—such as investment tax credits —should recognize that they may stimulate growth in energy demand as well. On the other hand, tax incentives may be used to encourage investments that increase energy use efficiency.
- Energy conservation will not disrupt the non-manufacturing sector of the economy, if government policies toward housing, transportation, R&D, and the environment are consistent with the objective of conservation, and allow lead time for adjustment.
- Because of the slowdown in population growth, growth in the labor force is expected to slacken after 1980.
- Energy, employment, and economic growth are

interdependent—but they are in no way linked inevitably to the patterns of the past. Sudden unexpected shortages can cause severe unemployment in any scenario. Contingency planning for shortages must make provision to minimize unemployment. But over the long run, the United States can grow and prosper, have plenty of jobs—and still conserve energy.

U.S. energy policy in the world context

- The dramatic price increases in the world oil market over the past four years are more than a temporary aberration on the supply and demand charts. According to our judgment, they represent a fundamental shift in the power relationships between the world's industrial powers and the oil exporting nations.

- For this reason, the international trade in oil can no longer be left to the marketplace—a handful of international oil companies dealing with governments of the exporting nations. Instead, there should be systematic, multilateral, government-to-government discussions between oil exporters and importers so that they can seek to identify their shared interests.

- Agreements should be sought in a number of areas in order to stabilize the impact of the oil trade upon the world economy. They should be sought, for instance, to avoid disruptive changes in oil supply and prices. Also, the oil importers and exporters should come to the immediate aid of poor nations by creating a special multi-billion dollar emergency fund. In addition, the oil importing countries should agree to sell food to the poor countries at greatly reduced prices, and the oil exporters should sell oil to the poor countries at comparably discounted prices.

- It is impossible to say whether oil prices will stay at their present high level between now and 1985. Certainly they could decline somewhat, especially if the United States, Western Europe and Japan reduced the growth in their oil import demands by conserving energy at home. We think that a drastic reduction in oil prices—say, a return to the levels of the early 1970s—is unlikely for several reasons: the nonsubstitutability of oil with alternative energy sources in the short run, as well as the relative financial strength, concentration of ownership, and common purpose of the OPEC cartel.

- Rising oil prices have seriously aggravated the already existent world economic woes—inflation and currency instability. Close multinational consultation is needed to avoid competitive exchange rate devaluations and restrictive trade measures (which will only make matters worse), as well as to encourage the free flow of the oil exporters' surplus financial reserves back into the importing nations, in the form of long term investments and purchase of exports.

• Among all the energy policy options explored in this book, the only one that seems to pose foreign policy problems that might be termed "insurmountable" is the High Import option (11 million barrels a day by 1985) under *Historical Growth.* Interestingly, this is the course which the United States was on before the 1973–74 Arab oil embargo.

• Energy conservation is the most effective unilateral tool available to the United States for coping with most of its energy-related foreign policy problems—vulnerability to politically motivated oil import cutoffs, international economic instability, and relations with Europe and Japan.

• As more and more nations acquire nuclear power capabilities, the problem of governments diverting nuclear material from civilian nuclear programs into military use becomes acute. In a first step toward trying to stem the further proliferation of nuclear weapons, we urge that the nuclear exporting nations (United States, Canada, Soviet Union, France, India, Great Britain, et al.) seek agreement not to export nuclear fuels and equipment, either civilian or military, to the Middle East, a strife-torn region so rich in other energy resources. Additional agreements should be sought to cover other regions of the world that are yet untouched by nuclear power.

At the same time, the nuclear nations should provide developing countries with technical and financial assistance so that they can develop more economical domestic resources—solar, hydroelectric, coal, organic wastes, geothermal, and so on. These multinational efforts to restrict nuclear exports should, of course, be part of a broader effort, within the nuclear nations, to disarm their nuclear arsenals and to create adequate safeguard systems to keep their own civilian nuclear materials out of the hands of criminals and terrorists.

Environment

• Air pollution is the most serious immediate, widespread danger to human health from the use of energy. Yet the expectations of rapid growth in energy are leading to pressures to postpone or suspend the implementation of air quality goals. Available scientific evidence indicates that there is no basis for relaxing present air quality standards. For the near term, air quality goals should be pursued both by curtailing growth through conservation policies and through regulations and pollution taxes to require that control technologies be installed when they are available.

Small particle air pollution, with its associated sulfates and toxic trace elements, is largely uncontrolled today and may prove to be the most damaging form of air pollution. Increased use of both coal and oil will worsen this problem. A control program and control technologies are urgently needed to deal with it.

• Nuclear fission is potentially a very large source of energy. Nuclear energy is free of air pollution, generally requires less land in providing energy and, in the long run, allows us to avoid some of the global climatic problems that may be associated with the burning of fossil fuels. But the problems of reactor safety, nuclear theft, the proliferation of nuclear weapons through diversion of fissionable materials, and ultimate disposal of nuclear wastes are, as yet, unresolved. Moreover, the problems of our institutional capabilities for dealing with these issues have not yet been squarely faced. Resolution of these problems should come before, not after, a high level of nuclear capacity is installed.

Nuclear power is currently growing at a tremendous rate. But the current projections are based on the historical rate of growth in energy, which is high. Our studies show that a much slower rate for nuclear power is adequate to meet energy needs, if the conservation oriented policy we recommend is implemented. We do not advocate an absolute ban on new nuclear plants because the problems posed by using fossil fuels instead are also serious. But a conservation oriented growth policy will provide breathing room so that we can gain a better understanding of nuclear power problems, and reach some better judgments before major new expansions of nuclear power are made.

• Initiating major new energy developments of the coal and shale in the West, or of the oil off the Atlantic and Pacific Coasts or in the Gulf of Alaska poses serious problems of secondary impact on the regions and adjoining coastal zones that have not even been thoroughly evaluated, much less resolved. The problems involve water supply, the disruption of traditional land use values, and region wide demographic and economic changes, as well as on-site environmental problems of land restoration, air pollution and the like. The energy conservation growth rates will permit thorough evaluation and orderly development when, and if, possible, without sacrificing important environmental values.

• The final choice on which supply options are environmentally "best" cannot be settled on the basis of scientific evidence alone. Fossil fuels present a clear and present danger of air pollution. Nuclear power presents small, but uncertain risks, of terrible accidents. Developments in the Rockies and near the coastal zones threaten irretrievable loss of the amenities that characterize these regions, thereby degrading the diversity of our land environment. These decisions depend on value judgments of people which, aided by weighing benefits and risks and determined ultimately by political decisions, will depend a great deal on the people's access to information and their interest in affecting the political process that makes those decisions.

Private enterprise and the public interest

• The U.S. energy supplying industry does not constitute a monopoly by economic standards, although there are indications of diminished competition in some areas. In particular, changes in leasing policy, antitrust enforcement and tax laws are necessary to improve the industry's competitive performance.

• The basic problem is that the energy industry —particularly the oil industry—possesses a unique combination of political advantages which has enabled it to exert considerable influence on public policy. This influence is manifested in a variety of energy policies that are highly favorable to the industry.

• A necessary first step in reforming energy policies is to remove the main sources of the oil industry's disproportionate political strength through strong campaign finance reform measures. This must be accompanied by consumers taking an active interest in energy policy issues at the ballot box, where their views can be most effective.

• By uncoupling economic and political power in this way, it would then be possible to implement effective reforms in areas of diminished competition and the other areas of reform set forth in this book.

• If the modest reforms to strengthen competition and create an environment in which the industry is more responsive to public concerns are insufficient, then it may be necessary for social control to be exercised by the government taking a more active role in the organization of energy supply. The federal chartering of companies appears to us to be a promising technique for doing so.

• No matter which route is taken, citizen participation is essential. Citizen interest must be connected to the decision making process on a continuous basis so that it becomes a lobby or a pressure group to be taken into account when decisions are made. Citizens must develop countervailing power at all levels of government.

Utilities

• Electricity is generally still priced to promote growth through discounts for greater use at a time when greater conservation is needed. We recommend a new method of pricing electricity that encourages conservation by charging more for electricity consumption at the time of heaviest system use, so as to reflect actual costs and encourage thrift.

• The current system of regulation of the electricity industry that relies primarily on state commissions is inadequate to cope with utilities that largely operate in regional power grids. Regional utility commissions could assure that utility expansion plans were integrated into regional grids so as to meet regional needs with maximum efficiency in the investment of capital and the use of land.

• Reforms in the structure of the industry to separate the generation and transmission of electricity from local distribution are desirable. This would facilitate the formation of regional companies, with public participation in their management, to achieve the economies in capital and land use that are possible in an integrated regional approach to power system expansions.

Federal resources

Over half the domestic fuel resources that remain in the ground are public property. Government decisions on their sale and use will be a critical force in determining patterns of future energy development. Conflicting social goals of fair value to the treasury, meeting energy needs, and environmental protection, are inherent in the government's resource management programs.

Past Interior Department resource management policies have failed to reconcile these conflicts and protect the public interest. There is no assurance that the resources are being developed at a time and price that correspond to national needs; vast amounts of the resource base have been released with a grossly inadequate return to the public treasury; and the environment has been poorly served by a lack of leadership commitment and inadequate protection measures.

For example:

- Over 22 billion tons of federal coal have been transferred to private hands at little or no cost, and with inadequate provisions to ensure that lands can be restored. Little of this coal has even been developed;
- The government has launched a program to lease ten million acres a year of federal offshore oil and gas lands with poor understanding of the extent of oil and gas to be sold, of the effects of such massive leasing on revenues returned to the treasury, and of the environmental and social costs associated with its development.

It is our conclusion that the public interest can best be served by truly competitive leasing under a reformed resource assessment, planning, and regulatory system. The key elements of this reform are:

- Immediate improvement of the resource data base, a greatly increased environmental study program, and an expanded data analysis capability;
- Creation by law and implementation of a three-tiered assessment and planning system (detailed in the body of this report), including an annually revised and planned energy resource schedule; regional resource assessment projects to address decisions on whether, where, and in what sequence, the resources of a region will be developed; planning for sales of leases to establish environmental controls at specific sites and to ensure a fair market value return to the treasury when the resources are sold;
- The resource management responsibilities should be consolidated in a new Energy Resource Administration within the Interior Department. Environmental and safety regulatory responsibilities should be transferred to the Environmental Protection Agency.

Until this new management system is implemented, there should be no leasing in as yet undeveloped regions of the Outer Continental Shelf; the existing de facto moratorium on further coal leasing should be continued; and other energy resource programs should be maintained at their present levels.

Research and development

- The present energy R&D efforts—both public and private—are concentrated in a very few supply oriented technologies. Allocation of funds is basically tilted toward supporting a high rate of energy growth. The nation is ignoring energy conservation technologies, neglecting environmental R&D, and failing even to achieve very much diversity in energy supply.
- There is a major need for redirecting federal energy R&D. The program should be goal oriented. Energy conservation, diversity of energy supplies, environmental protection, and health and safety should be major goals.
- A major new thrust in R&D addressed to energy conservation opportunities is urgently needed now to sustain the *Technical Fix* scenario beyond the 1990s and to implement the *Zero Energy Growth* scenario. Research of the institutional problems in introducing conservation technologies into the economy is relatively inexpensive, but important to the success of conservation oriented R&D.
- A major new thrust should also be initiated to pursue solar energy and other renewable energy sources.
- Environmental protection technology should be an integral part of energy R&D all the way to the marketplace. New technologies should not be considered commercial until they can satisfy health, safety, and environmental standards. Setting those standards and satisfying them should be part of the R&D process. We also need an urgent and aggressive R&D program to clean up existing technologies and to solve many health and safety problems.
- Government energy R&D programs should explore promising ideas, perform basic research, and advance the concomitant technology. The government should phase out its R&D commitment in the demonstration phase. A thorough assessment of the costs and benefits, as well as the health, safety, and environmental problems associated with the technology, should be undertaken by an independent technology assessment office before any funding for a demonstration project is authorized. Such assessments should be followed by full public debate under congressional auspices on the merits of proceeding with a demonstration

plant. Further evaluation should be undertaken, if necessary. If the decision to proceed is made by Congress, and the requisite industrial infrastructure exists, government funding for demonstration plants should be limited to 25 percent of the initial estimated cost. The remaining 75 percent and any cost overruns should be borne by the participating private industries. This procedure is not now a part of our energy R&D efforts and should be implemented immediately.

The breeder reactor program, to which we have committed a major portion of the federal R&D funds, is an outstanding example of the neglect of public participation as well as independent assessment, and of failure to protect the public treasury. We recommend that the present open-ended government funding commitment to the LMFBR demonstration project be terminated immediately. In addition, an independent assessment of the state of reactor technology and its associated health, safety, and environmental problems should be undertaken by the National Academy of Sciences on an urgent basis, so that the public may have the opportunity of debating the desirability of proceeding with the demonstration plant. When that desirability is established, the demonstration project should be funded along the lines discussed above.

Energy Policy Project Staff, Consultants and Advisory Board

PROFESSIONAL STAFF

S. David Freeman, Director
Monte Canfield, Jr., Deputy Director
Pamela L. Baldwin, Project Assistant
Steven C. Carhart, Technology Specialist
John Davidson, Physicist
Joy C. Dunkerley, Economist
Charles Eddy, Project Counsel
Jenrose Felmley, Librarian
Katherine B. Gillman, Writer
Lucille Larkin, Director of Public Information
Arjun B. Makhijani, Engineer
Kenneth J. Saulter, Economist
Majester L. Seals, Project Assistant
David Sheridan, Editor
Robert H. Williams, Physicist

SUPPORT STAFF

Shirley Cox
Edwin Farrell
Sharon Lynn
Valeria Moore
Billie Truesdell
Barbara J. Vance

FORMER PROFESSIONAL STAFF

Michael Fortune
Jeremy Main
Francisco X. Otero
Judith Pigossi
Shelley Prosser
Adam Sieminski
J. Frederick Weinhold

FORMER SUPPORT STAFF

Annette Burke
Evelyn Chisholm
Larry Dickter
Norma Dosky
Cheryl Garner
Dieu Granados
Patricia Lynch
Rebecca Jacobsen
Patricia Pierson
Judith Miller

CONSULTANTS

Saud Al-Sowayel—Independent Consultant
James P. Beirne—Independent Consultant
Daniel Bell—Department of Sociology, Harvard University
Manson Benedict—Department of Nuclear Engineering, M. I. T.
Kenneth E. Boulding—Department of Economics, University of Colorado
Harrison Brown—Division of the Humanities and Social Science, California Institute of Technology
Ronald Brown—Washington Bureau, National Urban League
Peggy Bruton—Independent Editorial Consultant
Barry Commoner—Center for the Biology of Natural Systems, Washington University
John Esposito, Lawyer and Freelance Writer
Frances E. Francis—Economist and Lawyer
Irene Gordon—Freelance Researcher and Writer
Perry R. Hagenstein—New England Natural Resources Center
Hendrik S. Houthakker—Department of Economics, Harvard University
William Iulo—Department of Economics, Washington State University*
Mark Levine—Stanford Research Institute, Menlo Park, California*
Walter J. Mead—Department of Economics, University of California, Santa Barbara*
James R. Murray—National Opinion Research Center, University of Chicago
John A. Neary—Freelance Writer
Donald E. Nicoll—Chairman, Joint Operations Committee, Land Grant Universities of New England
Alan Poole—Biologist
Marc J. Roberts—Department of Economics, Harvard University
William E. Shoupp—Former Senior Vice President, Westinghouse Electric

* Full-time staff members for 1972–1973 academic year.

Robert H. Socolow—Center for Environmental Studies, Princeton University

Ben J. Wattenberg—Writer, Editor, Publisher and Consultant

Richard Whalen—Independent Consultant and Journalist

Carroll L. Wilson—Alfred P. Sloan School of Management, M. I. T.

John W. Wilson—Independent Consultant

ADVISORY BOARD

Chairman—Gilbert White, Director, Institute for Behavioral Sciences, University of Colorado.

Dean Abrahamson, Professor, School of Public Affairs, University of Minnesota

Lee Botts, Executive Secretary, Lake Michigan Federation

Harvey Brooks, Dean, Division of Engineering and Applied Physics, Harvard University

Donald C. Burnham, Chairman, Westinghouse Electric Corporation

John J. Deutsch, Principal and Vice Chancellor, Queens University, Canada

Joseph L. Fisher, Former President, Resources for the Future, Inc.

Eli Goldston, President, Eastern Gas and Fuel Associates (Deceased)

John D. Harper, Chairman, Aluminum Company of America

Phillip S. Hughes, Assistant Comptroller General, General Accounting Office

Minor S. Jameson, Consultant (former Executive Vice President), Independent Petroleum Association of America

Carl Kaysen, Director, The Institute for Advanced Study

Michael McCloskey, Executive Director, Sierra Club

Norton Nelson, M.D., Director, Institute for Environmental Medicine, New York University

Alex Radin, General Manager, American Public Power Association

Joseph R. Rensch, President, Pacific Lighting Corporation

Charles R. Ross, Former Member, Federal Power Commission

Joseph L. Sax, Professor, School of Law, University of Michigan

Julius Stratton, President Emeritus, Massachusetts Institute of Technology

William P. Tavoulareas, President, Mobil Oil Corporation

J. Harris Ward, Director, Commonwealth Edison Company (Deceased)

Advisory Board Comments

Major issues of energy policy: Statement by the Advisory Board

Currently suffering from the absence of integrated policy in dealing with energy, the United States will have to make early and hard decisions on a series of complicated issues if it is to avoid greater distress in the future than afflicted the nation during 1973–1974. The lack of coherent policy and the major unresolved problems are outlined in this report by the staff of the Energy Policy Project. More detailed information on these problems, together with some differing views as to their solution, is contained in the set of supporting studies being published by the Project. The staff report and those documents warrant review by all who are concerned with the nation's energy situation.

The prescient decision of the Ford Foundation to launch the Energy Policy Project in early 1972 came at a time when there was doubt that the Project's findings on what then was seen as a momentous public question would command public attention when released. By the late autumn of 1973 the Arab oil embargo had helped plunge the country into what was widely regarded as a state of crisis. During that period any pronouncement on energy attracted notice, and piecemeal remedial proposals multiplied. By the summer of 1974 complacency again prevailed in many sectors of public discussion; interest in conservation of energy use had slackened, and confidence in supplies was fostered by ambitious schemes to become self-sufficient in oil, gas, coal, and nuclear fuels.

In our judgment this is no time for complacency. The decisions that must be made—if only by default—over the next few years are crucial to the welfare of the nation. The scope of possible public and individual action is immense. Much of the basic evidence has been marshalled by the Energy Policy Project. We believe that the full series of reports illuminates most of the critical issues. We differ among ourselves as to how well some of the issues are treated, and as to the wisdom of the recommendations for action.

Inasmuch as the Advisory Board was selected to reflect a broad range of individual outlooks, consensus among us on issues and remedies was neither expected nor sought. Our relationship with the Director and staff of the Project was one of offering advice on study plans, the selection of expert assistance and grantees, the quality of supporting materials, and the content of the preliminary and final reports. The staff and participating grantees brought together an important resource of experience and competence. We enjoyed candid exchanges of opinion on their products, recognizing that the Director could never fully respond to our variegated suggestions and that he was free to ignore all of them. The concluding report is the staff's, and this appraising word is ours.

In offering our opinions we recall with warm appreciation the contributions of two colleagues who died before the study was completed. Eli Goldston and J. Harris Ward were vigorous and original in their critiques of the plans and preliminary drafts. They enriched our thinking and graced our discourse as the study unfolded.

We regard the Project as a unique attempt to define the issues of national energy policy. It identifies numerous gaps in knowledge and fills some of them. It provides a useful framework for public discussion. By its recommendations as well as by its omissions and its assembly of evidence, it should stimulate more intelligent examination of the choices ahead. However, it should be studied as a beginning rather than as an end. In commending it as a point of departure for thoughtful appraisal of the nation's energy options we note a number of issues of unquestioned significance. Some of them are presented as our joint judgment. The others are emphasized as objects of disagreement to which we address ourselves in our statements of individual views.

It cannot be overstressed that the nation is in need of a genuinely integrated policy for promoting research and development on energy problems and for allocating responsibility for energy resources and demand. The emergency measures taken during the recent "crisis" do not meet this need.

Intelligent canvass of the components of an integrated policy is handicapped by lack of conclusive information at numerous points relating to environmental, demand and supply considerations. It should be noted that the Project's supporting studies were well underway when the embargo on overseas oil temporarily disrupted the national economy. They could not take advantage of analysis of new economic and political conditions that may persist for a long time. This fresh experience should be examined along with other investigations recommended in the report.

The three scenarios described in the report are in no sense the only options open to the American people. Many other assumptions and actions in addition to those outlined in the *Historical Growth, Technical Fix,* and *Zero Energy Growth* scenarios warrant attention. These three illustrate the ways in which the future opportunities may be examined, but should be taken as an invitation to explore alternative assumptions rather than as setting limits on possible action.

It is clear that there are no simple, viable choices at hand. Reliance on a massive program of research or concentration on a single fuel source or creation of an overarching federal agency cannot alone assure the country it is on the right road to avert future disruption of severe proportions.

There can be no doubt that more emphasis should be given at the national level to conservation while work continues on development of energy supplies. However, much remains to be learned in planning for such activities. Experience in influencing energy use is very limited: until recently the emphasis has been almost wholly on growth, and basic studies of methods and implications are lacking.

Whatever the course of action taken, the time is short in which to begin concerted programs. The report shows at numerous places the long lead times that are involved in technological innovation, in alterations in use patterns, and in restructuring of private or public organization. It will take five to ten years in most instances to translate a decision into significant changes in the life of the nation.

In looking to the future, the report gives inadequate attention to the question imposed by the finite nature of fossil fuels and what happens when the readily available supplies of oil and gas are exhausted. The approach of that situation in the short term of two or three decades is not presented as a sufficiently serious problem. We believe it deserves more thorough appraisal in terms of its implications for survival of industrial society, for international comity, and for the environmental and economic effects of transition processes that would be triggered long before the marginal resources are depleted. A position on this question is basic to much of the controversy that centers on the amount of attention to be given to development of alternative sources such as solar energy, or possible fusion to support a hydrogen-electric economy, or nuclear power.

The case for putting the federal house in order in managing federal oil, gas, coal, and other resources is persuasive. There is room for difference in judgment as to how to reorganize the federal agencies that now share responsibility in a weak fashion, but some kind of strengthened and unified management should be initiated promptly.

The report was not intended to provide a political guide for achieving the objectives we seek. Those steps will be intricate and will require a great deal of further attention if equity and sound national policy are to be assured. A problem is how to balance the effects of policy choices upon different regional and economic groups.

So much for issues that we can define with moderate agreement. Of larger import are the questions on which we are divided and which some of us regard as inadequately presented in the report. These are noted in capsule form and treated at more length in our individual views.

The promise and handicaps of nuclear power generation are sharply in dispute among us. The hazards attached to nuclear reactor accidents, nuclear theft, and nuclear waste disposal are viewed by some as being so severe as to warrant holding up nuclear energy programs. Others regard the threats as inflated and see the program as essential to a sound policy on energy supply.

There is strong difference in judgment as to the reliability of estimates of the volume of oil and gas resources.

The report considers environmental impacts wherever appropriate but is not able to mediate the argument of how much environmental quality should be traded against changes in prices of energy and its availability. The question of whether air pollution standards are to be for emissions as well as for ambient conditions plague a good deal of the debate over the effects of energy generation upon environmental quality. Depending upon the position taken, the costs of further development of fossil fuels may be greater or lesser, and opinions differ as to what is acceptable to our society. There is also major difference of judgment as to the degree to which patterns of energy consumption will be influenced by the combination of price changes and devices such as building codes or pricing schedules. Some feel these measures have been neglected; others regard them as overrated or misrepresented. An example is the report's stress upon eliminating promotional electricity rates for industry; it is argued that such schedules are in fact rarely used.

One of the more contentious issues raised by the report has to do with the role of government in coping with energy problems. The extent of the report's reliance upon new government organization at the federal or regional level to work out and execute far-seeing policies is regarded by some as essential and by others as naive and impracticable in the light of past performance. The making of energy allocations through the marketplace in the opinion of some does not receive sufficient prominence. They feel that big business is unjustly blamed for ills of energy production.

Those and other issues noted in the following statements must be resolved in some fashion as the nation forges an improved energy policy. We hope that the debate stimulated by the Energy Policy Project's materials will contribute to their more constructive resolution.

Gilbert F. White, *Chairman*
Dean E. Abrahamson
Lee Botts
Harvey Brooks
D. C. Burnham
John J. Deutsch
Joseph L. Fisher
John D. Harper
Phillip S. Hughes
Minor S. Jameson, Jr.
Carl Kaysen
Michael McCloskey
Norton Nelson, M.D.
Alex Radin
Joseph R. Rensch
Charles R. Ross
Joseph L. Sax
Julius A. Stratton
William P. Tavoulareas
J. Harris Ward*

* Harris Ward died at the time the final draft of the report was under review and had indicated his intention to his colleague Gordon Corey to submit individual views as appended.

Supplementary statement by Julius A. Stratton, Harvey Brooks, D. C. Burnham, John D. Harper, Phillip S. Hughes, Joseph L. Fisher, Alex Radin, and Joseph R. Rensch

Throughout this study there is a predominant emphasis upon the need to conserve energy and upon measures, both technical and social, that should be taken if substantial conservation is to be achieved. No one will challenge the vital importance of reducing waste, of increasing efficiency, and of promoting the conservation of energy in every rational way. In the short term such measures may well preserve the nation from the disruptive consequences of serious energy shortages.

But the country must recognize and bear constantly in mind that while conservation will buy time, even the most austere self-discipline will fail to resolve the very real and ultimate problem of supply. The fundamental issue that lies before us is not the simple one so commonly identified as the increase of energy for consumption by a greedy nation. Rather it is the question of whether this nation will be ready with new, practical, economical sources of energy to replace oil and gas when their availability begins to decline.

The year 2000 is only twenty-five years away. Despite any steps toward conservation taken in our own country, the global demand for energy is destined to mount steadily in these years. Without debating here the exact time span, we can be certain that in the foreseeable future the world's reserves of oil and gas are bound to diminish. The lead time for the development of basically new resources to the point of economical, practical utility is exceedingly long. Those new resources may include the conversion of shale and coal to hydrogen or hydrocarbons, the economic application of solar energy, the further advance of nuclear reactors, or the possible success of fusion to support a hydrogen-electric economy. The lead time involved reflects not alone the years of technological research and development, but also the time required for adequate assessment of environmental impact, for perfection of manufacturing methods, and for the lengthy and complicated process of social and political adaptation.

The hour is late. We doubt whether the public at large has any conception of the heritage of social and economic disaster that it may leave to a coming generation through its exhaustion of the resources of the earth if it fails to concentrate *now* on the development of alternative sources of energy supply. Somehow our plight must be brought forcefully to the attention of the public, and short-term measures must not be allowed to obscure the urgency of a long-term problem of the highest priority.

Supplementary statement by Dean E. Abrahamson, Lee Botts, John J. Deutsch, Joseph L. Fisher, Phillip S. Hughes, Michael McCloskey, Norton Nelson, Alex Radin, Charles R. Ross, Joseph L. Sax, and Gilbert F. White

This report presents a set of policy recommendations. As members of the Advisory Board we are in general agreement with most of these policy recommendations. Our individual differences about substantive matters are specified in individual comments below.

Individual views

Dean E. Abrahamson

This report does an excellent job of identifying and illuminating energy policy issues. It is neither elegant nor profound, but is superior as a general exposition of the elements of energy policy and will elevate the stature of debate over energy policy by treating energy as an explicit policy issue, and defining and supporting a consistent and broad set of policies.

There are some shortcomings. The report begins with a statement that, "Drift is surely the worst of the alternatives before us.", implying that there has been "drift" in the past in the determination of energy policy. This is naive; energy policy decisions have been and are influenced by careful, rational planning, but by the energy suppliers for their own vested interests rather than by the broader spectrum of those influenced by these policies taking into account the public interest.

The responsibility for other apparent deficiencies of the report must be shared by all who were associated with the Energy Policy Project. For example, the time horizon of the report, which is extremely short, was specified in advance. Also, the failure of the report to deal adequately with the interactions between energy and the production and availability of food was due to a general failure of Board and Staff alike in emphasizing the importance of these interactions.

Other issues are very important, but the available data simply do not permit specific recommendations. Some of these issues are mentioned in the statement of the Board as a whole but attention should also be drawn to the global climatic impacts of the use of fossil fuels. These impacts are possibly of paramount importance, yet the evidence available to date does not permit a definitive conclusion about their magnitude. The report suggests that these global climatic effects could limit the use of fossil fuels to a level only two or three times those of the present. But with the information currently available the report must stop with recommending major expansion of research bearing on these questions.

In other respects the report is timid. There is a growing body of evidence that solar energy can, within the time span considered by the report, begin to supply a significant portion of our energy needs. Yet solar energy is assigned a very minor role in meeting energy needs through the year 2000. Another example is the treatment of reducing growth rates in energy consumption. The suggestion that zero energy growth (ZEG) is an extreme case is ludicrous. Further, in discussing the possibilities of decreased levels of energy consumption, in contrast with limiting growth rates, the report is defensive and apologetic. Even were ZEG attained in the near future we would be faced with major energy supply problems and exacerbation of the present spectrum of environmental insults. It is also unrealistic to expect energy, at the per capita levels of ZEG in the United States, to be available globally. Do we then accept as inevitable a continuation of gross inequality in energy availability between the nations? It is disappointing to find these, and other fundamental considerations, not addressed by this pioneering study.

The generally excellent report is undercut by an apologetic and defensive tone which pervades some sections of the report, and by an unwillingness—for whatever reason—to draw definitive recommendations on many issues. One example is the treatment of nuclear power. The report spells out in a straightforward and realistic manner the hazards associated with nuclear power. In the discussions and recommendations, for example, that the nuclear exporting countries refrain from supplying reactors to the

Middle East and other regions of the world as yet untouched by nuclear power, it is tacitly recognized that effective international safeguards are very unlikely. Yet the report draws back from recommending a general abandonment of nuclear power.

In spite of these, and other shortcomings and compromises, this report with its set of policy recommendations provides a firm foundation and support for an elevation of the stature of the energy policy debate. It clearly demonstrates that to the extent definitive action is not taken, the nation will see its interests poorly served by having decisions made without public participation. We can influence our future or we can continue to be herded into the future on the basis of decisions made by the energy companies.

Lee Botts

The most important contribution of this report is to put before the American people the possibility and importance of choice in energy policy for the future. The fact that choice is possible is even more important now that the false promise of Project Independence is being promulgated.

Other events of the past year have reinforced the necessity for the American people and the Congress as their agency of decision making to realize that where energy comes from and how it is used can be a positive rather than a negative factor for future well-being in this country and the world. My city, Chicago, has for the first time had to call Ozone Watches to protect the health of residents. A nuclear plant in the region experienced minor sabotage by an employee, and a package containing low level radioactive wastes from another was lost en route to its destination. But the death rate on highways decreased substantially with a decrease in driving and a lower speed limit due to conservation measures. We can and we must exercise the opportunity for choice in both energy source and energy use now that the relations between these events can be better understood. The Energy Policy Project report can help.

There are three important areas in which the report provides a good starting point for debate but does not go far enough to provide a basis for final decision. I urge that the Congress address itself particularly to these issues in taking leadership for future national energy policy. The three issues are as follows:

1. How real costs of energy production can be attached to the cost of energy production and the extent of real environmental costs.
2. Why nuclear power should be rejected as the major source of energy for the remainder of this century until and unless positive protection can be provided against environmental contamination from this source.
3. Whether it would be possible to achieve Minus Energy Growth with even more positive long term benefits than those offered by Zero Energy Growth.

1. Real costs of energy production: The report suggests an energy tax as a means of using traditional market mechanisms to reduce consumption. This suggests that the costs of environmental degradation associated with energy conversion can be both measured and paid for in money. The suggestion in Chapter Eight that it is always possible to calculate in advance the cost of environmental degradation does not reflect present ecological understanding of potential destructive consequences of continued degradation. Nor does the report adequately reflect present knowledge of potential future environmental changes associated with high growth in energy consumption.

Some biologists believe that Lake Michigan has passed "a point of no return" in accelerated eutrophication. While it is still a comparatively clean lake in its open waters, away from the inshore sources of pollution, no amount of money could restore the life in the lake as it existed before man began to exploit its fishery and to use its waters for waste disposal. Now that we understand what happened and how, we can, if we will, avoid causing still more irreversible biological destruction.

Specifically, the Energy Policy Project report states that "there is a limit to what clean air is worth." Such a proposition could have seemed valid before we began to understand the full costs of our dependence on the automobile. It is difficult to accept literally in light of present scientific speculation about possible global climatic changes associated with growing accumulation of fossil fuel wastes in the atmosphere. The point is that, by the time global climatic effects are apparent it will be too late to debate what clean air is worth. *Now,* while it still may be possible to stop the Los Angelezation of Chicago, is the time to consider continued energy consumption in light of the potential long term consequences and to plan for a margin of safety for the future of life on our planet. The report acknowledges that the Clean Air Act had little such margins but finds much significance in the fact that achieving its goals will cost little in money. It does not emphasize enough that the choice may be survival rather than between levels of enjoyment of material comfort.

2. Need for positive protection against nuclear disaster: While urging that the nuclear industry assume full responsibility for insurance against nuclear accidents, the report ignores the fact that there would have been no nuclear program for power production if the industry had had to assume such liability heretofore. Nor does the report deal with the fact that the most serious consequence of dependence on nuclear power in the long run may be the genetic consequences of accumulated radioactivity in the environment. The accumulation may come as the result of accident in operation of a plant, to which the question of liability insurance is directly applicable. It could come as the result of unlawful diversion of nuclear materials, the possibility to which the Energy Policy Project has directed appropriate attention. It could also come simply with continued operation of many nuclear plants, a possibility to which the report, in my opinion, does not give enough consideration.

In stating that "Nuclear energy offers an escape from the global climatic problems of fossil fuels" the report suggests that nuclear power does not offer potential global environmental problems of its own. Yet the report's own identification of the unsolved problem of long term storage of nuclear wastes made necessary by operation of the plants even without accident seems to contradict the view that nuclear power is somehow potentially less dangerous than dependence on fossil fuel. More simply, nuclear power does not cause emphysema but it can cause leukemia.

The report does not state forcefully enough that the real situation faced by the American people is a question of choosing their own poison as long as continued energy use on a very high level is assumed. In my view, the report gives ample reason to call for rejection of nuclear power but stops short of doing so.

3. Potential for minus energy growth: The third way in which the report stops short of a conclusion to which its arguments lead is the question of potential benefits with decrease in absolute energy use in the future. While the report states in Chapter Eight that "a satisfactory economy after the turn of the century might be possible with appreciably less energy" than shown in the report's *Zero Energy Growth* scenario, such a possibility is not examined.

With all the reasons stated in the report why an actual decrease in energy use may offer the most attractive alternative of all for both economic and environmental reasons, even the suggestion of such a possibility may be the most important step toward real choice for the future that the Energy Policy Project has taken. Congress should move toward considering what improvement in the world's prospects a policy of Minus Energy Growth could offer.

The role of government: Finally, I must take issue with my colleagues on the Advisory Board of the Energy Policy Project who protest that the report's conclusions on the need for new energy policy place an undue and dangerous burden on government. Some comments suggest that the report calls for government, particularly the federal government, to take action in areas in which it has not taken action in the past.

In truth, in my view, it is the way the federal government has developed and implemented energy policy in the past that has brought the United States to its present dilemma. In the past, government has not acted in the interests of the public as a whole in relation to energy resources, but in response to special interests. It has not considered the long term—or even in most cases the short term—consequences of promotion of ever-increasing growth in energy use except as measured by economic profits.

While it is true that some consequences could not have been taken into account because they were not perceived, this is no longer true. We are well warned about potential nuclear contamination. Now we know that costs of energy production must be measured in terms of human health as well as convenience. The value of survival of life cannot be compared in dollars alone to the profits of exploitation of the earth's resources.

I am grateful to the Ford Foundation for giving the people of the United States a new view of what could happen so that government can be directed in deciding what ought to happen. With the Energy Policy Project report, the choice can no longer be seen simply as a choice between continued ever-increasing growth in energy consumption in spite of real costs or economic disaster. The choice can be for increased well-being for people as part of the natural world.

Harvey Brooks and Carl Kaysen

The major propositions which the report presents deserve wide attention. We believe that some of the deficiencies in the report—rhetorical excesses and disproportionate attention to certain matters of detail—unfortunately detract from its persuasiveness, and may divert the attention of readers away from the central questions which it fairly poses. The most important of these is the simple proposition that as a nation we are faced with a long-run energy problem with which we are unlikely to deal in a desirable way by relying chiefly on the workings of the market. This is what we have done until now, with sporadic, diffuse and uncoordinated interventions by government—sometimes to assist producers, sometimes to protect consumers, sometimes to preserve the environment, sometimes in the name of national security. Further, the Report argues—in our view, properly—that a wise policy will be one which includes a substantial decrease in the rate of growth of energy use in the United States as an essential element. It is a virtue of the report to make clear the potential gains in both security and in our capacity to control the environmental costs of energy use from such a decrease, and to show that it is both feasible and not so costly in terms of other values, including the values of prosperity and consumer satisfaction, as to be impossible of achievement.

Chapter Three on the *Technical Fix* scenario, and Chapters Six and Seven on energy, employment, and economic growth, and the world context of U.S. energy policy, make a clear and forceful argument for these propositions.

Further, the report's discussion of the possibility that after 1985 we should seek to order our affairs so that there is no further growth in the demand for energy is interesting and provocative in an important way. We think the discussion of this issue removes it from the realm of faddishness into that of serious discussion. While we ourselves are unpersuaded that this is a necessary goal, we certainly think it wise to give its possibility every detailed examination.

Unfortunately, the good features of the report are marred by many superficial defects. The level of sophistication with which political issues are presented is unfortunately low. This is especially true of those involving conflicts between consumers and producers, or between producers desirous of expanding output at minimum costs and citizens concerned about the impact of their activities on the environment, and other similar issues which form the substance of political conflict. The populist speech writer seems at times to have taken over from the analyst in these discussions.

Thus, we have serious reservations about the completeness or balance of the discussions in Chapters Nine and Ten on private enterprise and the public interest, and reforming electric utility regulations. Chapter Seven, on U.S. energy policy in the international context, contains much naiveté of the opposite variety, in looking for more international cooperation and harmony than is likely to exist.

Despite these reservations, we consider the report an important document and its major propositions worth the serious concern of the country.

D. C. Burnham

I would like to compliment the Project for its very comprehensive and useful studies on energy conservation, and I agree that improved efficiency should be a key component of our future energy strategy. However, I am greatly concerned that this final report of the Energy Policy Project will seriously impede the formulation of a sound energy policy for the United States. *It misleads the nation on two counts: first, that hard decisions on commercial development of additional energy resources can be delayed a decade or more; and second, that greatly reduced energy usage will not seriously affect our economic well-being. I strongly disagree with these viewpoints and conclusions,* and believe further that the report is in error on three other key issues.

1. The urgency of shifting from an oil-gas energy base was overlooked: The report fails to recognize the need and urgency of initiating a conscious policy of systematically shifting away from our present excessive reliance on oil and gas. A policy of major reliance on oil and gas, if continued until the time of near depletion, will produce a disastrous situation once the production of these vital fuels begins to decline sharply. A period of several decades will be required to effect an orderly transition away from reliance on oil and gas, and that transition should be underway now.

A comparison (see Fig. A) of the nation's ultimately recoverable resource base with our present pattern of con-

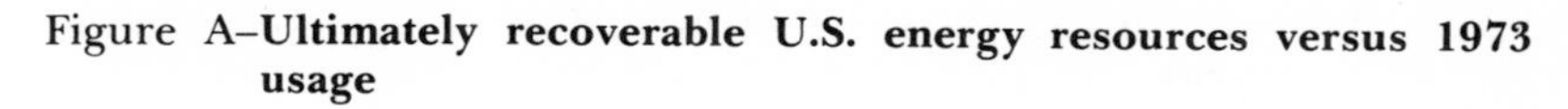

Figure A–**Ultimately recoverable U.S. energy resources versus 1973 usage**

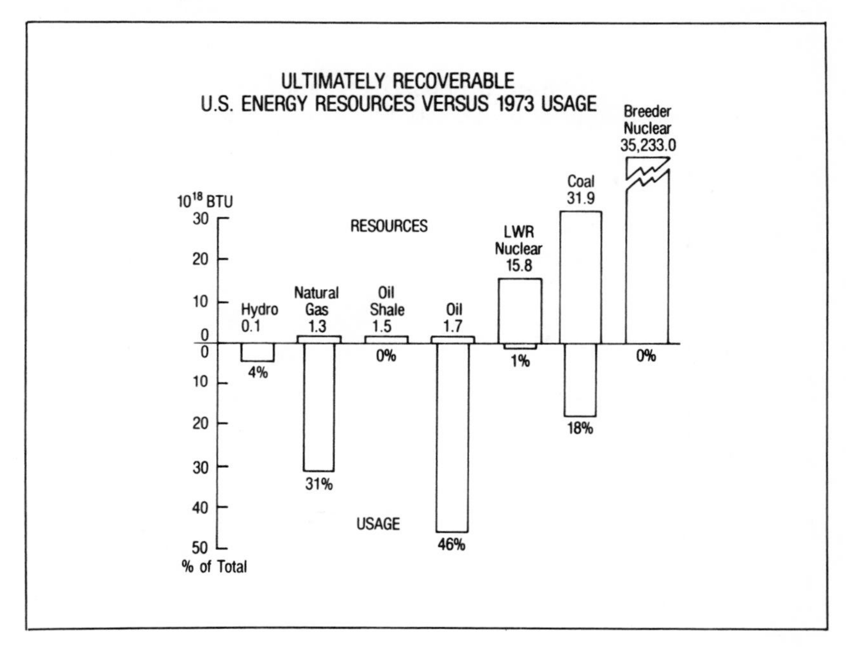

sumption makes the root of our energy problem dramatically clear. We are simply relying too heavily on oil and natural gas, our least plentiful energy resources, and neglecting our most abundant resources, coal and uranium. For example, the remaining recoverable resources of uranium for nonbreeder reactors are almost four times the remaining recoverable resources of petroleum, natural gas, and oil shale combined. With breeder reactors, our uranium resources will be increased over two thousand fold.

Changing this nation's energy base is far from a new concept (see Fig. B). We began the change from wood to coal 100 years ago, and from coal to petroleum and natural gas 50 years ago.

In making the change to a coal-nuclear energy base, increased reliance on electricity will be required. Electricity has great input energy source flexibility and can substitute for the direct burning of oil and gas in virtually every end use outside the transportation sector. At the point of use, electricity is the cleanest, most versatile, most efficient, most flexible, and most convenient energy form available to the ultimate consumer.

Figure B–**Historical and projected shifts in U.S. energy consumption patterns**

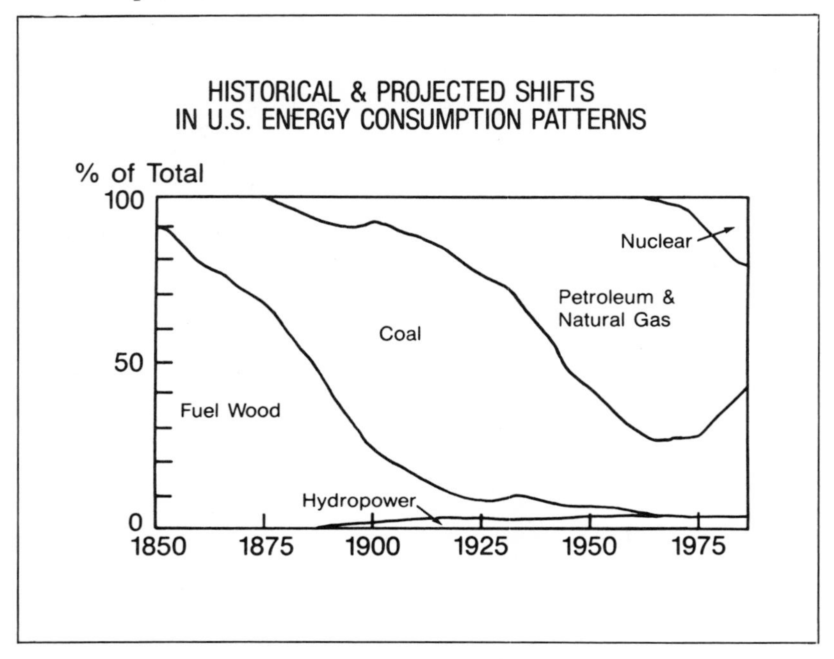

While some criticize electricity for the seeming waste involved in its production, they ignore its high end use efficiency, and the low end use efficiency of direct burning of oil and gas. The overall system efficiencies must be compared to give a true picture. Figure C provides a graphic illustration of the decidedly different picture one obtains by considering *total* system efficiency. In space heating, a gas furnace requires 2.2 units of energy out of the processing station for each unit of space heating provided. In contrast, the electric heat pump requires 1.6 units of nuclear fuel for the same unit of space heat provided. Note that this occurs even though two-thirds of the input energy to the electric utility plant is lost as waste heat. With direct burning of fuel, approximately one-half the energy is lost at the point of use. This example points out that criticism of electric heat being excessively wasteful is simply wrong.

Overriding the importance of system efficiency, though, is the much higher consideration of resource availability. The dwindling resources of oil and gas make it mandatory that we shift away from those fuels *regardless* of theoretical efficiency comparisons among systems. The Energy Policy Project's scenarios clearly illustrate the drastic

Figure C–**Comparison of gas space heating and electric heat pump**

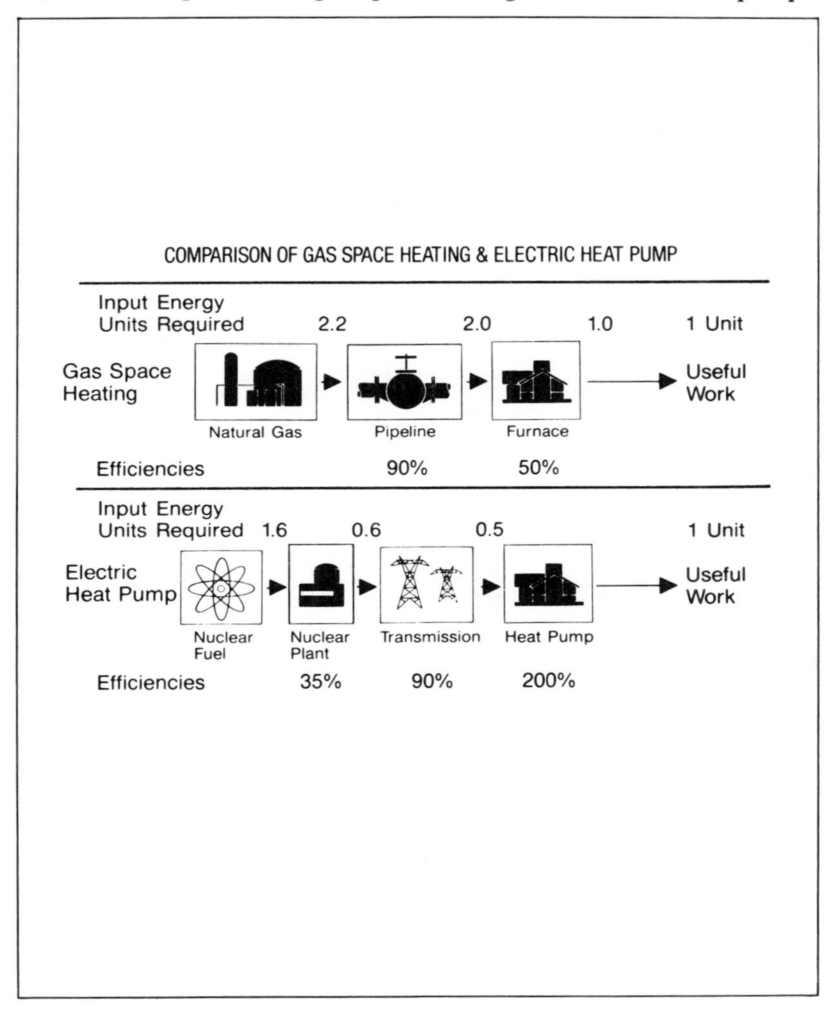

effects of a failure to shift from oil and gas to more abundant energy sources. A comparison of their projected demand (Fig. D) for domestic oil and gas with remaining recoverable domestic resources shows that complete exhaustion would occur in 2020 for the *Technical Fix* scenario, and 2030 for the *Zero Energy Growth* scenario. In marked contrast, a policy of gradually reserving oil and gas for critical nonsubstitutable end uses such as jet aircraft, large trucks, automobiles for long distance travel, drugs, fertilizers, and petrochemicals extends domestic resources to nearly the twenty-third century.

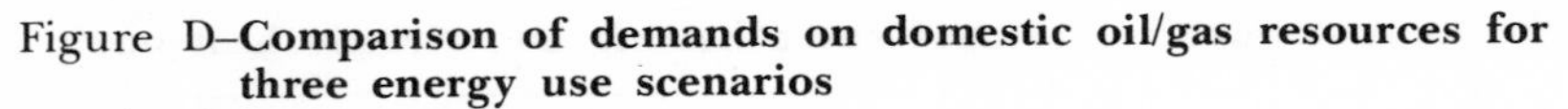

Figure D–**Comparison of demands on domestic oil/gas resources for three energy use scenarios**

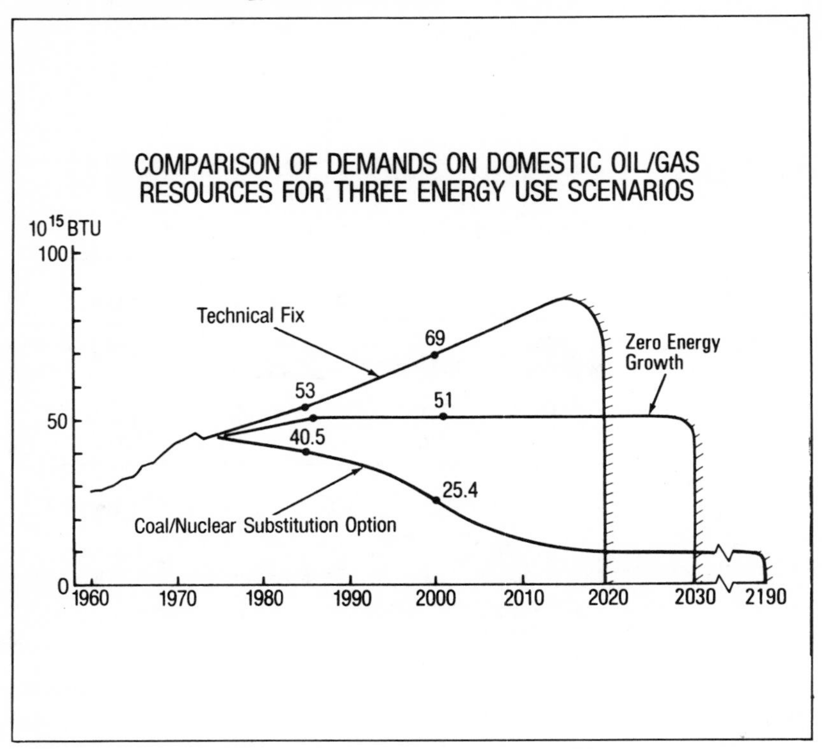

It is truly unfortunate that in their zeal to justify stopping the growth of nuclear power, electricity, and coal mining, the Energy Policy Project has adopted a strategy that will bring chaos to this nation's energy future. In emphasizing conservation alone, the Project has overlooked the fact that in the next few decades we truly have no practical option other than to shift from our present oil-gas energy economy to an electric energy economy based primarily on coal and nuclear. We must institute a policy *now* of substituting electricity for the direct combustion of oil and gas wherever this is technically and economically feasible.

2. *The future demand for energy has been severely underestimated:* The *Technical Fix* scenario appears to me to underestimate total energy requirements for the year 2000 by 25 percent, or 44 quadrillion Btu's. Our studies indicate that taking into account both conservation and technological changes in the way we use energy will require 168

quadrillion Btu's in the year 2000, not the 124 the Project indicates. There are three apparent reasons for this discrepancy.

First, the base case used a period in which significant structural changes to the energy supply were taking place; railroads shifted from coal to diesel, reheat turbines allowed dramatic increases in electric generation efficiency, coal was replaced by oil and gas for space heating, and glass wool insulation was introduced. These significant increases in end use efficiencies had the effect of masking the growth in the amount of useful work performed, and thus understate the true historic growth. This error can be eliminated by assuming the historical period to begin in 1960 rather than 1950. Then after correcting for the decrease in population growth rates in the future (U.S. Bureau of the Census, Series F), a base case requiring 210 quadrillion Btu's in the year 2000 is established. This is some 25 quadrillion Btu's above the Energy Policy Project's *Historical Growth* scenario.

Second, since energy utilization has historically included a constant stream of energy efficiency improvements, the failure to eliminate that portion of projected savings which would have occurred in due course, double counts the savings.

Third, inadequate provision was made for new energy using devices. There are no equivalents of jet planes, air conditioning, plastics, fertilizers, or any other similar significant new energy uses projected in the report.

In the *Technical Fix* and more especially in the *Zero Energy Growth* scenarios, the Energy Policy Project has assumed that the growth in energy usage and the growth in the nation's economy can be uncoupled. I consider their assertions to be totally unsupported and unsubstantiated by the facts. There is a wealth of data which substantiates the widely accepted contention that growth in energy usage and growth in the economy are inextricably linked (see Fig. E). The historic relationship between GNP growth and energy growth has continued during the first half of 1974 despite an increase of 75 percent in energy costs and a tripling of the rate of inflation.

Further, I have extreme difficulty in accepting the Project's contention that the *Technical Fix* scenario does not also require changes in life style, and that these changes are minor in the *Zero Energy Growth* scenario. I believe that the changes they propose would result in substantial social unheaval, as well as economic stagnation.

Figure E–**The interdependence of energy and economic growth**

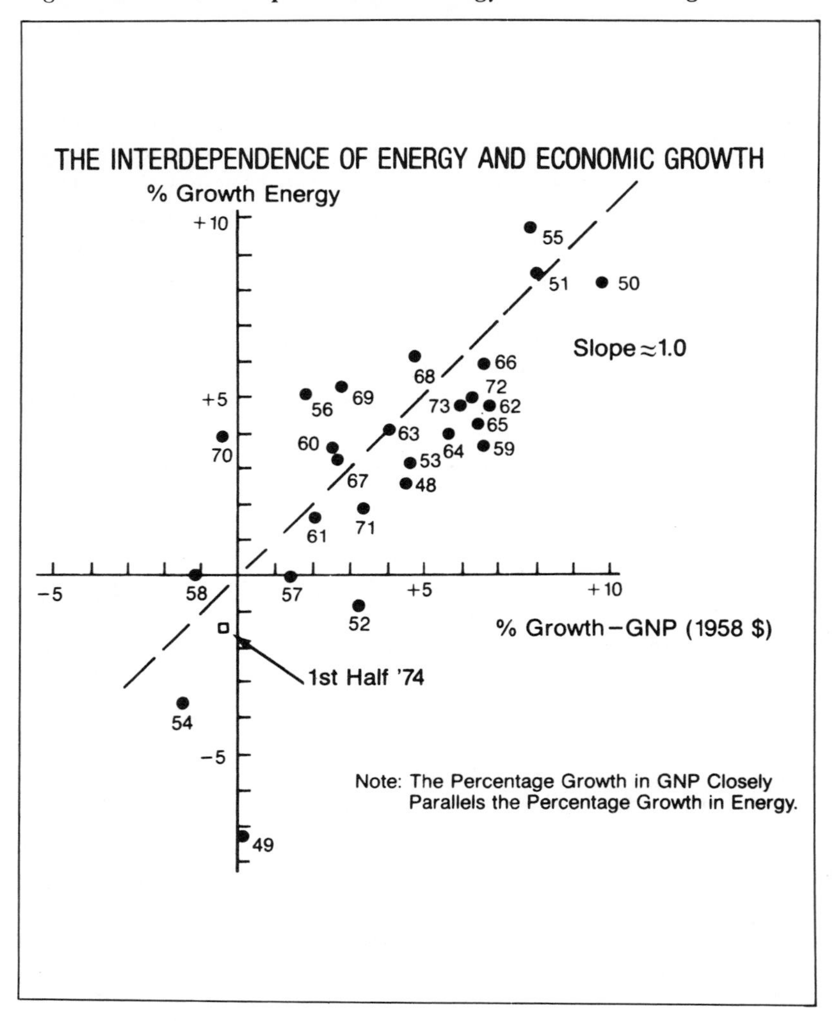

Following the Project's recommendation to defer decisions on developing new energy sources would have tragic consequences if their projection of future energy demand does indeed turn out to be seriously in error. Yet the question of what forecast is correct fails to change the essence of the basic problem. This nation *will* run out of oil and gas in the foreseeable future without a concerted effort to provide new energy sources. The magnitude of future energy requirements only shifts by a decade or so the point of their complete exhaustion. This adds more urgency to the need to develop additional energy sources, and to begin the shift from an oil-gas energy base. A failure to do so is too great a gamble to take with this nation's future.

3. The maturity and safety of nuclear technology was not fairly presented: No technology in the history of man has ever been so thoroughly researched or as stringently regulated as the nuclear industry, yet nuclear energy was discussed in the report as though it were a completely new technology just about to move from the research laboratory into its first prototyping. Presently, 227,000 megawatts of nuclear generation have been ordered by the electric utility industry, equivalent to half the total capacity existing in 1973. If these plants were to be closed down, as the Energy Policy Project seems to propose considering, the oil equivalent to generate the same amount of electricity over the 40-year plant life would be 93 billion barrels—almost ten times the amount of the Alaskan north slope reserves, and two and one-half times total U.S. known reserves.

Nuclear plants have proven to be as reliable—if not more reliable—than fossil fueled power plants. From 1960 to 1972 the Edison Electric Institute reported that for plants of 390 megawatts and above, availability averaged 80 percent for nuclear units and 78 percent for fossil units. The industry has, moreover, now accumulated over 183 reactor years of commercial power plant operations. In that cumulative experience, no member of the general public has ever been injured or killed through the operation of these plants.

In numerous places throughout the report, the danger of nuclear power plants has been discussed at length, with the implication being that the technology presents a grave danger to the nation. Nothing could be further from the truth. It has been statistically proven, for instance, that an installation of a coal fired plant in place of a 1000 MW nuclear plant increases the probability of an occupational fatality by a factor of 10. With respect to the general public, some of the most recently published analyses show that the danger of a person becoming a fatality as a result of all possible accidents at a nuclear power plant is about the same as the danger in smoking three cigarettes every year, being four ounces overweight, or, for a farmer, spending an extra twelve days of his life in the city. Surely such low risks are acceptable, given the benefits of nuclear power.

There is a danger involved in terrorists hijacking nuclear materials. This threat, like airplane hijacking, is sufficiently real to justify stringent security measures. But, like the airline situation, neither giving up airplanes nor nuclear power plants seems the logical solution. Similarly, nuclear waste can cause problems, yet these problems are easily controlled. The volume of nuclear wastes from commercial reactors through the year 2000 can be easily and safely stored in an area of less than two square miles. Further, this material constitutes "waste" only because we presently have no use for it. It is quite conceivable that in the future significant uses will be found for many of these waste materials, and they will be recycled back into the economy to serve useful purposes.

Recently, many strong foes of nuclear power have reevaluated their positions and become strong supporters of nuclear power. Responding to Ralph Nader's address in March to the joint session of the Massachusetts legislature, five PhD's in nuclear engineering from MIT protested Mr. Nader's emotionalist scare tactics as "blatant attempts to polarize the public around non-issues." Their conclusions about nuclear power included the following.

- "During normal operation, nuclear plants pose less risk to public health than coal or oil fired plants.
- "The risk to the public, for the worst hypothetical accidents for both nuclear and fossil plants, is less than most of the risks society has historically accepted.
- "The overall impact of nuclear plants on land, air, and water is far less than that of coal fired plants and comparable to that of oil fired plants.
- "The waste effluents from normal operation of a coal fired plant are more of a health hazard than the maximum credible accident of either the oil or nuclear-fueled plant."

The dangers of nuclear plants, like Mark Twain's premature obituary, have been slightly exaggerated.

In summary, I find that I am unable to accept the report's definitions of the nature of the United States and world energy problem, or the range of policy alternatives offered. The report's lack of attention to our excessive dependence on oil and gas and the potential exhaustion of these two vital fuels within one or two generations has caused it to completely overlook the critical need for a shift to more plentiful energy resources. This shift must rely on electricity produced from coal and nuclear fuels as the dominant energy form for the future. Nuclear power is a mature technology fully capable of providing the required energy—so long as unnecessary and unrealistic restraints are removed from its path. It is vitally important that we use our limited resources of oil and gas only for those uses to which they are uniquely suited. Conservation and efficiency improvements are necessary and should be pursued vigorously. However, they are not a means of insuring long term energy availability, but merely buy us the time in which to make an orderly transition to our most abundant fuels, coal and nuclear. The shift is inevitable; no viable alternative exists.

John J. Deutsch

I am in general agreement with the policy recommendations but I do not endorse all of the particular methods proposed for their attainment, nor can I agree with the tone of some of the discussion which, unfortunately, implies deficiencies and motivations which are either irrelevant or unproven. In choosing among the alternative methods available for the achievement of the basic purposes which many of us support, I favour the use of the market mechanism, as against direct controls, to a somewhat greater extent than do the authors of the report.

Joseph L. Fisher

I find the foregoing report of the EPP staff to be broad in scope, stimulating, and especially important because it makes a careful and respectable case for slowing down, even stopping, the growth of energy consumption. I have only a few comments in addition to those in the statement of the Advisory Board.

Of the three scenarios presented I think the one labeled *Technical Fix,* is preferable assuming it can be done without unacceptable environmental or social consequences. It would provide for modest growth in energy consumption, offer more hope to poor people with inadequate heat, electricity, and transportation, and stimulate innovation in the economy. The term "technical fix" is unfortunate; it connotes trickiness and artificial escape from the real problem. Technological progress would have been better.

In case the *Technical Fix* pathway turns out not to work entirely, then *Zero Energy Growth* will be the direction to move. Perhaps zero energy growth per capita would be more realistic.

Here and there the staff report falls into the trap of free market vs. government regulation as did the discussions of the Advisory Board. I regard this division as unproductive. No doubt United States energy policy will continue to be a mix of the two with both approaches being important. The challenge, in my opinion, is to specify the role of each in practical situations, avoiding ideological assertions of position.

I was pleased to see also that the report does not get hung up on organizational matters of the federal government as though the hard problems of energy could be solved by rearranging the boxes on the chart of organization. Where things are done is important but policy integration derives more from careful definition of the problems, analysis of alternative courses of action, and the process for selecting the best course. In regard to this last it is necessary, however difficult, to arrange things so that interested citizens have a chance to participate in the decisions in such a way that they can contribute without unduly delaying the decisions.

The report rightly emphasizes energy conservation from the mining of fuels through conversion to ultimate use. Conservation is good medicine at any time but especially when energy prices are high and the outlook for supply is uncertain.

The Advisory Board statement could with advantage have dealt briefly with Project Independence, which now seems to be the main objective and direction for action by the United States government. My own preference would have been for a short paragraph stating that some reduction in U.S. dependence on oil imports, now again running at one-third or so of total domestic consumption, is in order but that anything like self-sufficency would be prohibitively costly, if attainable at all, and also destructive of important elements in the country's foreign relations.

Finally I would have liked to see the following two paragraphs in the statement of the Advisory Board.

- It was generally agreed that an appropriate agency periodically prepare estimates of demand and supply for energy commodities under various assumptions regarding prices, possible substitutions, different end uses, and alternative policies. Then estimates should be presented by regions, should look ahead to both the short and the long run, should be both comprehensive and systematic, and should be reworked annually. Such an exercise would do much to establish a common framework within which consensus as to outlook and policy could be found.

- Major shifts in patterns of energy sources and uses are in the offing, we agreed. Such shifts will bring in their wake difficult economic problems: transitions in employment, growth of some industries and regions with decline of others, large adverse movements in the U.S. balance of international payments, substantially increased requirements for investment funds. These and related problems will challenge national and world policy leaders to new responses.

John D. Harper

I have made several suggestions which follow under specific headings of the Project report, but I would like to raise one objection concerning the entire report. It deals with two interrelated concepts of freedom—economic and political.

In my opinion, the report places far too much reliance on federal planning and economic regulation as the basic arbiter in our society. I am disturbed by a general lack of faith in the free market mechanism, which traditionally has allocated energy resources in a satisfactory and efficient manner.

The report's premise that government permanently project itself into virtually every phase of energy activity, including corporate decision making, is abhorrent to me and I am sure to most of the people in this nation. It is my personal belief that government intervention will lead to immediate and constant distortions of the pricing mechanism and of resource allocation and so weaken private enterprise that the whole business structure will come to depend on government control for its very survival. Such controls, no matter how well intentioned they may be, would cost Americans their economic freedom of choice and—inevitably—their political liberties as well.

Structuralist bias: There is an unmistakable orientation throughout this report that suggests something the authors may not have intended. I refer to a weighted bias, typical of some courses in economics, called the structuralist approach. In this report, the structuralist approach is characterized by the convenient ignoring of interindustry competition which—because it is a bulwark of private enterprise—cannot be ignored in any study of our economic system.

Most damaging, in my estimation, is that the structuralist approach in this report favors and encourages government intervention rather than the marketplace as the final arbiter of supply, demand, price, competition, and profit. One finds this orientation manifested many times throughout this report, but most notably in Chapters Nine and Thirteen.

In Chapter Nine for example, we find the crux of a pending complaint by the Federal Trade Commission against the nation's eight largest oil companies. An obvious prejudice against the complainants shows up throughout this chapter as well as in several of the conclusions drawn. Also in Chapter Nine the report reads, "Possibilities for collusion are greater when a few firms control a large share of an industry's activity. . . ." There is absolutely no consensus on this very dangerous assertion. In fact, there is considerable evidence to suggest that it is not true at all.

In Chapter Nine the report states ". . . but a rule of thumb has been formulated which says that monopoly power begins when the four largest firms account for more than 50 percent of the industry's output, or when the eight largest firms account for more than 70 percent." Yet no authority is given for this far-reaching "rule of thumb," nor is any evidence whatsoever given to substantiate the allegations made. Furthermore, there is a growing body of research at UCLA and other universities that indicates that high concentration ratios and high profitability do not necessarily coincide.

Also exhibited in Chapter Nine is the bias I mentioned earlier—in this case very definitely against the oil companies. The report, unfortunately, ignores the long profit performance of the oil companies (at or below the average of all manufacturing industries). It also takes no notice that energy prices in the United States have for many years been significantly lower than in other parts of the world.

I take issue with a sentence of Chapter Nine which reads, "Such windfall profit taking should be distinguished from monopoly profits, which are produced by industry through implicit or overt collusion." There is a clear implication here that profits that are not windfall profits are monopoly profits generated by implicit or overt industry collusion. This is an unsubstantiated and wholly false statement. It is, however, consistent with numerous, similar unsupported allegations throughout this report.

Another such statement appears in Chapter Nine: "This is the sort of intervention which is needed if government is to stop being the ally of the energy industry and become instead the ally of the citizen who elects and supports it." Such statements, fully thought out or taken out of context, are highly inflammatory and do no service to this report.

Econometric model accuracy: Chapter One of this report sets out several qualifications and disclaimers regarding the econometric models used to evaluate various economic aspects of the three scenarios.

Developers of econometric models are normally rather modest in their claims of reliability of results. In this report, however, it appears that an unusual degree of faith is placed on the accuracy of scenarios evolved. In nearly every chapter, the future of each scenario—even 25 years ahead—is referred to in absolute terms. Part of the problem stems from the questionable accuracy of the aggregate data used. Part is due to the near impossibility of encompassing in mathematical equations the infinite complexities of our dynamic market system. The aggregate numbers used in macroanalysis reflect only an overall, gross look at the economy; detailed industry components are difficult to evaluate within such a general framework.

The reader, therefore, gains the impression that the path of each scenario is pat and scientifically predictable. This is not so; some of the assumptions cited are questionable, and the implicit assumptions are not discussed at all. For example, in view of the nation's inflation problem and the changing structure of the labor force, the full employment assumption may not be realistic if it is defined in terms of 4 percent unemployment rate.

In the *Zero Energy Growth* scenario, it is difficult to rationalize both the substitution of more labor intensive processes in manufacturing and the growing employment in services with the projected productivity rate. It further appears in this scenario that prices of different forms of energy are changing in a vacuum, since little is said about relative changes in other product prices or overall inflation rates. Apparently, the report assumes *ceteris paribus,* other factors remaining constant, which of course is never the case in the real world.

Zero Energy Growth scenario: The report deliberately leads the reader to believe that the greatest hope for the future is to be found in the *Zero Energy Growth* scenario. This bias prevails despite what appears to be editorial attempts to balance the discussion with token references to the *Technical Fix* scenario.

I discussed earlier some of my doubts about the technical accuracy of the econometric model, with its explicit promise that we can continue to have a dynamic society with zero energy growth.

Almost every chapter of the report ends with an assurance that zero energy growth is the way to minimize the problems. Yet, it is totally unrealistic to suggest that energy growth can be controlled in a free society, any more than population growth can be controlled. It is likewise true that if people decide that they want big cars, frequent excursions to Europe, and that they don't care to invest their money in storm windows, there is no way that we can expect to have zero energy growth.

Leading people to believe that ZEG is the solution to our energy problems is absolutely irresponsible.

Energy and the environment: There are two references in this section to which I take exception. The first one deals with the report's conclusion that acceptable SO_2 scrubber technology is available and that it is not being installed because ". . . utility officials are reluctant to buy this expensive equipment as long as they can avoid it." I know for a fact that acceptable and reliable SO_2 technology is not available.

The second reference is in regard to the report's preoccupation with particle problems. The report suggests that much more is known about particulate matter and its effects than is told. This unexplained reference implies some impending danger that has not been substantiated. I suspect that the reference, like the discussion of nuclear accident possibilities, is used to promote the concept of zero energy growth.

Minor S. Jameson, Jr.

In my considered judgment, this report is seriously lacking in objectivity and is not an appropriate guidance tool for evaluating or determining public policies as to energy. The reasons for this conclusion may be summarized as follows.

1. Implementation of the report's conclusions and recommendations would discourage and retard the development and production of U. S. energy supplies and lessen competition, particularly in the case of domestic oil and natural gas.
2. Such implementation would involve a substantial departure from reliance upon the free enterprise system and the competitive marketplace, and a massive intervention by government in not only the energy industries but also other segments of the economy.
3. The general tone of the report, and certain sections in particular, reveal an unjustified bias against private business, with many statements that are misleading and some that are contrary to the facts.

For these reasons, the report does not serve the best interests of energy consumers, the economic welfare of the country and national security.

1. Energy supply: All three scenarios in the report, and all other authoritative energy studies, are in agreement that oil and natural gas necessarily will continue to be the primary source of energy for the United States over the next decade at least and probably well beyond. All three scenarios call for increases in the total supply of oil and natural gas during this time span.

The report correctly recognizes the danger and inadvisability of undue reliance on uncertain and cartel-controlled foreign sources of petroleum. Increases in *domestic* oil and gas production are projected in all three scenarios.

The report concludes that limitations on U. S. oil and gas availability are likely to stem from a combination of environmental, social, and political constraints on rates of development rather than a physical limit on the quantities in the ground. I am in agreement with that conclusion except for the fact that it should include economic constraints—prices, generation of capital, tax incentives, return on investment, etc. The report blandly dismisses economic constraints by asserting that *today's* price of crude oil provides not only adequate but excessive incentives to produce the required quantities of domestic oil and gas. Today's federally controlled price of natural gas is ignored, as are future price levels for both oil and gas.

The "Supply Actions" in the report's conclusions and recommendations call for only one specific governmental action affecting U. S. oil and gas production: the elimination of the percentage depletion provision of the federal income tax laws and the elimination of the provision permitting the expensing of intangible costs. Even reputable economists opposed to these tax provisions recognize that their practical effect is increased oil and gas production at prices lower than would prevail otherwise. Conversely, elimination of these tax provisions would mean less oil and gas at higher prices.

Elimination of these tax provisions would impact more on the smaller, nonintegrated independent producers of oil and gas than on the larger major oil companies. Many independents would be forced to sell their producing properties to the larger companies and discontinue exploration and development activities. As a result, competition in the industry would be lessened and the multiplicity of effort needed to increase oil and gas supplies would be diminished.

Another section of the conclusions and recommendations calls for an indefinite ban on leasing of the Outer Continental Shelf, one of the most promising potential sources for increased domestic oil and gas supplies.

Out of the many complexities of the energy problem, one indisputable fact emerges: domestic production of oil and natural gas must be increased. There is no alternative during the next ten to fifteen years. Otherwise, the country faces the continuing and menacing threat of embargoes on oil and gas imports, crippling shortages, and exorbitantly high prices. It follows, therefore, that governmental energy policies must be directed to encourage and make possible the needed increases in U. S. petroleum production. To that end, the EPP report provides no guidance whatsoever toward formulating policies. Instead, the report deludes the public into the false complacency that increasing supplies of U. S. oil and gas are assured. At the same time, the report recommends government actions that would discourage and depress domestic production. This is a cruel deception of the consuming public, counterproductive to solving the nation's serious energy problems.

2. *Free enterprise vs. governmental intervention:* As the report points out, "the United States has basically a private enterprise market economy." This serves as a departure point from which the report moves into governmental intervention on a massive scale. To bring the extent of this intervention into clear focus would require a document

almost as lengthy as the report itself. Hopefully, some awareness by the public and all concerned of the consequences may be furthered by the following examples.

According to the report, the major finding is that the United States should substantially reduce, or even bring to a halt, growth in the nation's use of energy. Such an objective obviously and clearly could be achieved only by governmental edict. In effect, the government would decree, for example, what kind of automobile the consumer could purchase and whether he could travel by air, mass transit, or private car. Home builders and industrial plants would be under government mandate in construction and the use of energy. To insure "social equity," the government would impose "income redistribution measures such as a guaranteed minimum income or a negative income tax."

The report endorses such actions and concludes that a substantial reduction, or halt, in the growth of energy use is desirable, technically feasible, and economical. I strongly disagree. Governmental actions required to achieve a no-growth objective would be distasteful to the public, injurious to economic progress, and ultimately fatal to the private enterprise market economy on which the American way of life has been built.

One additional point. The report looks favorably on greatly increased intervention and even active participation in the business of finding, producing, and making available crude oil, natural gas, and finished petroleum products. This would involve, among other things, government chartering of oil companies, government monitoring of operations, a TVA type of government owned oil company, with the clear implication that nationalization of the petroleum industry may prove to be necessary and desirable.

I know of no evidence in modern history that would support this conclusion that direct government participation in, or control over, the business of finding and producing oil and gas would provide additional supplies in the public interest. In addition, it would be the first step toward nationalization of other basic industries. All concerned with policy making not only in the field of energy but also in broad economic matters should reject this approach.

3. Anti-industry bias: Perhaps the most unfortunate aspect of the report is the obvious bias against private business, compounded by misleading and nonfactual statements. While this tone permeates the report and assumes an adversary relationship between government and industry, it is most evident in Chapter Nine, "Private Enterprise and the Public Interest."

In Chapter Nine, and in the report's conclusions and recommendations, the oil industry is portrayed as consisting of a handful of large companies whose economic and political power is so great that actions must be taken so that they "can no longer run the government to achieve the policies which suit them best, regardless of whether those policies suit the rest of the people." The report defines this quote as "fine words" and asks "How can they be implemented?" In this same vein, the report accuses the President of the United States of taking action "due in no small part to the influence of the oil industry."

This description of the industry, and these charges against both industry and government, reflect seriously biased thinking and are based on misleading analysis involving statements clearly in conflict with the facts. Evidence cited in the report in support of the charges that the "oil industry" can and does "run the government," include the imposing of state conservation laws more than 40 years ago and the maintenance of the government's program to limit oil imports more than fifteen years ago.

It is ironic indeed to attribute such policies to the political power of a handful of large oil companies. State conservation laws and the oil import program were initially advocated and supported by the smaller independent producers who otherwise would have been forced out of business. These independent producers, who play a vital role in finding and developing U. S. oil and gas resources (approximately 80 percent of all wells completed in the United States are drilled by independents—not the handful of large companies), are virtually ignored in this chapter of the report. The only two references to independent producers are patently untrue. Contrary to the suggestion in the report, the pipeline transport system is not being used to "squeeze out" independent producers. The extreme lack of understanding is the suggestion that percentage depletion "benefits the large firm more than the small independent producers." The truth is that elimination of percentage depletion and expensing of intangible costs would create a financial crisis for independent producers, many of whom would be unable to continue in business.

It is one thing to falsely accuse the large oil companies of running the government. It is something else again to infer that the government in general and high officials in particular have acted to benefit the oil industry "regardless of whether the policies suit the rest of the people." For example, the report condemns the President for acting on oil import policy in response to the influence of the oil industry in 1972. This implies equal condemnation of President Eisenhower, who established the oil import program in 1957, and President Kennedy, who made the program more restrictive in 1962. It follows, therefore, that such condemnation extends to Cabinet officers and members of Congress who have been involved in determining energy policies.

These are serious charges indeed. The failure to document and substantiate them discredits the report as an objective study.

Conclusions: Fundamental differences in philosophical viewpoints existed among Board members and between Board members and the Project's staff. Perhaps it was naive to hope that a balance would be struck in the final report. Unfortunately, the report is a far cry from such a balance. It has become necessary, therefore, for me to submit this vigorous and sharp dissent, with faith in the wisdom and integrity of our country's policy makers to act in the best interests of all concerned.

Michael McCloskey

It is a pity that the report of the Energy Policy Project was not available a couple of years ago to help guide us through the turmoil of the energy crisis. That crisis has already taught us the main message of the report: we do have choices to make about how much energy we want to consume in the future. Nonetheless, the report should be extremely useful in focusing future debate on the implications of various alternative energy policies.

In general, I agree with the conclusions of the report. It is unfortunate, though, that a decision to include conclusions and recommendations was made so late in the process of drafting the report. The discussion throughout the report does not always reflect the force of the final conclusions. Because of this, I feel that I should underscore some points which should emerge more strongly from the report, as well as note a few omissions in the discussion and points of disagreement.

The central tool of the report is the presentation of three alternative scenarios for energy growth. Initially, the purpose in focusing on this tool was to examine what would happen if the nation heeded the advice of various protagonists in shaping its energy policy: the energy industry, disinterested technicians, and environmentalists. However, as the scenarios were developed they came to embody a restatement of these positions in terms of what the authors thought was most reasonable and plausible. Thus, as the report emerged the scenarios have an uncertain and even arbitrary quality to them. They represent neither the positions of outside groups nor predictions of the most likely developments along certain pathways. As tools they are useful, but one could readily change some of the assumptions and alter the results.

This point can be seen in the way both the *Historical Growth* and *Zero Energy Growth* scenarios are handled. The *Historical Growth* scenario is certainly not presented in the way the energy industry would; it is closer to the medium growth projection of the National Petroleum Council rather than the high growth projection they prefer. So we lack an examination of the implications of following the outer limits of the energy industry's position. On the other hand, the discussion of the *Historical Growth* scenario vacillates between accepting it as a possible alternative and doubting its feasibility. It does not sufficiently emphasize the inner contradiction in this scenario: consumer prices must stay down to foster demand of this magnitude but raw material prices must go up to stimulate discovery and development of vast amounts of lower grade energy resources. Industry cannot have it both ways—unless it can induce the federal government to intervene even more on its behalf with subsidies to fill in the difference.

The report does not examine the likelihood of this happening at a time of record oil company profits. The taxpayer is not likely to want to bear a heavier personal tax load in order to liberalize depletion allowances and similar subsidies. As a result, the industry will probably try to get these subsidies less directly through such measures as rapid leasing of federal offshore oil, coal, oil shale, and geothermal fields. If these resources are leased at too rapid a rate, the prices are likely to be depressed and the public treasury deprived of a fair return. The result will still be that the rest of the taxpayers will bear a burden that could have been lighter. This kind of governmental intervention to give away public resources at unduly low prices, of course, is welcomed by industry while its intervention to protect the environment, the consumer, or the taxpayer is denounced as being "against the American way of life."

The discussion of the *Historical Growth* scenario also fails to fully examine the practical problems of continuing to push development at such a rapid pace. While it alludes to "bottlenecks of manpower, drilling equipment, and the like," it fails to elaborate. There are already shortages of steel for drilling operations, of enough trained manpower to keep up with the pace of offshore oil leasing; there are not enough experienced crews and knowledgeable managers to push oil shale development as fast as the Interior Department plans. The report fails to assess the difficulties of providing the infrastructure needed to support continued rapid expansion in energy supply.

Moreover, the report does not really examine the implications of tying up prodigious quantities of capital in supplying an exponentially increasing quantity of energy. The $1.7 trillion worth of capital needed to realize this scenario represents an increasing share of all capital investment. It is capital taken away from housing, from the production of consumer goods, and from investments in energy conservation. Diverting so much capital will provide a major stimulus to inflation because the price of money will have to rise to attract such an immense sum. It will mean higher interest rates for home owners, and higher rents, and more erosion in buying power of those on fixed incomes.

An ill-advised effort to perpetuate historical growth rates may well mean a lowered standard of living for most people, with an economy even more afflicted with inflation and high taxes. And it certainly will bring a reversal in our efforts to clean up the environment, with fouler air, poorer health, and a battered landscape.

Unfortunately, the report's presentation of the range of alternatives makes zero energy growth by the year 2000 appear to be one extreme in the spectrum. Actually that goal is one veering toward the middle range of alternatives. There are environmentalists who feel the net national welfare would improve if we consumed less total energy than we now do. We actually did consume less energy in the winter of 1973–74 than in the prior year, and in some ways we were better off. We suffered with less air pollution and endured fewer fatalities on the highways, though admittedly the suddenness of the embargo caused annoying lines at the gas stations. But the point is that reducing consumption is no longer a theoretical possibility; we recently did it. The study should have looked at a wider range of alternatives in curtailing consumption to examine the tradeoffs in public and private welfare. It should have delved more deeply into how other industrialized nations achieve high standards of living with less energy consumption than the United States. If it had done so, a policy of gradually reigning in growth rates by the year 2000 might appear to be a far less formidable undertaking.

When the Project began, the issue of the desirability and feasibility of national self-sufficiency in supplying our energy had not emerged as it now has. The report attempts to deal with this issue, which is symbolized by "Project Independence," in the context of variations in the scenarios. This manner of treatment, however, obscures some critical comparisons. For instance, while there are significant differences in the totals for projected consumption by 1985 under the three scenarios (116, 91, and 88 quadrillion Btu's respectively for *Historical Growth, Technical Fix,* and *Zero Energy Growth*), the amount of oil to be imported is the same for the self-sufficiency *Historical Growth* scenario and the *Zero Energy Growth* scenario—i.e., an amount equivalent to 9 quadrillion Btu's. The environmental protection variation in the *Technical Fix* scenario projects an amount only slightly higher for 1985, 12 quadrillion Btu's. Thus, this nation's exposure to the problems of embargoes on imported oil in the years immediately ahead is not significantly different under these variations in the three scenarios. Other variations do allow imports to fall by the year 2000, but the impact on our national security of depleting our domestic oil and gas supplies in the shortest possible time is not fully explored. As the Project Director once said, "Draining America First" hardly sounds like a prescription for national security. We would then really be prostrate before the Middle East.

In its recommendations, the report relies on both changes in market prices and governmental policy to turn us along the various pathways we can choose. Industrial critics keep reiterating that the government should stay out of the marketplace for energy. Their argument, of course, is selective. They want a host of special tax dispensations and the ready sale of huge quantities of cheap federal energy resources, with minimal constraints. But they oppose governmental intervention to foster energy conservation and to protect the environment, the consumer, and the poor. While environmentalists want the market to do its job in reflecting full economic, social, and environmental costs and thus in helping to allocate resources, there are a number of fundamental reasons the government must intervene to overcome deficiencies in the performance of the market.

First, existing public policy (at the behest of the industry) already unduly encourages demand and understates the price of energy through a maze of subsidies and incentives; these policies must be reversed or counterbalanced. Second, only the government can make decisions regarding tradeoffs between environmental, social, and supply questions to protect the public interest. So-called externalities will never be fully reflected in the price of energy. Third, only the government can deal with the public's long range interest in preventing the depletion of nonrenewable energy resources. The market discount rate does not adequately deal with resource use–timing questions of this sort. Fourth, most industries are under pressure to maximize their short term profits for stockholders. This interferes with efforts to moderate their present behavior in the interests of long range considerations. The recent policy of the oil companies in boosting demand to the brink of shortages illustrates the problem. Their policies and the nation's long range interests frequently diverge.

Industry, however, would have us believe that governmental intervention means some sort of officious meddling in our lives. This does not have to be the case. To rein in the growth rate in energy use in the next decade to two percent or less, the Project recommends public policies which will still allow the consumer freedom of choice. Cars with poor mileage could still be purchased, but new tax rates would encourage most of us to buy efficient vehicles. No one would have to insulate his home, but new special loans would be available to help us do it. Building codes might change to require less energy wastage in commercial structures, but these requirements would merely complement existing specifications on matters such as minimizing fire hazards. Utilities would have their rate structures modified to discourage excessive consumption, particularly at peak load periods and of expensive blocks of new power, but this would be brought about through the existing regulatory structure, and anyone could still elect to pay the price for power in expensive blocks and at expensive times. And the most important policy changes would involve a reduction of governmental intervention in the marketplace through an end to government subsidies to the energy industry. These policies would make it economically possible for us to do more to conserve energy, but anyone could still be prodigal if he wanted to.

It needs to be stressed, too, that the poor need not suffer under such policies. They will not be cut off from access to more energy. Even under the *Zero Energy Growth* scenario, there will be half again more total energy by the year 2000, and more per capita too. The poor would have better and more economical transportation opportunities, with more mass transit and more efficient and economical cars moving into the used car market. More insulated housing would gradually move into the market, to reduce heating costs for the poor as they come to occupy such housing as it ages. Less air pollution would affect their health in inner cities. More jobs would be preserved for those of limited skills as the trend to substitute energy for human labor is blunted. With low growth the poor will not suffer under the kind of inflationary pinch that high growth will engender.

While it is true that the conservation oriented scenarios contemplate the need for higher energy prices to moderate demand, some of these price rises have already taken place (e.g., in the case of post-embargo gasoline prices). Moreover, the inflationary impact of contrived efforts to perpetuate high historic growth rates may result in rises in the general price level that are far greater than conservationists advocate for energy. The energy industry shows no concern for the impact this inflation will have on the poor.

Many conservationists do favor measures to offset the impact on the poor of modest rises in the price of energy. They advocate preferential utility rates for those who consume small quantities of gas and electricity, with those in this category (mainly the poor) paying less than the full cost. Such underpricing might be offset by charging consumers of large quantities of power an additional amount. Moreover, they would support further measures to offset any regressive impact on the poor, such as a negative income tax or the issuance of "Energy Stamps" similar to food stamps. These could be used for the purchase of gasoline and other fuels. If the chances of enacting such offsetting measures are clouded, it is because they will surely be opposed by most industry lobbies, as they simultaneously decry the impact of conservation scenarios on the poor. In the meantime, they will revel in record profits and plead for more federal subsidies.

Environmental values fare well in the report, with them being treated seriously and sympathetically. However this treatment is flawed by the fact that some relevant environmental issues are not discussed at all and others are discussed in a conclusionary way that fails to explain much about how the conclusion was reached. Explanations are especially scanty on biological questions (the staff lacked background in this area). The unevenness in the discussion of environmental subjects stems from the fact that the Project only chose to contract for studies of problems about which little was felt to be known. The result, however, is that the final report fails to make a balanced assessment of the full range of environmental problems involved in energy use, including the presumably more well understood subjects.

For instance, the report says nothing about the problem of siting power plants and coping with the thermal pollution they cause; about the preemption of millions of acres of productive farms, forests, and ranges for power plants and transmission lines; about the role of energy in triggering visual blight and declining amenities; about the pressure to exploit our last wild rivers for hydroelectricity (and the illusory output of pumped storage plants); about the destruction of coastal marshes and estuaries by canals dredged for oil drilling; about the prodigious quantities of discarded crankcase oil flowing from our sewers into the seas; about the growing volume of oil that tankers spill into the oceans; and about the problem of building up so much heat in metropolitan centers from increasing energy use (they become giant "heat sinks") that local microclimates change.

Moreover, the report fails to fully comprehend the fact that the public may have a long-term interest in minimizing depletion of "stock" resources, such as fossil fuels (particularly oil), and in, instead, converting itself into a posture of reliance on "flow" resources, such as solar and geothermal power, which can be utilized indefinitely. The report does not look much beyond the year 2000, when we may begin to exhaust our oil reserves. There may be a long-term need to use oil for lubrication and to produce fertilizer, fiber, and plastics. Current generations may have no right to keep using up the options of future generations.

No environmental assessment can omit so much and begin to be adequate. If the environmental assessment had been fuller, the argument for controlling growth would have been even stronger. And as it stands, the case is stated in compelling terms.

Alex Radin

The Energy Policy Project has performed a valuable public service by defining some of the options available to the United States in determining energy policy. The Project's report should be useful to citizens, the Congress, and the Administration in arriving at a course of action in an area which has been characterized by uncertainty, drifting, and conflicting policies.

Perhaps the single most useful aspect of the study has been its focus on various ways by which conservation of energy can be achieved, without detriment to the nation's standard of living. Also to be commended are the emphasis on achieving social equities, the need for greater competition in the energy industry, the uncoupling of the political and economic power of the oil industry, the desirability of establishing an Energy Policy Council, the methods of achieving better management of public resources, the benefits of separating generation and transmission from the distribution of electric energy, and the importance of regional planning.

Despite these and other aspects of the report which I regard as favorable, there are some recommendations to which I take exception. The tone of the report also leaves some implications which I believe are undesirable.

More specifically, the report gives only faint recognition to the benefits that have accrued to society as a result of the availability of abundant sources of energy, while at the same time it emphasizes the dangers that result from energy production or conversion. For too long our society has utilized increasing amounts of energy without showing sufficient concern for the hazards of energy production. A righting of the balance is necessary. But the general tenor of the report, it seems to me, overcompensates in the direction of decrying the detriments of energy production, without giving sufficient recognition to the benefits that have been derived from the use of energy, and the importance of adequate sources of energy to the future of our society.

Similarly, the report seems to be optimistic with regard to the possibility of conserving energy, while being pessimistic about the potentialities for finding means of producing energy with minimal adverse impact on the environment.

In this connection, I question whether there is justification for the degree of optimism displayed in the report with respect to the potentiality for conserving energy by converting a large share of the nation's economy to a service orientation. The projections of the nation's ability to shift to a service economy are based on econometric models whose assumptions are open to question, and whose accuracy cannot be assured. Within the next decade and probably beyond we will have tremendous needs for new housing, new mass transportation, and other products of the economy which would not, it seems to me, permit such a large diversion to the service economy as is projected in the report.

Based on the assumption that there can be a substantial reduction in the rate of growth of energy demand, the report states that "a pause of several years in new power plant starts is possible for the nation as a whole."

It seems to me that this would be a risky course to follow. Projections of future demand for electric energy are highly uncertain at this time, because we have entered upon an unprecedented era of rising prices, environmental problems, and construction and financing difficulties. The desire for conservation and higher electric rates may indeed slow down the traditional 7 percent per year growth in demand for electricity. On the other hand, many utilities are reporting that because of uncertainties in the availability and price of alternate energy sources, there has been an upsurge in installation of electric heating in new homes as well as conversion to electric heating in older homes. Some industries also are converting to electricity because of concern about availability of fuel and the ability to meet environmental standards. It is also likely that electricity will be used more extensively for transportation in the future.

These new demands may well compensate (or more than compensate) for some reductions that could result from higher rates, or efforts to conserve electricity. If this is true, a "pause" of several years in construction of new power plants could result in shortages of supply that would in turn cause hardships to consumers and have an adverse effect on the economy.

Furthermore, in an effort to "catch up," the electric industry might be forced to build fossil-fired generating stations which could be put into service on a faster schedule than nuclear plants, but which would result in several adverse effects: (1) higher rates to consumers because of reliance on fuels more expensive than nuclear power; (2) a drain on coal and oil supplies; and (3) quite possibly, a greater dependence upon foreign oil.

Because of the long lead time needed to build new generating facilities (about ten years for nuclear power plants), and continuing uncertainties as to the future requirements for electric energy, I think it would be a mistake to suggest that a "pause of several years in new power plant starts is possible for the nation as a whole." Conservation and more efficient utilization of energy are essential, but if we are to err, there would appear to be less risk in erring in the direction of having an excess of generating capacity for a short time, rather than being caught short.

Another statement in the report (Chapter Eight) which I believe merits comment is the following: "Electricity is generally still priced to promote growth through discounts for greater use at a time when greater conservation is needed."

This and other similar statements leave the reader with the impression that electric utilities are still designing rates to promote greater use of electricity. There was a time when electric utilities did adopt "promotional" rate schedules which were intended to encourage consumption, but it is my impression that few, if any, utilities are doing so today.

It is important to make a distinction between rates which are designed to be "promotional," and rates which have the effect of charging large consumers less per kilowatt hour. The two concepts are not necessarily synonymous, but the distinction is not made clear in the report.

Most electric utilities charge lower rates today to those who consume more energy, because the cost of providing service is lower for large volume consumers. For example, distribution and administrative costs do not vary in the same proportion as increased consumption. The practice of charging lower unit costs for increased consumption is no different from that followed in other industries which charge lower prices to customers who buy in bulk quantity, and where the cost of providing service is lower.

Interestingly enough, the "fundamental" reform which the report advocates—peak load pricing—also would result in lower kilowatt hour costs to large consumers.

Chapter Ten of the report advocates basing "the price of electricity on a capacity charge for the power used on peak days plus a price for each kilowatt hour used which reflects its costs." Under this policy, large consumers, such as industries, would have a fixed capacity charge, but because they would consume a large number of kilowatt hours, their average cost per kilowatt hour would be lower than that of smaller consumers, such as residential users.

The report also tends to underestimate the cost of the rather sophisticated type of demand meter which is proposed, and some of the difficulties inherent in the use of such a meter. It should also be noted that experience in England has shown that whereas demand meters have tended to reduce peak demands—a desirable objective—they have not reduced total energy requirements.

In some cases, the report tends to attribute to "promotional rates" results which are more properly attributable to other causes. For example, Chapter Ten states that promotional rates have encouraged consumers to utilize electric house heating, but that the greater use of energy for this purpose has resulted in building new, higher cost generating capacity that pushed up the price of electricity. These consumers, the report contends, thus were "victimized" by high heating bills. In reality, these consumers were "victimized" by high fuel costs, not high capacity charges.

Peak load pricing, which is advocated by the report, has considerable merit. I have no objection to this concept, as long as such pricing is based on costs. However, I believe that the report oversells the benefits of peak load pricing, and leads the reader to anticipate results which might not be attainable.

The following are additional facets of the report with which I do not concur, or which merit comment.

1. The recommendation that a federal tax be placed on public power enterprises leaves the reader with the impression that there is a considerable disparity between publicly and privately owned utilities with regard to the payment of federal income taxes. The fact of the matter is that federal tax payments by privately owned companies have been declining steadily for the past decade or more, because of various tax benefits accorded utilities. According to a recent report of the Senate Government Operations Committee, federal income tax payments of the nation's Class A and B privately owned utilities dropped in 1973 to a record average low of 2.6 percent of their operating rev-

enues. Forty-nine of the nation's investor owned companies paid no federal income tax at all. It should be pointed out that public power agencies are accorded an exemption from federal income taxes not in order to lower the price of electricity, but because such exemption follows the traditional practice of immunity of all local governmental activities from federal taxation. An attempt to place a federal tax on the income of local public power systems would open up the question of whether or not such a tax also should be placed on other revenue producing functions of local government, such as water departments, parking meters, auditoriums, etc. Such a policy would raise serious Constitutional as well as policy questions.

2. I take exception to the recommendation (Chapter Eight) that government funding for demonstration plants, including the breeder reactor, be limited to 25 percent of the initial estimated cost. If this policy had been in effect, it is doubtful whether nuclear power would be available as a commercial option today. Other promising technologies which are in the national interest also might not be developed. Rather than setting some arbitrary limit on government funding, it would seem preferable to attempt to devise a policy whereby the federal government paid for that portion of the cost of new technology that is of national benefit, over and above the cost that should be paid by the affected industry, because of the direct benefits it might receive from the new technology.

3. The report recommends (Chapter Eight) the enactment of pollution taxes "supplementary to regulatory actions to reflect environmental costs of fuels extraction and energy operations." If such taxes are imposed, the proceeds should be specifically earmarked for research and development and other steps needed to abate the pollution effects of energy production. Otherwise, such taxes would become a burden on the consumer, without necessarily being used to correct the problem for which the tax is imposed.

Although there are several other comments to which I would take exception, the above are the principal items with which I would disagree.

I reiterate that, on the whole, the report makes a significant contribution, and its recommendations deserve serious consideration. The Ford Foundation is to be commended for initiating a project which is so timely and which can be so significant to the future welfare of the United States.

Joseph R. Rensch

I appreciate having had the opportunity to work with the capable Project Director and others involved in the preparation of this thought-provoking document. I also appreciate having the right to dissent from some of the report's statements and conclusions. In the interest of space I will comment briefly only on those general policy areas of the report with which I have fundamental disagreements.

I believe the report may work a disservice to the American public as a whole because it lacks objectivity and is not sufficiently comprehensive in dealing with critical energy matters. Some special interest groups will find the report most satisfying because it champions their particular causes—for example, the arbitrary restriction of energy production, or the theme of providing more energy functions through big government. Yet there are important reasons to reject these and other of the report's key conclusions. Three basic positions of the report give me particular concern.

First, the assumptions adopted early in the report can mislead the concerned citizen into thinking that merely moderating growth in energy demand will solve difficult and complex energy supply problems. Overemphasis of the contribution of conservation can cause us to move too slowly and inadequately in developing new energy sources which require long lead times. This, in turn, can force the nation into an unnecessary long term energy supply crisis, recovery from which would be at a severe cost to the economy and our nation's standard of living.

My second concern is the report's failure to grasp the significance of a basic new energy form for the nation's future energy economy. Adding hydrogen to our future energy supply promises significantly lower costs and more efficient energy utilization, yet the authors concentrated only on what was familiar to them—electricity. To ignore the hydrogen economy contribution is a critical omission.

My final basic concern is the populist approach taken in the latter part of the report, particularly those chapters dealing with the operation and structure of the energy industry. The report sharply and critically focuses on the shortcomings of the existing energy industry. The authors recite faults they say now exist or might occur in the future, even if they do not exist now. They suggest changes to avoid these faults. Unfortunately, the report does not balance its outlook by presenting both sides of the picture. There is no showing of the solid results obtained with the existing energy industry structure. There is no recognition of the serious deficiencies likely to occur with the proposed changes.

Conservation alone will not be adequate to remedy the nation's growing energy imbalance: The report's introductory and scenario chapters on energy growth stress the value of conservation. Everyone agrees with the importance of this. The impression presented, however, is that by conserving and possibly achieving zero energy growth, we will have largely solved the nation's energy supply problem. This is misleading, not only because the assumptions regarding our ability to limit energy demand are far too optimistic, but because it can also create a false sense of complacency by offering too easy a choice. The report states that a computer model was used to test the scenarios. Unfortunately, this was not the critical type of testing required—it could really be described as an effort to support, not test, the author's position. Other models, or indeed that same model structure with more realistic assumptions, would have produced totally different results.

There is general agreement across the nation on the need for an all-out conservation ethic and programs to reduce wasteful uses of energy. However, important as this conservation step is, it alone will not be sufficient because new energy sources must be developed to fill in for the depletion of existing primary energy sources—and this is an immediate problem. Long before the turn of the century, substantial new oil, gas, coal, and uranium supplies must be found and developed in our nation to offset declining production from existing low cost sources that are being consumed. The report's position is that adequate supplies will be forthcoming, even when we forego the development of major new energy sources, especially the presently undeveloped offshore areas, western coal and shale, and nuclear power.

All evidence is to the contrary. Recent experience from efforts at secondary and tertiary production from existing oil fields, the accelerated development of new but lean oil and gas field prospects, and the opening of new underground coal mines shows that the nation will fall far short of the necessary production requirements even with present, high energy prices. To obtain additional production from existing sources will require substantial price increases, or the extensive relaxation of environmental and safety standards.

This is an unnecessary price for the nation to pay just to limit energy production arbitrarily to existing sources. All of the nation's areas should be considered as supply sources for needed energy, including the undeveloped offshore areas, western coal and shale, and nuclear power. The type of energy, its location, and rate of development would be based on the overall economic and environmental attractiveness to the nation. Our latest safety and environmental standards must of course be met, and the costs for these would be a factor in evaluating the acceptability of each project. The report's suggestion that developing these now undeveloped areas threatens serious environmental damage and irretrievable loss of amenities appeals to all of us who want to preserve our precious environment, but it is an argument without real foundation. First of all, none of the oil, gas, coal, shale, or nuclear development would occur without meeting the nation's latest and constantly improving safety and environmental standards. Secondly, wells are drilled one at a time; mines and plants are also developed and built one at a time. The progress of each is visible and any changes required to overcome any unforeseen problem can be made as it arises or the system can be shut down until the problem is solved.

Failure to consider hydrogen in the nation's future energy economy is a critical omission: In looking to the nation's long term energy future, the authors properly included the essentially limitless sources: solar, controlled nuclear fusion, and geothermal. But they took a narrow approach in considering the application of solar and nuclear technology to produce only electricity. While nuclear electricity is familiar and technically feasible, electricity is significantly more expensive than it direct use counterpart, gas. The authors may have concluded that since the supply of natural gas is finite, there was no reason to consider it in the long term future. But production of synthetic gas from coal is now underway (and shale can also be utilized) to extend our nation's gas supplies by making use of these abundant resources. Then later on, around the turn of the century, hydrogen can be used to supplement natural and synthetic gas. It would be produced by separating water into its component parts, hydrogen and oxygen, by utilizing solar or nuclear energy sources.

There was no need before now to consider the production of synthetic hydrocarbons and hydrogen, because natural resources were available at less cost than synthetics. Now, extensive research efforts are underway to develop optimum means of producing low cost hydrogen. Several high efficiency processes are being developed and tested in laboratories. Hydrogen is the perfect environmental fuel. It can be distributed through the most efficient and esthetically attractive energy transportation system—the same type of underground pipeline system now distributing gas—and it can supply both direct combustion and the fuel cell which produces electricity on site without pollution. The prime advantages of hydrogen are that this form of energy can be produced more efficiently and at lower costs than electricity. It will have a major future role, even though the report fails to recognize its importance.

The suggested restructuring of the energy industry would be a step backward: In dealing with the energy industry, especially the later chapters on Electric Utilities, Private Enterprise, and Energy Resources, the report takes a populist approach that the nation's energy problems are largely the fault of the existing industry structure and therefore changes are required. The report fails to mention that what the nation considers as its major energy problems; namely, rising prices and an imbalance between energy demand and supply, would have occurred regardless of the industry structure because available, low-cost domestic resources were being used up and inexpensive foreign crude became a memory when the exporting nations raised their prices. Ignoring these facts of the physical or resource situation, the authors conclude that greater public participation in energy must be achieved, and that individual government agencies must be consolidated. They suggest that the existing private industry should be reorganized and restructured with government taking a much larger direct role.

While our nation's energy industry is less than perfect, the same is true of government and all organizations run by human beings. The report's proposals for greater direct government and public involvement in the energy industry are not new, either in this country or the rest of the world. Many of these schemes have been proposed over the past 30 to 40 years, and an examination of the record will show that government operation of business is no panacea. In fact it can be argued that, the greater the government's involvement, the poorer the performance. This point is well illustrated in our country, and we are fortunate to be able to look now at many other nations around the globe to see how poorly government manages business and hopefully, to learn our lesson before it's too late.

By any standard the nation's energy industry has performed well—real energy prices, that is those with inflation effects removed, declined over the period from 1950 to 1970. During this period energy supplies expanded significantly while employment in the energy industry grew at a much slower rate, or did not grow at all, indicating the industry was improving its productivity. By contrast, most other industries were expanding their employment at a relatively much greater rate, and the government sector of the economy was growing fastest of all.

The reader is told briefly in an earlier chapter that the American standard of living is the highest in the world and that energy plays a key role in achieving that standard. The report then abandons the conclusion to be drawn from those facts.

While presenting their material as a comprehensive review and analysis of the energy industry, its problems and their solutions, the authors largely cover only one side of many important issues. For example, in the supporting study attacking the electric utility industry the authors of that study accurately admitted that their biases would be apparent throughout. The Project report carries this bias intact into the final edited version. In another area the authors based their work on an encyclopedic review of technical and legislative literature, selecting that material which supported their position. Many of the references cited the horrors from past experiences in air pollution, oil spills, and surface mining which occurred before the nation considered environmental quality important. More recent material which shows substantial progress to date in air pollution, oil spill prevention and clean up, and surface mine reclamation, was largely ignored. In the areas dealing with competition in the energy industry, the authors cite theorists who suggest competition could be threatened if the industry is allowed to operate as it has to date; that is, with some large integrated private firms. The authors ignore the view of other distinguished economists who base their conclusion on observed facts and dispute the contention that the present energy industry is anticompetitive. In the areas dealing with life styles, the authors make a point of the hardship that increased energy prices will bring, and in areas dealing with energy consumption the authors stress the need for conservation. Yet later on the report suggests that the nation move to the all-electric economy which is both more expensive and less energy efficient than the direct use of fuels in appliances.

The report's failure to look at both sides of the issues, and to consider all points which would contradict the positions taken, is a critical omission. This need for thoroughness and objectivity was recently well expressed by Richard Feynman, Nobel Laureate and Professor of Theoretical Physics at California Institute of Technology. Professor Feynman reminded the Cal Tech graduating class that it is not sufficient merely to conduct research that supports one's theories. The true scientist must also consider, and present for others to see, all of the evidence that might prove his theory wrong. He must literally lean over backwards in this effort. Otherwise he can fool himself and others.

It would be a great tragedy for the nation if our future energy policies were based on invalid presumptions and incomplete analysis. And it could be equally disastrous if basic changes in industry structure are mandated without taking into consideration the probable adverse consequences of such changes.

Joseph L. Sax

This is an important document by the only true test of importance; it requires us to alter the questions we ask about energy problems. No one who reads this report will ever again be able to view energy policy as merely a matter of developing supplies to meet demands projected into the future from unquestioned historical patterns of use. Because it describes specifically a number of ways in which historical patterns of demand can be substantially modified without sharply disappointing consumer expectations, it is also a document that will obtain the serious consideration of those who formulate and execute our public policies.

In two other respects, this report deserves high praise.

It puts to rest the threadbare argument that debates over energy policy are in essence controversies over how much or how little government intervention is appropriate in our society; the excellent Chapter Eleven, for example, in which it is noted that more than half our fossil fuel energy resources are found on publicly owned land, strikingly demonstrates that energy policy is in very large part necessarily governmental policy. Moreover, in detailing the serious hazards associated with increasing energy development, the report persuasively demonstrates the inadequacy of the view—so assiduously promoted in industry advertising —that the only way to insure the well-being of the society is to embark on all-out development of our energy resources. The report shows that a determined effort to reduce the historical rate of growth of demand is an essential element of insuring against grave hazards that full scale development itself produces.

The failures of the report are many, but they do not detract from its substantial achievement, for they are basically failures of stopping short rather than of affirmative error. The most important of these is the refusal of the authors of the report to identify priorities among the developmental hazards that reduced demand will permit us to defer or to avoid altogether. Nuclear development, surface mining in the arid West, and some offshore development of oil resources are all described as serious hazards, but to the policy maker who will doubtless have to choose which are the most serious among them, and thus to be averted first, the report gives no guidance. This is a failure of nerve on the part of the authors.

The report also fails to be specific enough in areas where specificity and persuasive detailed explication were clearly needed. We are told that a transition to a less energy intensive society, in which more durable goods will be produced, will not in the long run reduce employment opportunities. But the painful short-term problems of transitional employment, the problems that necessarily most trouble the labor segment of the society, are tossed off in a few pages of generalities. This is a failure of responsibility.

Finally, for all its seeming venture, the report never finds itself able to look beyond the present high consumption economy to which we have all become addicted. In essence it recommends nothing more than that current levels of consumption can be made less energy intensive. Doubtless this perspective gives the report a quality of contemporary realism that will attract the attention of elected officials. But nowhere does the report come to terms with the problems other than energy use that a gluttonously consumptive society faces—problems of congestion, of land use, of water pollution, and of the dissemination of toxic substances. This is a failure of vision.

William P. Tavoulareas

Background: When the Energy Policy Project was first organized in 1972, I was assured that the purpose of the Project was to achieve a balanced view of the future of energy supply and demand through the use of research by numerous consultants, augmented by that of the Project staff, in order to provide information to facilitate a discussion of this vital area of national concern.

I was attracted to the concept of a study which would draw on all sides of opinion and expertise to bring together a balanced treatment of this very complicated topic. It was primarily for this reason that I accepted a position on the Advisory Board. It was made clear to all of us on the Board that the Director was responsible for the content of the report, and that the Board's function was only one of advice. Nevertheless, I accepted the position on the Board upon assurance from the Foundation that my own point of view could be made available at the same time as the report was issued.

By the fall of 1972 it became apparent that the great majority of the consultants retained by the Project had already taken well-known positions with respect to controversial aspects of the energy problem and that these positions were not evenly distributed with regard to these controversial aspects but were extremely biased to one side. In fact one of the first consulting grants was for the purpose of assessing energy decision making in the U.S. government. This grant was made to an organization which had just sued the U.S. government to prevent the continued leasing of offshore acreage in the Gulf of Mexico. The fact that many of the consultants had already adopted positions on the subject was of particular importance in light of the fact that many of the individual consulting grants were not funded to a degree which would permit any amount of meaningful original research; therefore the best that could be expected in many cases would be a search of the existing literature on the subject or a reassembly of previously completed research. (I do not by any means want to say that every grant should be characterized in this fashion, and indeed some have done a good deal of creditable research.)

While I expected and fully accepted that the Advisory Board should represent a wide range of opinion, I did expect and, in fact, was assured by the Foundation that the Director was approaching the subject with an open mind. I was therefore shocked to learn of a major policy speech which the Director made on January 25, 1973, in which he stated in a public forum the broad outline of the program which is now embodied in the final report of the Project. These conclusions were expressed before a single one of the consultants had given the Project the benefit of their advice, and without any advice from the Advisory Board. (Indeed, the Advisory Board was not even informed of the pendency of the speech.)

Despite my consistent comments to the Project and the Foundation that the results were largely preordained by the Director's public statements and by the sources of the advice being received, efforts to obtain research in areas affecting the other side of these controversial issues were almost totally absent. In those few cases where such advice was received, it has been largely ignored. It is therefore no surprise that a lack of balance is evident in the final report. The basic thesis of the Project becomes: "The search for energy and the use of energy is bad while energy conservation is good." On this base all else rests. There has been no effort to deal with the advantages of energy use in the same social and economic areas where the disadvantages arising from the search for and use of energy—both real and imaginary—are so meticulously catalogued.

My comments on the preliminary report (*Exploring Energy Choices,* published earlier in 1974) have produced almost no change in the direction of the final report. Although longer than the earlier report, its views are basically the same. I had hoped for a better result. We are thus left at the end of two years with a result that was predictable almost from the beginning, and in which the only significant added ingredient is the expenditure of time and over $4 million.

Specific comments: At the last meeting of the Advisory Board, the Board decided to impose restrictions on the length of separate comments by Board members, even though the Ford Foundation has assured me that I would retain my right for comment and dissent. It was never mentioned or even hinted that this right could in effect be frustrated by unreasonably curtailing the length of the space I would be allowed. The Chairman of the Advisory

Board has kindly undertaken to relax this restriction somewhat; but it still remains a particular burden to me. Over three-fourths of the energy in the United States is being supplied by oil and gas; and it is therefore no surprise that the bulk of the report deals directly or indirectly with the oil and gas industry. Since I am the only oil company executive on the Advisory Board, it is incumbent on me to attempt to answer at least the most important distortions in the report. In the space available to me there is no way that I can deal even with the most important of these distortions. Indeed, a serious consideration of Chapter Nine would involve the entire space available to me, and yet there are other points which also require treatment. I must therefore comment briefly at the risk that points will not be fully developed, and at the further risk that some may falsely assume that I have agreed with statements which are omitted merely by reason of space limitations.

• *Campaign finance reform measures should be adopted to remove the oil industry's disproportionate political strength.* Campaign financing reform is, if anything, a larger subject than energy policy. This conclusion of the Project is a perfect example not only of the imbalance of the report but of the superficiality of its treatment of the most complex subjects. Finally, on a personal note, I should say that if the oil industry has disproportionate political strength today, one might well conclude, in the face of the punitive anti-oil legislation now pending in Congress, that the disproportion is on the low rather than the high side.

• *Building codes would be updated to make energy conservation a priority objective.* This assumption, which is very briefly stated, ignores the well-known difficulties in coordinating even relatively minor changes in the building codes administered by over 8,000 jurisdictions in the United States. To suppose that changes of the magnitude assumed in the report, even if they were desirable, could be made within the time frame assumed by the Project is highly questionable. Moreover, the concept does not take account of the real advantages which exist today for differing building code standards for, say, a summer home in the north, versus a year-round residence in the same area. Finally, this area is an example of the ambivalence of the report towards public participation in decision-making. "Public participation" is generally hailed as being desirable. Presumably today's hodge-podge of building codes reflects the widest sort of "local option" expression, yet the report finds this diversity undesirable.

The report repeatedly uses the need for public participation to support its objective of slowing down the development of additional energy resources; but when it comes to the report's prime objective of a drastic curtailment in energy use, public participation is ignored. A governmental imposed change in life styles is the solution offered.

• *Revamp the railroad tariff regulations to provide for flexible rate making.* Anyone familiar with the labyrinth of railroad tariffs cannot be sanguine that this subject can be attacked with dispatch.

• *Policy could include specific subsidies to the poor.* This recommendation provides a good example of the way in which the really difficult problems are dealt with in the report. Because of the great uncertainty that the consumer will use less energy even if he is paying the full cost of it (including social costs), the report recommends that a gradually increasing energy tax be enacted. Naturally this tax would be regressive. The regressivity problem is deftly solved by the simple assumption that income could be redistributed through subsidy programs for the poor. Considering our lack of success in dealing with welfare programs and such problems in this country involving more than energy use, it is startling that the authors of the report would consciously recommend increasing the financial burdens of the poor, and then assume that these added burdens could be readily alleviated.

• *Establishing new communities of 55,000 population at the rate of 20 per year.* These 200 new communities per decade involve enormous commitments for new infrastructure (and energy use), none of which is costed out in the study. Jobs and services have to be relocated if the new cities are to be viable economic entities. Considering the decades of dislocation and hardship which have followed much less traumatic economic transfers in the past, this goal must be, at least, wildly optimistic.

• *Establish automobile efficiency standards.* This recommendation is one of several involving mandatory technological change. We have had painful experience with the efforts of Congress to set technological standards for automobiles which go beyond what can be accomplished with existing technology. But even assuming that there will be new technological developments, the efficiency standards will in essence mandate the size, weight, and relative safety of the automobile. Again, the image is one of regimentation rather than free choice. More importantly, it will deny a person the choice of a larger, more comfortable, and safer automobile, which might even be driven fewer miles (because of the cost), rather than the model which was mandated by law.

• *Create a federal "yardstick" corporation as a benchmark for costs and prices.* This suggestion is one of several for the establishment of a government role in commercial transactions. Ironically, while the report is highly critical of government agencies which today have knowledge of oil and gas (the Interior Dept. and the FPC), it makes the suggestion for a federal corporation on the apparent assumption that such a corporation would not only operate as efficiently as the private companies, but at an even lower cost. The matter would be a legitimate subject for debate were there not such a plethora of examples of government oil corporations around the world which are highly inefficient. The reader must seriously ask himself whether the services he now receives from the federal government and from the various federal, state, and city government agencies are delivered in an efficient and inexpensive way, and whether he could expect a better record from a federal oil and gas corporation. Many of us are old enough to remember when the New York subways were taken over to save the 5¢ fare. At present the operating cost per passenger of the system is far greater than the subsidized fare of 35¢.

• *Change treatment of income tax payments to foreign governments.* This recommendation is made without an understanding of the need for U.S. corporations to be competitive abroad. If a discriminatory action taken by the U.S. government against its own corporations operating abroad causes them to lose their competitive posture, versus the corporations of other consuming nations, the result will not be additional revenue for the U.S. Treasury. Instead it will constitute a gift of these foreign businesses to competitors in the various foreign countries. These competitors are willing and able to pick up any portion of the foreign oil business which the U.S. is preparing to relinquish.

The final report characterized: There are basically four key assumptions which underlie the report recommendations that the United States adopt a policy of low energy growth—and ultimately no energy growth. These assumptions are:

1. It will be technically, economically, and socially feasible to reduce energy consumption drastically and quickly to a level (that is, approximately zero growth) consumption pattern.
2. It will be possible to set up a sophisticated workable government control mechanism to direct public consumption and living patterns in a way consistent with the national policy and to redistribute wealth so as to overcome inequities to the poor created by the economic distortions caused by the policy.
3. Increased energy supplies will always be "expensive" (in a total sense) and environmentally unsatisfactory; the use of more energy will yield no advantages which will offset these disadvantages.
4. Our experience with the private sector is such that we must as a policy matter replace the use of market mechanisms by explicit, highly tuned government controls.

The report places great emphasis on the need for public participation in policy decisions. Yet, if the recommendations of the report are followed, the decisions in which the public could participate are sharply hemmed in.

The report expresses great fear that the public would not make a "right" decision with respect to energy use if the supplies were available at a cost the consumer was prepared to pay. As a consequence, the report really involves a complex plan of delay in the development of new supplies. Each decision to force reduced consumption is to be taken early and without public debate. Each decision to develop additional supplies is to receive the most careful consideration, with every level of the public to be afforded the opportunity of a veto. While masquerading under the banner of public participation, the report, in essence, advocates just the reverse since it carefully manipulates the result and then asks for a public rubber stamp of approval. In short, the report does not contribute to a debate of all aspects of the issues. Instead, it is an unabashed primer for regimentation.

The report justifies a number of its sweeping assumptions by asserting that no real change in life style is involved. But is this really so? Is it really not a change of life style if we require the American public to substitute bicycles and walking for their automobiles (to the degree assumed in the report)? Is it really not a change in life style to require permanent accommodation to cooler indoor temperatures in winter and warmer temperatures in summer? Is it really not a change to require multifamily housing even where a preference exists today for single family housing? Will the American public really agree that if they are forced to take vacations near home, this is not a change of life style?

The report does acknowledge that its goals will be unattainable without a certain degree of government control. Thus certain government measures are suggested in the achievement of these goals. Again, it is assumed that the suggested government action can and will be taken readily and without public resistance. Yet, there is no way to guarantee that energy conservation would rank quite so high in the public view of national priorities, unless national policy—or the lack of it—artificially limits available supplies. Moreover, the enactment of that part of the recommended legislation which falls within state and local governmental jurisdiction would be most difficult to achieve.

Throughout the report, the low / zero growth scenarios are based on energy scarcity and high energy prices. The recommended supply policies are designed to keep energy scarce and therefore expensive. Yet, the report admits (on the basis of its own consultants' studies) that the physical energy resources of all types are adequate to support much higher growth rates. Indeed, it accepts the need ultimately to develop these resources, since even a zero-growth case requires considerable energy consumption. It is at this point that the report most grievously fails in its obligation to inform the public. Having roundly criticized every energy source which is available to us in quantity today, while at the same time admitting the need for energy production, the report then steers clear of any indication of priority among sources in terms of the total impact on the environment and the economic system.

The reader of the report is left with the uncomfortable feeling that he must have energy. But all the Energy Policy Project can tell him is that all of the sources are bad, and he can nowhere find any guidance as to which alternative he should at least temporarily select among those available.

Since I believe many energy sources including oil can be developed with acceptable risk to the environment, I would actively pursue the development of a variety of sources so as to meet the needs of continued economic growth and less dependency on foreign sources.

A viable alternative: In contrast to strategy in the report, I would not want to be so sanguine that all of the assumptions in the report will develop into reality. Let me at the outset agree that we should continue the national dialogue with respect to the desirable level of consumption. Let me also agree that we must make every effort to squeeze waste energy usage out of the system. But to postpone resource development until after the demand situation has been worked out involves too great a risk. I would like to suggest some elements of a viable alternative strategy.

- *First,* we should have the objective to eliminate government controls which unnecessarily interfere with the development of additional supplies.

- *Secondly,* we should go forward with the orderly development of supplies, even to the point of creating an energy surplus again. If it appears desirable the entire development scheme can later be modified at any stage in its implementation. We should recognize that all the decisions will not be taken at one time. Coal mines will be opened one at a time. Oil wells offshore will be drilled one at a time. Refineries will be built one at a time.

- *Thirdly,* the timetable on environmental objectives should be carefully reviewed in relation to the energy needs. Here I particularly emphasize I am referring to the timetable and not to the objectives themselves. I continue to believe that the advance of technology and the development of clean energy sources will permit us to realize our environmental objectives. I ask only that the two programs be viewed as part of a single problem allowing for the tradeoffs between them.

- *Fourthly,* we must encourage energy research so that the problems that we have experienced in the 1970s will not again become problems in the 1980s and 1990s. Energy resources are abundant; and if we have the technology to utilize them in an optimal fashion, we need have no concern for future energy growth.

• *Finally,* we must deal with the social costs of higher-priced energy. The appearance of higher energy costs in the economy will create dislocations. The extent of these dislocations is at present unclear. However, arbitrary controls which delay the development of additional supplies only aggravate the problems of the poor.

This solution to the energy problem would involve less controls than the report implies; would involve a return to a surplus of energy as a means of keeping prices down; would involve reasonable preservation of our environmental objectives; and would involve explicit attention to the problems of the poor.

The issue of public participation in decision-making is one which we have mentioned a number of times in these comments. If we are to have an honest and consistent presentation to the public, we must recognize that the decisions we are facing are not easy, nor will we find unanimity in reaching them. It is entirely possible that we will have to make the very difficult decision to sacrifice the comfort and esthetic sensitivities (but not the health) of the few for the advantages to be gained by the many. To me this is the essence of the democratic process as it is practiced in the United States. Not everyone can have a veto; otherwise public policy could never go forward.

Again, as I stated in my remarks on the preliminary report, there are essentially two alternatives in dealing with the energy problem. The first would delay the development of new supplies on the assumption that energy usage can easily be reduced enough to bring supply and demand into balance. This is the case which the report implicitly adopts. The second alternative, not covered in the report, would increase supplies, eliminate waste usage, and examine all implications of further energy reductions which may have an impact on life styles. We should ask ourselves which course carries the greater risk. If the assumptions behind the low growth cases are wrong the result will be energy scarcity, high energy prices, unemployment, and other economic and social dislocations. On the other hand, if the assumptions supporting the case for increased supplies are wrong we will have energy surplus and low prices. It seems clear to me that this latter risk is the more tolerable one.

J. Harris Ward

Nuclear power: In a number of places the report suggests that nuclear power is an emerging science—risky and only partially understood. The impression is also given that its contribution to power production in this country and elsewhere in the world is insignificant. Roughly one-third of Chicago and northern Illinois' electricity is now made with nuclear energy, and the oldest atomic unit has been operating for almost fourteen years. The state electric monopoly in France has decided that all new power stations will be nuclear. Twenty nations other than the United States, including all major industrial powers, are moving ahead on the nuclear power front.

A proposal to defer nuclear development in this country is dangerous to the national security and the national welfare for a number of reasons. In the first place, nuclear power is one of the best means of solving our energy shortage and our environmental problems. Secondly, a nation which hesitates to keep up technologically today must make double or triple the effort to catch up tomorrow. Thirdly, nuclear power has a fine safety record. If it proves to have hazards which we are unable to manage, with existing methods, we can and must learn to handle the hazards with other methods—not abandon the technology. Finally, nuclear waste management methods are new but they are not beyond our knowledge or capacity. If they should turn out to be inadequate, they too can be corrected.

Nevertheless social activists and a small minority of scientists, representing what they consider to be the public interest, have succeeded in extending the construction time of most recent nuclear units on the basis of environmental and safety concerns. This report considers and almost suggests a moratorium on nuclear power development until disagreements among scientists have been settled. The cost of avoiding decision until scientists are unanimous will, in my judgment, place too great a burden on the American people.

Zero energy growth: ZEG, zero energy growth, is discussed in some detail in the preliminary report. ZEG may reduce energy problems somewhat but its long term social, economic, and international effects will be both massive and unpredictable.

The rising standard of living in the world and in the United States is related very directly to the substitution of other forms of energy for human sweat. Progress in this respect has increased geometrically as the ox and the mule have been replaced by wood, coal, oil, gas, and uranium in an ever-increasing supply of energy units. Fission is here and fusion is on the way. Neither the minds nor the data are available today to tell us the effects of additions to or changes in the energy mix, nor the growth rate of total energy use.

The foregoing statement by Mr. Ward was prepared for the preliminary report, Exploring Energy Choices, *which was published early this year. At the time of his death on July 18, Mr. Ward was planning to submit a similar statement. While I concur in the foregoing statement, I have additional reservations about the report's ingenuous assumption as to the ease of attaining zero energy growth without economic and political disruption, apparently through substitution of government decision-making for individual choice. However, my tenure on the Advisory Board since Mr. Ward's death has been so brief that a more extended statement on my part is unwarranted.*

Gordon R. Corey
Vice-Chairman, Commonwealth Edison Company

Gilbert F. White

The report moves public thinking about energy problems in a sound direction and puts forward healthily provocative proposals. However, if the American people are to avoid rude surprises in carrying out a national energy policy they need to be sensitive to three considerations which are mentioned but not stressed in the report or by my colleagues. The recommendations would benefit from a leavening of imaginative caution as to basic assumptions, from a more quizzical view of the state of the world, and from concrete examination of regional differences in the United States. When taken into account those considerations add weight to arguments for encouraging large flexibility in national options.

First, in the preparation of the three scenarios it was necessary to make a series of assumptions as to how technology will unfold, how people will act, and how society will respond to new information and conditions. In the circumstances these are useful for purposes of projection. However, experience would suggest that some of the assumptions will turn out to be wrong and that at least a few may be terribly wrong. The assumption that consumer preferences reflected in demand curves and life styles will remain the same may grossly underestimate the capacity of Americans to change: how many reports in the early 1960s predicted the shift in values that marked the environmental movement at the end of the decade? Projections of the year 2000 population may be wide of reality. And so on.

To make such assumptions is necessary to the analysis. To challenge them and suggest disturbing alternatives calls for imagination and a stubborn willingness to confuse the calculations with doubts and possibly fanciful observations. While the scenarios are a good start, people should be encouraged to stretch their minds to explore the effects of still different views. Just as the scenarios break out of the rather slavish linear extrapolations which characterized electricity growth projections for so long, the new figures deserve fresh and continuing appraisal.

Second, the report's treatment of the global setting in which U.S. policy takes shape is commendable in its departure from a narrow national view that afflicted so many studies in past. It nevertheless seems unduly sanguine as to a number of factors affecting the welfare of the human family. The widespread high rate of population growth, the distressingly slow improvement in economic conditions in several score developing countries, the dwindling global stocks of food, the new views of rights to natural resources, the prospects for serious shortages in several minerals, the increasing vulnerability of industrial society to catastrophic disruption, and the rapidly changing political alignments of nations producing raw materials heighten instability. These would be disturbing even were there no doubts as to possible climatic change or new sources of environmental deterioration.

In these circumstances the United States should seriously consider the ways in which it could cope with far more stringent impediments to international materials flow and processing. For example, a scenario for a declining level of energy consumption should not be dismissed as impossible. Hopefully, the United States will play a constructive part in bringing about a more stable and equitable sharing of the earth's resources, and energy policy will be a key element in that long term effort.

Third, although the report mentions areas which will be affected by proposed energy developments, it does not specify how development would occur in different mixes in different regions. In the formation of balanced national policy it is essential that the people of each region understand what a particular set of programs would mean to them and their children, and what shifts might occur when another program is adopted elsewhere. Deep tanker ports on the New Jersey coast, offshore drilling on the New England shelf, air pollution regulations affecting shopping center locations in Illinois cities, and oil shale development on the western slope of Colorado not only have distinctive local impacts but affect needs and activities in the other areas. Careful studies of spatial linkages within and between geographic areas will be in order if intelligent public choices are to be made.

Chapter notes

Chapter Two–The Historical Growth scenario

1. Data Resources Inc., "Energy Projections: An Economic Model," a draft report to the Energy Policy Project, May 1974.
2. Resources for the Future, "Toward Self-Sufficiency in Energy Supply," a draft report to the Energy Policy Project, September 1973.
3. Resources for the Future, "Toward Self-Sufficiency" (Note 2).
4. Jerome E. Hass, Edward J. Mitchell and Bernell K. Stone, *Financing the Energy Industry,* a Report to the Energy Policy Project, Cambridge, Mass.: Ballinger, 1974.

Chapter Three–The Technical Fix scenario

1. E. P. Gyftopoulus, Lazaros Lazaridis and Thomas F. Widmer, *Potential Fuel Effectiveness in Industry,* a Report to the Energy Policy Project, Cambridge, Mass.: Ballinger, 1974.
2. Data Resources, Inc., "Energy Projection: An Economic Model," a draft report to the Energy Policy Project, May 1974.
3. American Institute of Architects, "Energy Conservation in Building Design," a draft report to the Energy Policy Project, May 1974.
4. Jerome Weingart, Richard Shoen et al., "Institutional Problems of the Application of New Community Energy System Technologies," a Caltech Environmental Quality Laboratory draft report to the Energy Policy Project, November 1973.

5. C.N. Cochran, "Aluminum—Villain or Hero in Energy Crisis," Alcoa Research Laboratories, Pittsburgh, April 1973.
6. J.P. Dekaney and T.C. Austin, "Automobile Emissions and Energy Consumption," U.S. Environmental Protection Agency, Ann Arbor, May 1974.
7. Task Force on Railroad Productivity, "Improving Railroad Productivity," final report to the National Commission on Productivity and the Council of Economic Advisors, Washington, D.C., November 1973.
8. J. J. Mutch, "The Potential for Energy Conservation in Commercial Air Transport," RAND Report R-1360-NSF, Santa Monica, October 1973.
9. The Conference Board, *Energy Consumption in Manufacturing,* a Report to the Energy Policy Project, Cambridge, Mass.: Ballinger, 1974.
10. Midwest Research Institute, "Potential for Recycling Metal from Urban Solid Wastes," a draft report to the Energy Policy Project, April 1974.
11. Calculation based on data from Foster Associates, Inc., *Energy Prices 1960-1973,* a Report to the Energy Policy Project, Cambridge, Mass.: Ballinger, 1974.
12. J. Herbert Holloman, et al., "Energy R&D Policy Proposals," a draft report to the Energy Policy Project, July 1974.
13. Jerome E. Hass et al., *Financing the Energy Industry,* a Report to the Energy Policy Project, Cambridge, Mass.: Ballinger, 1974.
14. Bruce Hannon, "System Energy and Recycling," Center for Advanced Computation Report, University of Illinois at Urbana-Champaign, January 1972.

Chapter Four–**A Zero Energy Growth scenario**

1. For an overview of the arguments for limiting growth, the following readings are recommended: "The No-Growth Society," *Daedalus* (Journal of the American Academy of Arts and Sciences) 102: 4, Fall 1973; H. E. Daly, ed., *Toward a Steady State Economy,* San Francisco: W. H. Freeman, 1973; D. L. Meadows, et al., *The Limits to Growth,* New York: Universe Books, 1972; and "Blueprint for Survival," *The Ecologist,* January 1972.
2. W. J. Campbell and S. Martin, "Oil and Ice in the Arctic Ocean: Possible Large Scale Interactions," *Science,* July 6, 1973, p. 56.
3. For a discussion of this point of view see J. Kenneth Galbraith, *Economics and the Public Purpose,* Boston: Houghton Mifflin, 1973; and Richard N. Goodwin, *The American Condition,* Garden City, N.Y.: Doubleday, 1974.
4. There is, of course, a prolific literature of contemporary U.S. social criticism. See, for example: Theodore Roszak, *Where the Wasteland Ends,* Garden City, N.Y.: Doubleday, 1972; Lewis Mumford, *The Pentagon of Power,* New York: Harcourt Brace Jovanovich, 1968; and Abraham Maslow, *Toward a Psychology of Being,* New York: Van Nostrand, 1968.

5. Maslow, *Toward a Psychology of Being,* (Note 4).
6. Data Resources, Inc., "Energy Projections: An Economic Model," a draft report to the Energy Policy Project, May 1974.
7. For a detailed discussion of alternative policy instruments, see J. H. Holloman et al., "Energy R&D Policy Proposals," a draft report to the Energy Policy Project, July 1974.
8. Alan Poole, "Potential Energy Recovery from Organic Wastes," a draft report to the Energy Policy Project, June 1974.

Chapter Five–**The American energy consumer: rich, poor, and in-between**

1. U.S. Bureau of the Census, *Current Population Reports,* Series P 60, Number 90, "Money Income in 1972 of Families and Persons in the United States," December 1973.
2. Paul M. Tyler, "Home Insulation, An Effective Conservation and National Defense Measure," U.S. Department of the Interior, Bureau of Mines Information Circular 7166, April 1941.
3. "Michigan Pushes Home Insulation," *New York Times*, August 26, 1973.
4. A. B. Makhijani and A. J. Lichtenberg, *An Assessment of Residential Energy Utilization in the U.S.*, Electronics Research Laboratory, University of California at Berkeley, January 1973.
5. U.S. Environmental Protection Agency, "EPA's Position in the Energy Crisis," *Environmental News*, January 1974.
6. U.S. Bureau of the Census, *1970 Census of Housing*; *Detailed Housing Characteristics* (U.S. Summary), July 1972.
7. U.S. Bureau of the Census, *Current Population Reports*, Series P20, Number 246, "Household and Family Characteristics: March 1972."
8. A. B. Makhijani and A. J. Lichtenberg, *An Assessment of Energy and Materials Utilization in the U.S.A.*, Electronics Research Laboratory, University of California at Berkeley, September 1971.
9. Eric Hirst, *Energy Use for Food in the United States*, Oak Ridge National Laboratory, October 1973.
10. David Pimentel et al., "Food Production and the Energy Crisis," *Science*, November 2, 1973.
11. John S. and Carol E. Steinhart, "Energy Use in the U.S. Food System," unpublished paper, 1974.
12. U.S. Department of Agriculture, Agricultural Research Service, *Food Consumption of Households in the United States*, Spring 1965.
13. Eric Hirst, *Direct and Indirect Energy Requirements for Automobiles*, Oak Ridge National Laboratory, February 1974.
14. Makhijani and Lichtenberg, *Energy and Materials Utilization,* (Note 8).
15. Bruce Hannon, Robert Herendeen, and Anthony Sebald, "The Energy Content of Certain Consumer Products," a draft report to the Energy Policy Project, Energy Research Group, Center for Advanced Computation, University of Illinois, July 1973.
16. R. W. Sullivan et al., *A Brief Overview of Energy Requirements for the*

Department of Defense, Columbus, Ohio: Batelle Columbus Laboratories, August 1972.

17. U.S. Department of Interior, Bureau of Mines, *Minerals Yearbook*, 1972.

Chapter Six–**Energy, employment, and economic growth**

1. The most current and detailed annotated bibliography of energy related research, including energy conservation studies, is published monthly and available free of charge from Oak Ridge National Laboratory, Oak Ridge, Tennessee 37830. Edited by Miriam P. Guthrie, it is titled *NSF-RANN Energy Abstracts: A Monthly Abstract Journal of Energy Research.*
2. For example: Joel Darmstader, "Appendix; Energy Consumption: Trends and Patterns," in Sam H. Schurr, ed., *Energy, Economic Growth, and the Environment*, Resources for the Future, Baltimore: Johns Hopkins Press, 1972; and National Economic Research Associates, Inc., "Appendix; the Energy/GNP Ratio," in *Fuels for the Electric Utility Industry 1971–1985*, a report of National Economic Research Associates, Inc. to the Edison Electric Institute, EEI Publication No. 72-27, New York, 1972.
3. Anne P. Carter, ed., *Structural Interdependence, Energy and the Environment*, Hanover, N.H.: University Press of New England, forthcoming.
4. The Conference Board, *Energy Consumption in Manufacturing*, a Report to the Energy Policy Project, Cambridge, Mass.: Ballinger, 1974, summary section.
5. H. S. Houthakker and Dale W. Jorgenson, "Energy Resources and Economic Growth" (Chapter 3), a draft report to the Energy Policy Project, September 1973.
6. See Appendix F, Economic Analysis of Alternative Energy Growth Patterns, 1975–2000. This is a report based on the model developed for the Project by Dale Jorgenson and Edward Hudson of Data Resources, Inc., henceforth referred to here as the "DRI model."
7. The Conference Board, *Energy Consumption in Manufacturing,* (Note 4); and E. P. Gyftopoulos, Lazaros J. Lazaridis and Thomas F. Widmer, *Potential Fuel Effectiveness in Industry,* a Report to the Energy Policy Project, Cambridge, Mass.: Ballinger, 1974.
8. The Conference Board, *Energy Consumption,* summary section, (Note 4).
9. A. P. Carter, ed., *Structural Interdependence*, Chapter 1, (Note 3).
10. AFL-CIO, *The National Economy 1973*, Washington, D.C., 1974.
11. U.S. Department of Labor, Bureau of Labor Statistics, Announcements of Unemployment Statistics, January 1974.
12. *The Data Resources Review*, Vol. II, No. 10, Lexington, Mass.: Data Resources, Inc., December 1973.
13. A survey conducted by the Conference Board and detailed in the summary section of their report, *Energy Consumption* (Note 4),

suggests firms are now employing more man-hours to monitor and conserve energy in plant operation.

14. Appendix A, Energy Requirements for Scenarios.
15. The Commission on Population Growth and the American Future, *Population and the American Future*, Chapter 4, "The Economy," Washington, D.C., GPO No. 5258-0002, 1972.
16. U.S. Department of Commerce, Bureau of the Census, "Projections of the Population of the United States, by Age and Sex: 1972 to 2020," Series P25, No. 493, Washington, D. C.: U.S. Government Printing Office, 1972.
17. *Population and the American Future,* Chapter 4, (Note 15).
18. U.S. Department of Labor, "The United States Economy in 1985: An Overview of BLS Projections," *Monthly Labor Review*, December 1973.
19. Stephen Enke, "Population Growth and Economic Growth," *The Public Interest* 32, Sumrner 1973.
20. Enke, "Population Growth," (Note 19).

Chapter Seven–**U. S. energy policy in the world context**

1. Barabara Ward, "First, Second, Third and Fourth Worlds," *The Economist*, May 18, 1974.
2. Leonard Mosley, *Power Play: Oil in the Middle East*, New York: Random House, 1973.
3. M. A. Adelman, "Is the Oil Shortage Real? Oil Companies as OPEC Tax-Collectors," *Foreign Policy*, Winter 1972–73.
4. M. A. Adelman, *The World Petroleum Market*, published for Resources for the Future, Baltimore: Johns Hopkins Press, 1972.
5. *BP Statistical Review of the World Oil Industry, 1973*, London: The British Petroleum Co. Ltd., 1973.
6. *BP Statistical Review, 1973,* (Note 5).
7. "A Talk with the Shah of Iran," *Time*, April 1, 1974.
8. Federal Trade Commission, *The International Petroleum Cartel*, staff report to the F.T.C., submitted to the Senate Subcommittee on Monopoly, U.S. Senate Select Committee on Small Business, August 22, 1952.
9. Brookings Institution, "Energy and U.S. Foreign Policy," a draft report to the Energy Policy Project, June 1974.
10. *International Petroleum Encyclopedia*, Tulsa, Oklahoma: Petroleum Publishing Co., 1973.
11. Guy de Carmoy et al., *Cooperative Approaches to World Energy Problems,* Washington, D.C.: The Brookings Institution, 1974.
12. Ward, "Worlds," *The Economist*, May 18, 1974, (Note 1).
13. World Bank, "Additional External Capital Requirements of Developing Countries," unpublished report, March 1974.
14. Brookings, "Energy" (Note 9).
15. Brookings, "Energy" (Note 9).

16. Mason Willrich and Theodore B. Taylor, *Nuclear Theft: Risks and Safeguards*, a Report to the Energy Policy Project, Cambridge, Mass.: Ballinger, 1974.
17. The President's Materials Policy Commission (William S. Paley, Chairman), *Resources for Freedom*, a Report to the President, Washington, D. C.: U. S. Government Printing Office, 1952.

Chapter Eight–**Energy and the environment**

1. U. S. Department of the Interior, Office of Research and Development, "Energy Research Program of the U. S. Department of the Interior," March 1974.
2. U. S. Department of the Interior, Office of the Secretary, "Final Environmental Statement for the Prototype Oil Shale Leasing Program," August 1973.
3. Appalachian Regional Commission, "Acid Mine Drainage in Appalachia," June 1969.
4. U. S. Department of the Interior, Bureau of Mines, "Environmental Effects of Underground Mining and Mineral Processing," unpublished.
5. "Industry's safety record improves; more changes on the way," *Coal Age* 79: 2, February 1974, p. 80.
6. U. S. Department of the Interior, Mining Enforcement and Safety Administration, personal communication.
7. *Statistical Abstract of the United States*, U. S. Department of Commerce, 1972.
8. "Vigilance and New Equipment Help British Coal Mines Maintain Low Accident and Fatality Rates," *Coal Age* 78: 8, July 1973, p. 106.
9. National Science Foundation, Science and Technology Policy Office, *Chemicals and Health*, report of the Panel on Chemicals and Health of the President's Science Advisory Committee, September 1973.
10. Statement of the Conservation Foundation submitted by Malcolm Baldwin, in *Regulation of Surface Mining*, Hearings before the Subcommittee on Mines and Mining of the Committee on Interior and Insular Affairs, U.S. House of Representatives, Serial No. 92-26, 1971.
11. Statement of Dr. John Moore, Department of Economics, University of Tennessee, and Dr. Schmidt-Bleek, Director, Appalachian Resources Project, University of Tennessee, in *Regulation of Surface Mining*, Hearings before the Subcommittee on the Environment and Subcommittee on Mines and Mining of the Committee on Interior and Insular Affairs, U.S. House of Representatives, Serial No. 93-11, April 1973.
12. Illinois Institute for Environmental Quality, *Strip Mine Reclamation in Illinois*, December 1973.
13. E. A. Nephew, *Surface Mining and Land Reclamation in Germany*, Oak Ridge National Laboratory, ORNL-NSF-EP-16, May 1972.
14. National Academy of Sciences, *Rehabilitation Potential of Western Coal*

Lands, a Report to the Energy Policy Project, Cambridge, Mass.: Ballinger, 1974.

15. U.S. Department of the Interior, Bureau of Reclamation, "Montana-Wyoming Aqueduct Study," 1972.
16. D. E. Kash et al., *Energy Under the Ocean*, Norman: University of Oklahoma Press, 1973.
17. D. F. Boesch, C. H. Hershner and J. H. Milgrim, *Oil Spills and the Marine Environment*, a Report to the Energy Policy Project, Cambridge, Mass.: Ballinger, 1974.
18. S. J. Williamson, *Fundamentals of Air Pollution*, Reading, Mass.: Addison-Wesley, 1973.
19. U.S. Environmental Protection Agency, National Environmental Research Center, "The Economic Damages of Air Pollution," May 1974.
20. Council on Environmental Quality, *Environmental Quality,* the Fourth Annual Report of the Council on Environmental Quality, Washington, D.C.: September 1973.
21. Associated Universities, Inc., *Reference Energy Systems and Resource Data for Use in the Assessment of Energy Technologies*, submitted to the Office of Science and Technology, Executive Office of the President, Washington, D.C.: April 1972.
22. Department of Health, Education, and Welfare, National Air Pollution Control Administration, *Air Quality Criteria for Photochemical Oxidants*, AP-63, March 1970.
23. S. M. Horvath, "Effects of Carbon Monoxide on Human Behavior," published in *Proceedings of the Conference on Health Effects of Air Pollutants*, Assembly of Life Sciences, National Academy of Sciences—National Research Council, prepared for the Committee on Public Works, U.S. Senate, Serial No. 93-15, November 1973.
24. Medical College of Wisconsin, Department of Environmental Medicine, *"Normal" Carboxyhemoglobin Levels of Blood Donors in the United States*, Report No. ENVIR-MED-MCW-CRC-COHB 73-1, May 1973.
25. U.S. Environmental Protection Agency, Office of the Administrator, "Questions Submitted by Chairman Rodgers and EPA's Answers," in *Clean Air Act Oversight—1973*, Hearings before the Subcommittee on Public Health and Environment of the Committee on Interstate and Foreign Commerce, U.S. House of Representatives, Serial No. 93-62, September 1973, pp. 174–207.
26. Testimony of Hon. John A. Quarles, Acting Administrator, U. S. Environmental Protection Agency, in *Clean Air Act Oversight—1973*, Hearings before the Subcommittee on Public Health and Environment of the Committee on Interstate and Foreign Commerce, U.S. House of Representatives, Serial No. 93-62, September 1973.
27. National Academy of Sciences, *Report by the Committee on Motor Vehicle Emissions*, February 1973.
28. Arthur D. Little, Inc., *A Study of Technological Improvements in Automobile Fuel Consumption*, preliminary evaluation, October 1973.
29. R. Frank et al., "Sulfur Oxides and Particles; Effects on Pulmonary Physiology in Man and Animals," published in *Proceedings of the Conference on Health Effects of Air Pollutants*, Assembly of Life

Sciences—National Research Council, prepared for the Committee on Public Works, U.S. Senate, Serial No. 93-15, November 1973.

30. D. P. Rall, Director, National Institute of Environmental Health Sciences, National Institutes of Health, Department of Health, Education, and Welfare, "A Review of the Health Effects of Sulfur Oxides," October 1973.

31. J. Finklea et al., National Environmental Research Center, Office of Research and Development, U.S. Environmental Protection Agency, "Status Report on Sulfur Oxides," April 1974.

32. G. E. Likens et al., "Acid Rain," *Environment* 14: 2, March 1972, p. 33.

33. National Coal Association, *Coal News*, No. 4199, February 8, 1974.

34. U.S. Environmental Protection Agency, "National Public Hearings on Power Plant Compliance with Sulfur Oxide Air Pollution Regulations," January 1974.

35. S. K. Friedlander, "Small Particles in Air Pose a Big Control Problem," *Environmental Science and Technology*, Vol. 7, No. 12, December 1973, p. 1115.

36. J. R. Goldsmith, "Effects of Air Pollution on Human Health," published in *Air Pollution*, Vol. 1, edited by A. C. Stern, New York: Academic Press, 1968, p. 547.

37. J. F. Finklea, "Conceptual Basis for Establishing Standards," published in *Proceedings of the Conference on Health Effects of Air Pollutants*, Assembly of Life Sciences, National Academy of Sciences —National Research Council, prepared for the Committee on Public Works, U. S. Senate, Serial No. 93-15, November 1973.

38. D. Natusch et al., "Toxic Trace Elements: Preferential Concentration in Respirable Particles," *Science*, January 18, 1974, p. 202.

39. Vernon E. Swanson, U.S. Geological Survey, personal communication, 1974.

40. C. E. Capes et al., "Rejection of Trace Metals from Coal During Beneficiation by Agglomeration," *Environmental Science and Technology* 8:1, January 1974, p. 35.

41. A. P. Altshuller, "Atmospheric Sulfur Dioxide and Sulfate—Distribution of Concentration at Urban and Nonurban Sites in the United States," *Environmental Science and Technology* 7:8, August 1973.

42. S. Manabe, "Carbon Dioxide and Atmospheric Heating," in W. H. Matthews et al., eds., *Man's Impact on the Climate*, Cambridge, Mass.: MIT Press, 1971, p. 247.

43. *Inadvertant Climate Modification: Report of the Study of Man's Impact on Climate (SMIC)*, Cambridge, Mass.: MIT Press, 1971.

44. J. T. Peterson and C. E. Junge, "Sources of Particulate Matter in the Atmosphere," in W. H. Matthews et al., *Man's Impact on the Climate*, p. 310, (Note 42).

45. R. A. Bryson, "A Perspective on Climatic Change," *Science*, May 17, 1974, p. 753.

46. G. M. Brannon, *Energy Taxes and Subsidies*, a Report to the Energy Policy Project, Cambridge, Mass.: Ballinger, 1974.

47. "1974 Annual Statistical Report," *Electrical World*, March 15, 1974, p. 51.

48. T. H. Pigford et al., *Fuel Cycles for Electrical Power Generation. Phase I. Toward Comprehensive Standards: The Electric Power Case*, prepared for the Environmental Protection Agency by Teknekron, Inc., EPA No. 68-01-0561, January 1973.
49. United States Atomic Energy Commission, "The Safety of Nuclear Power Reactors and Related Facilities," WASH-1250, July 1973.
50. D. F. Paddleford, "Analysis of Public Safety Risks Associated with Low Probability Nuclear Power Plant Accidents," Westinghouse Electric Corporation, Nuclear Energy Systems, September 1973.
51. D. F. Ford and H. W. Kendall, "Catastrophic Nuclear Accidents" in *The Nuclear Fuel Cycle: A Survey of the Public Health, Environmental, and National Security Effects of Nuclear Fuels*, Union of Concerned Scientists, October 1973.
52. A. B. Lovins and W. C. Patterson, "The CEGB's Proposal for a Programme of Ordering Light Water Reactors," in *Appendices to the Minutes of Evidence: The Choice of a Reactor System*, Hearings before the Select Committee on Science and Technology (Energy Resources Subcommittee), British House of Commons, London, January 30, 1974.
53. Committee on Science and Technology, *Report on the Choice of a Reactor System*, British House of Commons, London, January 29, 1974.
54. Based on data in National Academy of Sciences/National Research Council, *The Effects on Populations of Exposures to Low Levels of Ionizing Radiation*, the Report to the Advisory Committee on the Biological Effects of Ionizing Radiation (BEIR Report), November 1972.
55. "Dixie Lee Ray Provides Some Details of Rasmussen Study on Nuclear Accidents," *Weekly Energy Report*, January 28, 1974.
56. Testimony of William Bryan (aerospace safety analyst) to the Subcommittee on State Energy Policy, Committee on Planning, Land Use and Energy, of the California State Assembly, February 1, 1974.
57. USAEC, "Everything You Always Wanted to Know About Shipping High Level Nuclear Wastes," WASH-1264, September 1973.
58. M. Ross, "The Possibility of Release of Cesium in a Spent Fuel Transportation Accident," University of Michigan Physics Department, January 1974.
59. Dudley Thompson et al., "Summary of Abnormal Occurrences Reported to the Atomic Energy Commission During 1973," report of the Office of Operations Evaluation, USAEC, May, 1974.
60. "Eight New QA Initiatives," *Nuclear Industry*, May 1974, p. 10.
61. A. M. Weinberg, "Social Institutions and Nuclear Energy," *Science*, July 7, 1972, p. 27.
62. H. Alfven, "Energy and Environment," *Bulletin of the Atomic Scientists*, May 1972, p. 5.
63. J. O. Bloemecke et al., "Managing Radioactive Wastes," *Physics Today*, August 1973, p. 36.
64. E. J. Zeller et al., "Putting Radioactive Wastes on Ice," *Bulletin of the Atomic Scientists*, January 1973, p. 4.
65. National Aeronautics and Space Administration, "Feasibility of

Space Disposal of Radioactive Nuclear Waste," Cleveland: Lewis Research Center, December 1973.

66. J. L. Cohen et al., "Use of a Deep Nuclear Chimney for the In-Situ Incorporation of Nuclear Fuel Reprocessing Waste in Molten Silicate Rock," UCLRL-51044, Lawrence Livermore Laboratory, USAEC, May 4, 1974.
67. A. S. Kubo and D. J. Rose, "Nuclear Waste Disposal," *Science*, December 21, 1973, p. 1205.
68. C. T. Hollocher, "Storage and Disposal of High Level Wastes," in *The Nuclear Fuel Cycle* (Note 51).
69. "Pittman Reports on Commercial Waste Management Status," *Nuclear Industry*, August 1973, p. 20.
70. Mason Willrich and T. B. Taylor, *Nuclear Theft: Risks and Safeguards*, a Report to the Energy Policy Project, Cambridge, Mass.: Ballinger, 1974.
71. D. P. Geesaman, "An Analysis of the Carcinogenic Risk from an Insoluble Alpha-Emitting Aerosol Deposited in Deep Respiratory Tissue," UCRL-50387, Feb. 1968; and UCRL-50387 Addendum, Oct. 1968, Lawrence Radiation Laboratory, Livermore, Cal.
72. A. R. Tamplin and T. B. Cochran, "A Report on the Inadequacy of Existing Radiation Protection Standards Related to Internal Exposure of Man to Insoluble Particles of Plutonium and Other Alpha-Emitting Particles," Natural Resources Defense Council, Washington, D.C., February 14, 1974.
73. E.A. Martell, "Comments on the Plutonium Toxicity Sections of the AEC Draft Environmental Impact Statement, LMFBR Program WASH-1535," Washington D.C., April 19, 1974.
74. T. H. Pigford, "Radioactivity in Plutonium, Americium, and Curium in Nuclear Reactor Fuel," a draft report to the Energy Policy Project, June 1974.
75. Center for Short-Lived Phenomena, "Miamisburg Plutonium Leak," Event Notification Report, Smithsonian Institution, May 15, 1974.
76. L. D. NeNike, "Radioactive Malevolence," *Science and Public Affairs*, February 1974, p. 16.
77. USAEC, "Draft Environmental Impact Statement, LMFBR," WASH-1535, Washington, D.C., March 1974.
78. T. B. Cochran, *The Liquid Metal Fast Breeder Reactor: An Environmental and Economic Critique*, Resources for the Future, Washington, D.C., 1974.
79. USAEC, "Potential Nuclear Power Growth Patterns," WASH-1098, Washington, D.C., December 1970.
80. J. P. Holdren, "Uranium Availability and the Breeder Decision," Environmental Quality Laboratory Memorandum No. 8, California Institute of Technology, January 1974.
81. USAEC, "The Special Safeguards Study," prepared for the Director of Licensing; released by Sen. Ribicoff, April 26, 1974.
82. ACRS letter of September 10, 1973, to Dixie Lee Ray, Chairperson of the Atomic Energy Commission.
83. Alvin Weinberg, "How Can Man Live With Fission?" presented at a Woodrow Wilson International Center for Scholars Conference, Washington, D.C., June 18, 1973.

84. H. P. Green, "Nuclear Power: Risk, Liability, and Indemnity," *Michigan Law Review*, January 1973, p. 503.
85. Joint Committee on Atomic Energy, *Selected Materials on Atomic Energy Indemnity and Insurance Legislation*, Congress of the United States, Washington, D.C., March 1974.
86. Directorate of Licensing, USAEC, "Draft Environmental Statement Related to the Proposed Legislation to Amend the Price Anderson Act," Washington, D.C., May 1974.
87. USAEC, "The Nuclear Industry 1973," WASH-1174-73, Washington, D.C., 1974.
88. T. S. Eliot, "The Wasteland," *Collected Poems*, New York: Harcourt Brace Jovanovich, Inc., 1963.
89. Letter from James Madison to W. T. Barry, August 4, 1882.

Chapter Nine–Private enterprise and the public interest

1. "United States of America Before Federal Trade Commission," (Hearing record), Docket #8934.
2. U.S. Department of the Treasury, Staff Analysis of the Office of the Energy Advisor on the Federal Trade Comission's July 2, 1973 "Preliminary Federal Trade Commission Staff Report on Its Investigation of the Petroleum Industry," unpublished paper, August 27, 1973.
3. The basic document is "Competition in the Energy Industry," a draft report to the Energy Policy Project by Thomas D. Duchesneau, which is based upon the research of the author but also incorporates the findings of the following additional draft studies commissioned by the Project:

   Thomas F. Hogarty, "The Geographic Scope of Energy Markets: Oil, Gas, and Coal," 1974.

   John H. Lichtblau, "The Outlook for Independent Domestic Refiners to the Early 1980s," 1974.

   Edwin Mansfield, "Firm Size and Technological Change in the Petroleum and Bituminous Coal Industries," 1973.

   Thomas Gale Moore, "Economies of Scale and Firms Engaged in Oil and Coal Production," 1973.

   Thomas Gale Moore, "Potential Competition in Uranium Enriching," 1973.

   Reed Moyer, "Price-Output Behavior in the Coal Industry," 1973.

   Lester M. Salamon and John J. Siegfried, "The Relationship Between Economic Structure and Political Power: the Energy Industry," 1973.
4. Arthur D. Little, Inc., "Competition in the Nuclear Power Supply Industry," a report to the U.S. Atomic Energy Commission, and the U.S. Department of Justice, December 1968.
5. "Investigation of the Petroleum Industry," Permanent Subcommittee on Investigations of the Committee on Government Opera-

tions, United States Senate, 93rd Congress, 1st Session, July 12, 1973, p. 34.

6. Walter J. Mead, "Competition in the Energy Industry," Energy Policy Project Staff Paper (unpublished).
7. Marine Engineers Beneficial Association, *The American Oil Industry: a Failure of Anti-Trust Policy*, New York, 1973.
8. Mead, "Competition in the Energy Industry."
9. Much of the ensuing analysis of the relationship between political and economic power is taken from Salamon and Siegfried, "The Relationship Between Economic Structure and Political Power," (Note 3).
10. Corwin D. Edwards, "Conglomerate Bigness as a Souce of Power," *Business Concentration and Price Policy*, Princeton: Princeton University Press, 1955, pp. 331–360.
11. *Census of Business*, Vol. 1, 1967.
12. *Fortune Magazine*, May 1974.
13. *Congressional Quarterly*, Vol. XXXI, January 6, 1973, pp. 24–25.
14. cf. Council on Environmental Quality, *Coal Surface Mining and Reclamation*, prepared at the request of Henry M. Jackson, Chairman, Committee on Interior and Insular Affairs, U. S. Senate, 93rd Congress, 1st Session, Washington, D. C.: U.S. Government Printing Office, March 1973, p. 10.
15. W. J. (Jack) Crawford, tax administrator of Humble Oil, quoted in Ronnie Drugger, "Oil and Politics," *Atlantic Monthly,* September 1969.
16. "Investigation of the Petroleum Industry," (Note 5).
17. This section draws heavily on the following two books:

    Gerard M. Brannon, *Energy Taxes and Subsidies,* a Report to the Energy Policy Project, Cambridge, Mass.: Ballinger, 1974; and Gerard M. Brannon, Ed., *Studies in Energy Tax Policy*, a Report to the Energy Policy Project, Cambridge, Mass.: Ballinger, 1974.
18. "Investigation of the Petroleum Industry," (Note 5).
19. Ralph Nader, "The Case for Federal Chartering," in Ralph Nader and Mark J. Green, eds., *Corporate Power in America*, New York: Grossman, 1973, p. 79.

Chapter Ten–**Reforming electric utility regulation**

1. Foster Associates, Inc., *Energy Prices 1960–73*, a Report to the Energy Policy Project, Cambridge, Mass.: Ballinger, 1974.
2. Foster Associates, *Energy Prices,* (Note 1).
3. Foster Associates, *Energy Prices,* (Note 1).
4. Edward Berlin, Charles Cicchetti and William J. Gillen, *Perspective on Power*, a Report to the Energy Policy Project, Cambridge, Mass.: Ballinger, 1974.
5. Federal Power Commission, *National Power Survey*, Washington, D. C.: Government Printing Office, October 1964, p. 196.
6. Berlin et al., *Perspective on Power*, (Note 4).

7. Berlin et al., *Perspective on Power,* (Note 4).
8. David A. Andelman, "Con Edison's Money Problems Are Serious And May Get Worse," *New York Times*, March 31, 1974.

Chapter Eleven–Federal energy resources: protecting the public trust

General: Much of the material in this chapter is based on extensive working conferences with U.S. Interior Department officials, other government officials, and unpublished Interior Department documents. These are not specifically cited in the text.

1. Public Land Law Review Commission, *One-Third of the Nation's Land*, Washington, D. C.: U.S. Government Printing Office, 1970.
2. U.S. Bureau of Land Managment, *Public Land Statistics 1972*, Washington, D. C.: U.S. Government Printing Office, 1973.
3. National Academy of Sciences, *Rehabilitation Potential of Western Coal Lands*, a Report to the Energy Policy Project, Cambridge, Mass.: Ballinger, 1974.
4. U.S. Bureau of Reclamation, "Appraisal Report on Montana-Wyoming Aqueducts," April 1972.
5. Montana Coal Task Force, "Coal Development in Eastern Montana," January 1973.
6. Basic resource laws discussed in the text include the Mineral Leasing Act of 1920, 30 U.S.C. 181 et seq.; the Outer Continental Shelf Lands Act of 1953, 43 U.S.C. 1331 et seq.; the Mineral Location Law of 1872, 30 U.S.C. 21 et seq.; and the National Environmental Policy Act of 1969, 42 U.S.C. 4321 et seq. These will not be referenced further.
7. Natural Resources Defense Council, "Energy Decision Making in the Interior Department," unpublished report to the Energy Policy Project, May 1973.
8. U.S. Department of Interior, "Interior Department Administration of the OCS," unpublished paper, 1971.
9. Walter Depree, Jr. and James A. West, "United States Energy Through the Year 2000," U.S. Department of Interior, December 1972.
10. *Natural Resources Defense Council, et al v. Morton.* (1971); see, e.g., U.S. Department of Interior, "Environmental Impact Statement for Proposed East Texas OCS Oil and Gas Lease Sale," Washington, D. C.: Government Printing Office, 1973.
11. National Petroleum Council, "U.S. Energy Outlook," Washington, D. C.: Government Printing Office, December 1972.
12. Based on reserve and revenue figures in "U.S. Department of Interior Draft Environmental Impact Statement for the Proposed Coal Leasing Program," May 1974.
13. Public Land Law Review Commission, "OCS Study," Washington, D. C., 1970.
14. Comptroller General of the United States, "Improvements Needed in Administration of Federal Coal Leasing Program," U.S. General Accounting Office, Washington, D. C., March 1972.

15. Public Land Law Review Commission, "OCS Study," (Note 13).
16. Don E. Kash et al., *Energy Under the Oceans*, Norman: University of Oklahoma Press, 1973; National Academy of Engineering, "Outer Continental Shelf Resource Development Safety," Washington, D. C., December 1972.
17. Donald F. Boesch, et al., *Oil Spills and the Marine Environment*, a Report to the Energy Policy Project, Cambridge, Mass.: Ballinger, 1974.
18. N.A.S., *Rehabilitation Potential,* (Note 3).
19. Comptroller General of the United States, "Administration of Regulations for Surface Exploration, Mining, and Reclamation of Public and Indian Coal Lands," Washington, D. C., August 1972.

Chapter Twelve–**Energy research and development**

1. Herbert Holloman et al, "Energy R&D Policy Proposals," a draft report to the Energy Policy Project, July 1974.
2. U.S. Atomic Energy Commission, Oak Ridge National Laboratory, *Inventory of Current Energy Research and Development*, Vol. 1, Prepared for the Subcommittee on Energy of the Committee on Science and Astronautics, U.S. House of Representatives, 93rd Congress, 1st Session, January 1974, Washington, D. C.: U.S. Government Printing Office, 1974.
3. "An Assessment of Solar Energy as a National Resource," NSF/NASA Solar Energy Panel, University of Maryland, 1972.
4. L. John Fry, "Methane Digestors for Fuel Gas and Fertilizer," New Alchemy Institute, *Newsletter* No. 3, Woods Hole, Mass., 1973.
5. M. A. Adelman et al., "Energy Self Sufficiency: An Economic Evaluation," *Technology Review*, 76: 6, May 1974, pp. 44–47 and 56–58.
6. U.S. Department of Interior, *Final Environmental Statement for the Prototype Oil Shale Leasing Program*, Washington, D. C.: U.S. Government Printing Office, 1973.
7. "Energy from Oil Shale," report to the House Committee on Science and Astronautics, Washington, D. C.: U.S. Government Printing Office, November 1973.
8. *Oil Shale Utilization, Progress and Prospects*, United Nations Publication 67.II.B.20, UN, New York, 1967.
9. Michel Grenon, "Energy R&D in the Industrialized Countries," a draft report to the Energy Policy Project, 1974.
10. "International Economic Report of the President," Washington, D. C.: U.S. Government Printing Office, February 1974.
11. Thomas B. Cochran, *The Liquid Metal Fast Breeder Reactor*, Resources for the Future, Inc., Baltimore: Johns Hopkins University Press, 1974.
12. Dixie Lee Ray, "The Nation's Energy Future," Washington, D. C.: U.S. Government Printing Office, 1973.
13. W. Seifert et al., eds., *Energy and Development—A Case Study*, MIT Report No. 25, Cambridge, Mass.: MIT Press, 1973.

14. Jerome Weingart, "Community Energy Systems," a draft report to the Energy Policy Project, 1974.

15. Clarence Zener, "Solar Sea Power," *Physics Today*, January 1973, pp. 48–53; National Science Foundation, "Ocean Thermal Energy Conversion," program solicitation, 1974.

16. *Energy Information,* Robert Morey Associates, Dana Point, Cal., Vol. 9, No. 12, June 1974, p. 2.

17. G. E. Branvold, "Solar Total-Energy Community Project," Sandia Laboratories Energy Report, Albuquerque, New Mexico, March 1974.

18. G. A. Samara and D. G. Schueler, "Solar Energy: Sandia's Photovoltaic Research Program," Sandia Laboratories Energy Report, Albuquerque, New Mexico, May 1974.

19. Alan Poole, "Potential Energy Recovery from Organic Wastes," a draft report to the Energy Policy Project, 1974.

20. Paul Kruger and Carel Otte, eds., *Geothermal Energy,* Stanford University Press, Stanford, Cal., 1973.

21. "Report of the Cornell Workshops on the Major Issues of a National Energy Research and Development Program," Cornell University, December 1973.

22. S. Schurr and B. Netschert, *Energy in the American Economy 1850–1975*, Baltimore: Johns Hopkins Press, 1960.

23. Robert Morgan, "Sea Horse," *Alternative Sources of Energy*, 10, March 1973.

24. R. F. Post, S. F. Post, "Flywheels," *Scientific American*, 229: 6, December 1973.

25. E. P. Gyftopoulos, Lazaros J. Lazarides and Thomas F. Widmer, *Potential Fuel Effectiveness in Industry*, a Report to the Energy Policy Project, Cambridge, Mass.: Ballinger, 1974.

Appendixes

Contents

Appendixes

Appendix A–Energy requirements for scenarios

RESIDENTIAL ENERGY

We use the same population[1] and housing projections[2] for all scenarios. Total population and housing projections are shown below.

Year	*Population (millions)*	*Occupied housing units (millions)*
1970	205	63
1985	236	80
2000	265	99

[1] U.S. Census Bureau's Series E projections, as published in late 1972.

[2] From Vary T. Coates "A Workbook on Alternative Future Lifestyles Related to Energy Demand," unpublished report to the Energy Policy Project, August 1973.

Space conditioning

The largest end use for energy in the residential sector is space conditioning (heating and cooling), which offers the greatest potential for energy conservation with presently available technology. We shall explain in detail here the space conditioning calculations; the bases for the calculations for other household energy uses are explained in the notes to each table.

Heat losses (in the winter), heat gains (in the summer), heating system efficiency, and air conditioner efficiency are the parameters that determine household space conditioning requirements. Heat losses and gains are determined by house size, location (weather), orientation, building type, construction materials, and construction quality. Heating (or cooling) system efficiency depends on the device(s) used, ventilation system design, location of the heater (air conditioner), system maintenance, and the weather. Fuel or electricity for the winter (or summer) requirements are determined by the formula:

$$\text{Annual fuel or electricity for heating (cooling)} = \frac{\text{annual heat loss (gain)}}{\text{average annual heating (cooling) system efficiency}}$$

Heating and cooling energy requirements are similarly calculated. We discuss only heating, but the discussion is equally valid for cooling (except of course that heat gain and cooling system efficiency determine the energy requirements rather than heat loss and heating system efficiency).

From the above equation, we see that there are two complementary ways to reduce heating energy requirements: (1) decrease heat loss and (2) increase heating system efficiency. Heat loss can be reduced by insulating ceiling, walls and floors, using storm windows and doors, weatherstripping doors and windows, by a tight fit between the house and window frame and proper shading and orientation of windows. Efficient heating devices reduce the amount of fuel required to provide a given amount of heat to the house. Good design of the flue and the ventilation system both reduces heat loss and increases the average efficiency of fossil fuel heating systems by cutting down on the number of times the unit starts up and shuts off.

There are no national, statistically significant field measurments on heating system efficiencies, and this should be an area for further detailed investigation. However, a few field tests and computer simulations yield results that indicate heating system efficiency (the heat deposited in the house divided by the heat value of the fuel) of about 60 percent for gas fired burners, with an air circulation system. Here we assume the 60 percent heating system efficiency for oil and gas fired systems (air circulation and steam).

The situation with regard to heat losses is even more complex. Information gathered from the building industry indicates that houses built before 1965 were, in general, poorly insulated—typically 1½" of insulation in the ceiling, none in the walls, none in the floors, and plain windows. About 50 percent of single-family homes are backfitted with storm windows.[3] In the 1965–1972 period, increasing amounts of insula-

---

[3] Dorothy K. Newman and Dawn Day Wachtel, Washington Center for Metropolitan Studies, "Energy in People's Lives," a draft report to the Energy Policy Project, April 1974.

tion have been installed. The present practice appears to be 2″ to 3″ in the walls, and 4″ to 6″ in the ceilings, with the higher figures generally applying to electrically heated homes. The windows are usually plain, though sometimes they are fitted with storm windows by the owner. The tightness of fit of doors and windows has an important bearing on heat losses—in a well-insulated house, it is the dominant mode of heat loss. The impression among some building contractors and people in the insulation manufacturing and installing industry is that construction practices in this regard have probably deteriorated in the past decade. Thus, the effect of better insulation may have been partially offset by leaky doors and windows. With the help of the above data, and existing computer studies on heat losses in homes, we arrive at an estimate for the average annual heat loss per housing unit of 14,000 to 15,000 Btu's per degree day.[4]

The average winter climate (weighted by population) is about 5,000 degree days, giving an annual heat loss of 70 to 75 million Btu's per housing unit per year. Thus, for an average housing unit with oil or gas heat, the annual fuel consumption would be about 125 million Btu's per year; this would increase to about 140 million Btu's per year when 10 percent losses in fuel processing are taken into account. For a house with electric resistance heat, the electricity requirements would be 70 to 75 million Btu's; this would increase to about 250 million Btu's of fuel when generating, transmission and distribution losses are taken into account. An electric heat pump would provide the same heating, using between 120 million and 170 million Btu's per year, depending on the design of the heat pump.

Heat losses and gains also depend on the kind of structure that houses the residence—single family detached, multifamily low rise, multifamily high rise, and mobile homes. In general, for a given housing unit size, insulation level, and climate, the losses from single family detached houses are considerably larger than those from multifamily dwellings.

The present stock of housing (1970 Census data) is predominantly made up of single family homes. Current trends (shown below) are toward various forms of multi-family housing and increased use of prefabricated or mobile homes.

Percent distribution of housing units, by type

	1970	*1985*	*2000*
Single family, detached	66	55	50
2 to 4 unit structures	16	17	15
5 or more unit structures[a]	15	22	26
Mobile homes	3	6	9

[a] Includes garden-type apartments and high rises.

Mobile homes are usually not moved once fixed on a site. They have floor, wall and ceiling insulation, and are generally smaller than single family homes (600 to 700 square feet compared to 1200 to 1400 square feet). Thus, the increasing trend toward multifamily structures

[4] A degree day is the equivalent of a 1° F temperature difference between 65° F and the outside of the house for 24 hours.

and mobile homes will tend to decrease energy requirements for space heating and cooling.

A new westward migration from the North-East and North-Central regions will further lessen space heating requirements. This latter effect is small, however (2 to 3 percent). The Series E Census projection shows the share of the population in the North-East and North-Central regions declining from 52.5 percent in 1970 to 46.4 percent in 2000, while the West's share rises from 17.7 percent in 1970 to 22.8 percent in 2000. The South's share stays at 30.8 percent.

Since multifamily dwellings (especially high rise buildings) have a smaller number of exposed surfaces, smaller window area, and often a smaller floor area, the heat losses and gains per dwelling are considerably lower than for detached single family dwellings with similar insulation. While the net effect cannot be calculated with precision, the combined effect of westward migration and a larger proportion of multifamily dwellings would be to reduce energy requirements between 5 and 10 percent by 1985, and 10 to 15 percent by 2000. The heating energy requirements for the *Historical Growth* scenario will be further reduced, somewhat, if the current trends toward higher insulation in the ceiling and walls (but plain windows) persist. Part of the decrease might be offset by an increase in the area of housing units; we have allowed a 10 percent increase in the average area of a housing unit.

The heat losses and heating system efficiency for new construction assumed for the calculations are shown below. In all of our calculations for space heating, requirements for the *Technical Fix* and *ZEG* scenarios are taken to be the same.

	Pre-1975 housing units		*1975-1985*		*1985-2000*	
	HG[a]	*TF and ZEG*	*HG*	*TF and ZEG*	*HG*	*TF and ZEG*
Heat losses per year (million Btu's/yr)	75	75	60	50	65[b]	55[b]
Average fossil fuel furnace efficiency	0.6	0.6	0.6	0.65	0.6	0.7
Heat pump coefficient of performance	—	1.5-2	—	2.0	—	2.2

[a]*Historical Growth* and *Technical Fix* will be abbreviated *HG* and *TF* throughout this appendix, where it is convenient.

[b]Due to the assumed 10 percent increase in area per housing unit between 1985 and 2000.

The various types of heating systems assumed for the *HG* and *TF* scenarios are shown below.

Electric heating saturation is assumed to level off at about 35 percent in the residential and commercial sectors, since a higher saturation would lead to a serious imbalance in the summer and winter space conditioning loads, especially in the *Historical Growth* scenario.

The rate of retirement of existing housing plus the new demand for housing determine the rate at which opportunities for energy conservation via better building and thermal insulation can be achieved. The table on page 435 shows the new (post-1975) and old housing in 1985 and 2000. These assumptions are common to all scenarios.

Number of heating systems for the HG and TF scenarios
(millions)

	1970	*1975*		*1985*		*2000*	
		HG	*TF and ZEG*	*HG*	*TF and ZEG*	*HG*	*TF and ZEG*
Electric resistance	4.4	7.5	7.5	20.0	5.0	35.0	—
Heat pump	0.5	0.5	0.5	—	7.0	—	35.0
Gas[a] and oil[b]	55.3	58.0	58.0	60.0	68.0	65.0	55.0
Coal, wood, other	2.8	2.0	2.0	—	—	—	—
Solar	—	—	—	—	small	—	10.0
None	0.4	—	—	—	—	—	—
Total	63.4	68.0	68.0	80.0	80.0	100	100

[a] Includes bottled gas and liquid petroleum gases.
[b] Includes kerosene.

Age of housing units
(millions)

	1975	*1985*	*2000*
1975 and pre-1975 housing	68	60	40
Post-1975 housing	—	20	60
Total	68	80	100

The retirement rate of housing has been taken as 1 percent of the standing stock of housing retired per year. This is based on 1960–1970 rates, derived from U.S. Census data. The number of persons per occupied housing unit is assumed to decline according to Census E projections from about 3.2 in 1970 to about 2.7 in 2000.

Appliance saturation levels

The saturation levels of various appliances are assumed to be the same in the *HG* and *TF* scenarios through 2000. We have presented the *ZEG* scenario as one in which the emphasis on gadgets is lower than in the *HG* and *TF* scenarios and, therefore, we have assumed lower energy use in the *ZEG* case for appliances as yet unknown. In *ZEG*, the appliance saturation levels for all presently known appliances are the same as the other scenarios through 1985. After 1985, we assume for *ZEG* that the miscellaneous category of appliances will be saturated, and will increase only in proportion to the number of households. Table A-1 shows the levels of saturation of appliances in 1985 and 2000 that form the basis of the residential sector energy calculations. Table A-2 shows the annual energy consumption for these appliances for *HG* and *TF* scenarios for the year 2000.

Residential energy requirements are summarized in Table A-3.

ZEG planned communities and residential energy savings

Variation in the number of planned communities (new towns and reconstructed center cities) is not expected to make a substantial differ-

Table A-1–**Appliance saturation levels**
(in percent unless otherwise specified)

Appliance	1970[a]	1985 HG	1985 TF	1985 ZEG	2000 HG	2000 TF	2000 ZEG
1. Space heat	100	100	100	100	100	100	100
2. Air conditioning (total)	35	100[b]	100[b]	100[b]	100	100	100
(a) Room	25	50	50	50	0	0	0
(b) Central	10	50	50	50	100	100	100
3. Water heat	100	100	100	100	100	100	100
4. Refrigerators (total)	100	100	100	100	100	100	100
(a) Regular		0	0	0	0	0	0
(b) Frost free		100	100	100	100	100	100
5. Lighting	100	100	100	100	100	100	100
6. Cooking ranges (total)	100	100	100	100	100	100	100
(a) Fossil fuel	50	40	40	40	30	30	30
(b) Electric	50	60	60	60	70	70	70
7. Dishwashers	25	100	100	100	100	100	100
8. Clothes dryer[c]	40	50	50	50	60	60	60
9. Clothes washers	60	80	80	80	100	100	100
10. Freezers	30	60	60	60	100	100	100
11. Portable appliances[d] (relative units)	1	2	2	2	3	3	2
12. Unknown appliances (relative units)	0	1	1	0.5	2	2	1

[a] Saturation numbers for 1970 rounded to the nearest 5 percent.

[b] Half the households in 1985 have room air conditioners, and the other half central air conditioning. Of the 40 million households with room air conditioners, 20 million are assumed to have one, and 20 million have two. Approximate average size of the room air conditioner is 10,000 Btu's per hr cooling capacity. In 2000, all households are assumed to be centrally air conditioned.

[c] Saturation is not 100 percent due to increased trend to multiple unit housing structures with common drying facilities.

[d] Portable appliances consist of things like TV, vacuum cleaners, electric irons, toasters, electric shavers, etc. The saturation of these appliances varies a great deal from very low to near 100 percent. Overall electricity use for portable appliances grew at about 7 percent per year in the 1960s and we extrapolated this trend. Since many low saturation appliances came into widespread use in the sixties, the unknown and low saturation category is to some extent included in the portable appliances category. The relative units in which saturation is measured reflect portable appliance electricity consumption.

ence in residential energy use among scenarios between 1975 and 2000. Savings, if any, would come from the transportation sector. The potential for energy savings in the 1975–2000 period is limited because the additional number of planned communities that can be built is probably not great. Whatever savings can be made in the residential sector will come from greater efficiencies in space heating and cooling, which have been assumed for the *Technical Fix* and *ZEG* scenarios, whether or not new building takes place in planned communities.

The rate of developing new communities in the *HG* and *TF* cases is

Table A-2–**Annual energy consumption for electrical appliances (2000)**

Appliance	*kwh(e)/yr.*[b] *HG*	*TF*
1. Room air conditioners	1500	1050
2. Central air conditioning	3000	2000
3. Water heater	5000	5000
4. Refrigerator (frost free)	1800	1200
5. Lighting	1200	900
6. Cooking range	1300	1300
7. Freezer	1000	800
8. Dishwasher[a]	250	250
9. Clothes washer[a]	100	100
10. Clothes dryer	900	900

[a] Hot water energy for dishwashers and clothes washers is included in water heating energy requirements.

[b] kwh(e) per year is rounded to the nearest 50 kwh(e) per year.

about eight new communities per year of 50,000 people each.[5] This gives a total population of 10 million in new communities by the year 2000. If the rate of construction of new communities were doubled, and an additional saving in space conditioning of 20 percent per housing unit were assumed due to more multifamily housing in the new communities, this yields a saving of 0.2 quadrillion Btu's per year by the year 2000. Some of the saving would perhaps be offset by the energy required to build the infrastructure (roads, sewer systems, etc.) for the new communities. On the other hand, fuel savings would result from the use of solar rooftop and integrated utility systems (total energy systems) in these planned communities. But even these additional savings are not likely to exceed 1 to 2 quadrillion Btu's per year.

In the long run (beyond 2000), a gradual program of sound energy planning built into land use planning and development of housing and commerce could yield much larger benefits. These benefits would accrue over the long period—50 to 100 years—that it takes to turn over most of the nation's stock of housing.

ENERGY IN THE COMMERCIAL SECTOR

The commerical sector calculations are based on the present patterns of energy use in this sector.[6] The energy requirements for various end uses are enlarged in accord with increases in the commercial sector area. This area is assumed to be proportional to the number of service sector employees.

Tables A-4, A-5, and A-6 show the basis for energy calculations in

[5] V.T. Coates, "A Workbook on Future Lifestyles," op. cit. (Note 2).

[6] The 1970 energy requirements for various end uses are based on data presented in the report "Patterns of Energy Consumption in the United States," prepared for the office of Science and Technology by Stanford Research Institute, January 1972.

Table A-3–**Residential sector energy requirements** (quadrillion Btu's per year)

	1985									2000								
	HG			*TF*			*ZEG*			*HG*			*TF*			*ZEG*		
Item	*Fuel*	*Elec.*	*Total*	*Fuel*	*Elec.*	*Total*	*Fuel*	*Elec.*	*Total*	*Fuel*	*Elec.*	*Total*	*Fuel*	*Elec.*	*Total*	*Fuel*	*Elec.*	*Total*
Space heat	7.2	1.2	10.7	6.5	0.5	8.0	6.5	0.5	8.0	7.4	2.2	13.3	4.4	0.7	6.3	4.4	0.7	6.3
Air conditioning	—	0.7	2.0	—	0.4	1.2	—	0.4	1.2	—	0.9	2.4	—	0.4	1.1	—	0.4	1.1
Water heat	1.5	0.5	3.0	1.8	0.2	2.4	1.8	0.2	2.4	1.4	0.9	3.8	2.0	0.1	2.3	2.0	0.1	2.3
Refrigerators	—	0.5	1.5	—	0.3	0.9	—	0.3	0.9	—	0.6	1.6	—	0.4	1.1	—	0.4	1.1
Lighting	—	0.3	0.9	—	0.3	0.9	—	0.3	0.9	—	0.4	1.1	—	0.3	0.8	—	0.3	0.8
Miscellaneous appliances	0.2	1.0	3.1	0.2	1.0	3.1	0.2	1.0	3.1	0.2	1.6	4.5	0.3	1.5	4.3	0.3	1.2	3.5
Subtotal	8.9	4.2	21.2	8.5	2.7	16.5	8.5	2.7	16.5	9.0	6.6	26.7	6.7	3.4	15.9	6.7	3.1	15.1
Presently unknown appliances and low saturation uses	0.8	0.3	1.7	0.8	0.3	1.7	0.4	0.2	1.0	1.0	0.9	3.4	1.0	0.9	3.4	0.5	0.5	1.9
Total	9.7	4.5	22.9	9.3	3.0	18.2	8.9	2.9	17.5	10.0	7.5	30.1	7.7	4.3	19.3	7.2	3.6	17.0

Notes: Electricity columns denote electricity at 3413 Btu's per kwh(e).
Total columns include electricity generation and transmission losses of 1.9 Btu's per Btu(e) in 1985 and 1.7 Btu's per Btu(e) in 2000. Other fuel processing losses are not included.

Table A-4.–**Commercial area projections**

	1970	1985	2000: HG and TF	2000: ZEG
Service sector employment (millions)	51	70	87	90+
Commercial area, relative units (1970=1)	1	1.35	1.8	2.0

Notes: Commercial area is assumed to be proportional to service sector employment between 1970 and 1985. In the 1985–2000 period, a 10 percent increase in per employee area is incorporated in the area projection for the *HG* and *TF* scenarios.

Since the service sector is assumed to expand faster in *ZEG*, a 10 percent increase in commercial area over *HG* has been assumed for *ZEG* for the year 2000.

the commercial sector. Commercial energy requirements are summarized in Table A-7.

Table A-5–**Basis for commercial sector energy calculations**

	Energy use per unit area (1970 area = 1), quadrillion Btu's						
		HG		TF		ZEG	
Item	*1970*	*1985*	*2000*	*1985*	*2000*	*1985*	*2000*
1. Space heat	4.5	5.6	6.4	4.5	3.6	4.5	3.6
2. Air conditioning	1.6	1.6	1.3	1.3	0.9	1.3	0.9
3. Lighting	0.9	0.9	0.9	0.7	0.7	0.7	0.7
4. Road oil and asphalt	1.0	1.0	1.0	1.0	1.0	0.9[a]	0.9[a]
5. Miscellaneous fossil fuels	0.8	0.8	0.8	0.8	0.8	0.8	0.8
6. Miscellaneous electricity	1.3	1.3	1.3	1.3	1.3	1.3	1.3

[a] Lower requirements due to the use of fewer (and smaller) cars.

Notes: Electricity production and transmission efficiency: 1970 = 29 percent, 1985 = 32 percent, 2000 = 35 percent. Losses are included.

Fossil fuel space heating efficiency assumed 70 percent for all cases.

Heat pump COP = 1.75.

For *TF* and *ZEG*, the space heating requirements on new construction are assumed to be reduced by 30 percent due to better building design and more insulation.

Source for 1970 data: Stanford Research Institute, "Patterns of Energy Consumption in the United States."

Table A-6–**Energy systems in the commercial sector, relative area (1970 area = 1.0)**

Item	1975	1985 HG	1985 TF	1985 ZEG	2000 HG	2000 TF	2000 ZEG
1. Total area	1.1	1.35	1.35	1.35	1.8	1.8	2.0
2. Pre 1975 area	1.1	1.0	1.0	1.0	0.8	0.8	0.8
3. 1975–1985 area	—	0.35	0.35	0.35	0.35	0.35	0.35
4. 1985–2000 area	—	—	—	—	0.65	0.65	0.85
5. Electric resistance heat area	small	0.2	—	—	0.6	—	—
6. Heat pumps	—	—	0.2	0.2	—	0.6	0.65
7. Total energy systems	small	small	0.1	0.1	small	0.25	0.3
8. Fossil fuel heating							
pre-1975 area	1.1	1.0	1.0	1.0	0.8	0.8	0.8
post-1975 area	—	0.15	0.05	0.05	0.4	0.15	0.25
9. Air conditioned area	0.7	1.0	1.0	1.0	1.6	1.6	1.8

Note: Area serviced by total energy systems: 30 percent of 1975–85 construction plus 25 percent of 1985–2000 construction.

TRANSPORTATION ENERGY

Automotive energy use

Historical Growth scenario

The following data provide a basis for projecting automotive energy requirements in the *HG* scenario:

Year	*Population*[a]	*Number of people per registered auto*[b]	*Fuel economy*[c]	*Fraction of vehicle miles (vm) driven in urban areas*[b]
1970	205 million	2.28	13.6 mpg	.515
1975	215	2.12	12.6	.535
1985	236	1.99	12.0	.585
2000	265	1.91	11.4	.605

[a] U.S. Census Bureau's Series E projections, as published in late 1972.

[b] From A. French et al. "Highway Travel Forecasts Related to Energy Requirements," Federal Highway Administration, December 1972.

[c] Declining fuel economy reflects the historical trend toward bigger cars with more energy consuming features.

Besides these assumptions for the *HG* scenario, all scenarios make use of the following assumptions:

1. The heating value of gasoline is 125,000 Btu's per gallon.
2. Cars are driven, on the average, 10,000 miles per year. This is an average figure for 1970 and has changed little over the years.
3. The urban load factor is 1.4 passenger miles per vehicle mile

Table A-7–**Commercial sector energy requirements**
(quadrillion Btu's per year)

	1985									*2000*								
	HG			*TF*			*ZEG*			*HG*			*TF*			*ZEG*		
Item	*Fuel*	*Elec.*	*Total*	*Fuel*	*Elec.*	*Total*	*Fuel*	*Elec.*	*Total*	*Fuel*	*Elec.*	*Total*	*Fuel*	*Elec.*	*Total*	*Fuel*	*Elec.*	*Total*
Space heat	5.2	0.6	6.9	4.7	0.3	5.6	5.0	0.3	5.9	5.3	2.0	10.7	4.8	0.4	5.9	5.3	0.4	6.4
Total energy systems	—	—	—	0.6	—	0.6	0.6	—	0.6	—	—	—	1.6	—	1.6	1.8	—	1.8
Air conditioning	—	0.5	1.5	—	0.4	1.2	—	0.4	1.2	—	0.7	1.9	—	0.5	1.4	—	0.6	1.6
Lighting	—	0.4	1.2	—	0.3	0.9	—	0.3	0.9	—	0.6	1.6	—	0.4	1.1	—	0.4	1.1
Road oil and asphalt	1.4	—	1.4	1.4	—	1.4	1.3	—	1.3	2.0	—	2.0	2.0	—	2.0	1.7	—	1.7
Miscellaneous	1.0	0.6	2.7	1.0	0.6	2.7	1.1	0.6	2.8	1.2	0.9	3.6	1.2	0.8	3.4	1.3	0.9	3.7
Subtotal	7.6	2.1	13.7	7.7	1.6	12.4	8.0	1.6	12.7	8.5	4.2	19.8	9.6	2.1	15.4	10.1	2.3	16.3
Presently unknown uses	0.8	0.2	1.4	0.8	0.2	1.4	1.0	0.3	1.9	1.0	0.2	1.5	1.0	0.2	1.5	1.4	0.4	2.5
Total	8.4	2.3	15.1	8.5	1.8	13.8	9.0	1.9	14.6	9.5	4.4	21.3	10.6	2.3	16.9	11.5	2.7	18.8

See notes to Table A-3

(pm/vm); the rural value is 2.4 pm/vm.[7]

4. The average rural fuel economy $(FE)_r$ is assumed to be 1.3 times the urban fuel economy $(FE)_u$. Both are thus related to the average fuel economy $(FE)_{av}$ by

$$(FE)_r = (1 + 0.3\, f_u)\,(FE)_{av}$$
$$(FE)_u = \frac{(1 + 0.3\, f_u)\,(FE)_{av}}{1.3}$$

where f_u is the fraction of vm driven in urban areas.

On the basis of these data we obtain the following for direct auto energy use in the HG scenario:

		Urban			*Rural*			
Year	*Number of cars (millions)*	*Vehicle miles (billions)*	*Fuel economy (mpg)*	*Energy (quadrillion Btu's)*	*Vehicle miles (billions)*	*Fuel economy (mvg)*	*Energy (quadrillion Btu's)*	*Total Energy (quadrillion Btu's)*
1970	89.3	460	12.1	4.8	433	15.7	3.4	8.2
1975	101	540	11.2	6.0	470	14.6	4.0	10.0
1985	119	696	10.8	8.1	494	14.1	4.4	12.5
2000	139	841	10.4	10.1	549	13.5	5.1	15.2

Technical Fix scenario

In the *TF* scenario we consider energy savings from improvements in auto fuel economy, to 20 mpg by 1985 and to 25 mpg by 2000. To achieve an average fuel economy of 20 mpg by 1985, fuel economy in new cars must be improved according to an ordered schedule. The following schedule results in an average fuel economy of 20 mpg for all cars in 1985. It is assumed that the age distribution of cars on the road and the variation in miles driven with age is the same in 1985 as in 1970.

Model year	*Miles driven annually per car (thousands)*	*Fraction of total cars*	*Fuel economy*
1985	17.5	.083	22
1984	16.1	.122	22
1983	13.2	.109	22
1982	11.4	.115	22
1981	11.7	.121	22
1980	10.0	.096	20
1979	10.3	.087	20
1978	8.6	.073	18
1977	10.9	.045	15
1976	8.0	.041	15
up to 1975	6.5	.108	12

Source: From the "National Personal Transportation Study Report #2: Annual Miles of Automobile Travel," U.S. Department of Transportation, April 1972.

[7] From the "National Personal Transportation Study Report #1:Auto Occupancy" U.S. Department of Transportation, April 1972.

The same number of cars would be in use in the *TF* scenario as in the *HG* scenario. Energy requirements (in quadrillion Btu's) would be:

Year	*Energy*	*Savings*
1970	8.2	—
1975	10.0	—
1985	7.5	5.0
2000	6.8	8.4

ZEG scenario

We consider four conservation measures for *ZEG:* improved auto fuel economy (33 mpg by 2000); a shift of urban traffic to buses (10 percent by 1985, 25 percent by 2000); development of new communities; development of urban bikeways and walkways.

Improved fuel economy: The savings relative to the *TF* scenario amount to 1.7 quadrillion Btu's in the year 2000.

Shift to buses: We assume urban bus energy requirements are 2600 Btu's per passenger mile,[8] compared to 5000 Btu's per pm (1985) and 3000 Btu's per pm (2000) for urban autos in the *ZEG* scenario. The energy changes (in quadrillion Btu's) are thus:

Year	*Reduced energy for autos*	*Increased energy for buses*	*Savings*
1985	0.5	0.3	0.2
2000	0.9	0.8	0.1

We see that the shift to buses from efficient cars does not produce substantial energy savings. If the shift is pursued, it should be done primarily as a means of improving urban living conditions (better air quality, reduced congestion, more mobility for those without cars—the poor, the young, and the old.)

The use of buses would reduce the need for autos below *HG* levels by 7.0 million in 1985 and by 21.0 million in 2000 in the *ZEG* scenario.

Development of New Communities: On the basis of estimates provided to EPP by Vary Coates,[9] we assume that new community development are pursued at a maximum rate (20 communities of 55,000 people each per year, compared to 7.5 for the other scenarios). Thus by 2000,

$$[(20 - 7.5) \text{ communities/year}] \times [25 \text{ yrs.}] \times [55{,}000 \text{ people}] = 17 \text{ million people (6 percent of population)}$$

could be living in new communities. We assume that in these communities total auto use would be cut in half. Thus the energy savings would be,

$$[\tfrac{1}{2} \times .06] \times [(139\text{–}21) \text{ million cars}] \times [10^4 \text{ miles/car}] \times \left[\frac{125{,}000 \text{ Btu's/gal}}{33 \text{ mpg}}\right] = 0.14 \text{ quadrillion Btu's}$$

---

[8] For diesel urban buses, as estimated by Richard Rice in "Toward More Transportation with Less Energy," *Technology Review*, February 1974, p. 45.

[9] V.T. Coates, "A Workbook on Future Life Styles," op. cit. (Note 2).

The need for autos would be reduced by an additional 4 million cars with new communities.

Bikeways and Walkways: If we eliminate by 2000 10 percent of the traffic in existing communities through use of bikeways and walkways, the energy savings would be,

$$0.1 \times [(841 - 210 - 23) \text{ billion vm}] \times \left[\frac{125{,}000 \text{ Btu's/gal}}{29 \text{ mpg}} \right] = 0.26 \text{ quadrillion Btu's}$$

This would further reduce the need for autos by 6 million.

The total energy savings (in quadrillion Btu's) for the *ZEG* scenario (relative to *TF*) are:

Year	*Improved fuel economy*	*Shift to buses*	*New communities*	*Bikeways and walkways*	*Total*
1985	—	0.2	—	—	0.2
2000	1.7	0.1	0.1	0.3	2.2

Comparison of auto use in scenarios:

Passenger miles per capita per year

	Urban		*Rural*		*Total*	
Year	*HG and TF*	*ZEG*	*HG and TF*	*ZEG*	*HG and TF*	*ZEG*
1970	3140	3140	5070	5070	8210	8210
1985	4130	3720	5020	5020	9150	8740
2000	4440	2750	4970	4820	9410	7570

Number of cars
(millions)

Year	*HG and TF*	*ZEG*
1970	89	89
1985	119	112
2000	139	108

Auto energy requirements
(quadrillion Btu's)

Year	*HG*	*TF*	*ZEG*
1970	8.2	8.2	8.2
1985	12.5	7.5	7.0
2000	15.2	6.8	3.8

Note that in 2000 auto energy use in *ZEG* is reduced over 70 percent while auto miles per capita are reduced only about 20 percent.

Bus energy use

Bus traffic is made up of commercial (urban and intercity) and school buses. School bus energy requirements and intercity bus energy requirements are taken to be the same in all scenarios. But urban buses are used more in the *ZEG* scenario, where a shift from autos to buses is assumed to take place (10 percent by 1985 and 25 percent by 2000):

$$.1 \times (696 \times 10^9 \text{ vm}) \times (1.4 \text{ pm/vm}) = 97.4 \times 10^9 \text{ pm by 1985}$$
$$.25 \times (841 \times 10^9 \text{vm}) \times (1.4 \text{ pm/vm}) = 294 \times 10^9 \text{ pm by 2000}$$

The travel by intercity and school buses is expected to be:[10]

Year	*Intercity buses (billion pm)*	*School buses (billion pm)*
1970	25	40
1975	26	40
1985	27	40
2000	28	40

Similarly, we obtain projections for urban buses in the *HG* and *TF* scenarios, while we add the above modal shift estimates for *ZEG*.

Year	*HG and TF (billion pm)*	*ZEG (billion pm)*
1970	25	25
1975	26	26
1985	30	127
2000	39	333

To convert these travel estimates to energy we use the following specific energy requirements:

3700 Btu's/pm for urban buses in *HG* and *TF* scenarios[11]
2600 Btu's/pm for urban buses in *ZEG* after 1980
1600 Btu's/pm for intercity buses[11]
1100 Btu's/pm for school buses[11]

and thereby obtain the following energy requirements:

Year	*HG and TF (quadrillion Btu's)*	*ZEG (quadrillion Btu's)*
1970	0.2	0.2
1975	0.2	0.2
1985	0.2	0.5
2000	0.2	1.0

A substantial number of new buses is required for the *ZEG* scenario. To estimate the number required, we note that in 1969 urban buses travelled an average of 25 thousand miles per year with an average

[10] Based on projections in the *1972 National Transportation Report,* Government Printing Office, Washington, U.S. Department of Transportation.

[11] From Eric Hirst, "Energy Intensiveness of Passenger and Freight Transport Modes, 1950–1970," Oak Ridge National Laboratory/National Science Foundation Report, April 1973.

load factor of 15 pm/vm[12] (i.e., 375 thousand pm per bus per year), so that the number of new buses required in *ZEG* would be:

260 thousand by 1985

780 thousand by 2000

Since there would be only about 70 thousand buses in 1975, the number of buses in use would have to grow at 13 percent per year till 1985, and at 7 percent per year from 1985 to 2000.

Finally, per capita bus use in the three scenarios would be:

Passenger miles per capita per year

		Urban	
Year	*Intercity*	*HG and TF*	*ZEG*
1970	122	122	122
1975	121	121	121
1985	114	127	538
2000	106	147	1257

Air Transport Energy

Historical Growth scenario

Various projections of future air transport demand have been made. Until recently most forecasts have assumed a continuation of the nearly explosive growth that characterized the 1960s. More recent estimates reflect the view that passenger air transportation is maturing while air freight is only beginning to grow. For the *HG* scenario we adopt the upper limit of a recent industry forecast for the period out to 1985. We extend the 1980–85 growth rates from this source to the period 1985–90, and project growth for 1990–2000 at a somewhat slower rate, reflecting further maturation of the industry. Thus we assume the following:

Percent growth per year

	Passenger miles			*Ton miles of freight*		
Year	*Domestic*	*International*	*Total*	*Domestic*	*National*	*Total*
1970–75	n/a	7.5	7.9	10.7	18.5	n/a
1975–80	n/a	10.2	9.5	10.1	12.5	n/a
1980–90	n/a	6.1	6.0	9.8	10.7	n/a
1990–2000	n/a	5.0	5.0	7.0	7.0	n/a

Note: Projections to 1985 are taken from "Dimensions of Airline Growth," prepared by Market Research Unit, Boeing Commercial Airplane Company, February 1974.

Using these growth rates we obtain the following projections for air transportation:

[12] A. French et al., "Highway Travel Forecasts Related to Energy Requirements," Federal Highway Administration, December 1972.

Freight
(billions of ton miles)

Year	*Domestic*	*International*
1970	3.3	2.4
1975	5.5	5.6
1985	14.2	16.8
2000	47.0	58.0

Passengers
(billions of passenger miles)

Year	*Domestic*	*International*
1970	110	50
1975	163	72
1985	336	158
2000	752	358

To calculate energy requirements corresponding to these demand projections and potential savings, we adopt the approach formulated by James Mutch[13] and assume the following:

1. Aircraft remain at the 1971 level of technology.

2. The distribution of traffic over route length remains stable at the values given in the 1970 Domestic Passenger Origin-Destination Survey.[14]

3. The 1971 Civil Aeronautics Board passenger load factor standards remain in effect and are met by the airlines. These standards call for an average passenger load factor (percent of available seating actually sold and used) of 54.1 percent.

4. Specific energy requirements are assumed to be as follows:

Freight		*Passengers*	
Domestic	*International*	*Domestic*	*International*
62,000 Btu's/tm	43,000 Btu's/tm	6200 Btu's/pm	4300 Btu's/pm

On the basis of these assumptions we obtain the following annual energy requirements (in quadrillion Btu's):

	Freight		*Passengers*		
Year	*Domestic*	*International*	*Domestic*	*International*	*Total*
1970	0.20	0.10	0.68	0.22	1.2
1975	0.34	0.24	1.01	0.31	1.9
1985	0.88	0.72	2.08	0.68	4.4
2000	2.91	2.49	4.66	1.54	11.6

[13] James J. Mutch, "The Potential for Energy Conservation in Commercial Air Transport," RAND report R-1360-NSF, October 1973.

[14] U.S. Civil Aeronautics Board, *Handbook of Airline Statistics*, Government Printing Office, Washington, 1971.

Technical Fix scenario

The foregoing projections of passenger miles and ton miles for air transport in the *HG* scenario are used as a basis for calculating potential savings in the *TF* scenario.

Short-term measures (for 1985): 1. Increase the average ton load factor to 58 percent. This corresponds to a passenger load factor of 67 percent, which probably involves no significant loss of service quality. With this load factor, as Mutch shows, the chances of a passenger not being seated are only one in a thousand. The higher load factor leads to a 28 percent fuel saving for domestic flights and an 8 percent saving for international flights. Thus the energy savings in 1985 amount to $0.28 \times (0.88 + 2.08) + 0.08 \times (0.72 + 0.68) = 0.95$ quadrillion Btu's.

2. Reduce flight speed to the speed corresponding to minimum fuel consumption; this would result in a 4.5 percent reduction in fuel use and a 6 percent increase in flight time. In the past airlines have travelled at higher than optimal speeds from an energy point of view because minimizing flight time has been more efficient in minimizing operating costs than conserving fuel: fuel costs have accounted for only one quarter of total operating costs. The potential saving is

$.045 \times (4.4 - 0.95) = 0.16$ quadrillion Btu's.

Thus the total potential savings in 1985 amount to 1.1 quadrillion Btu's or 25 percent of the total air transport energy use in the *HG* scenario.

Long-term measures (for 2000): For the long term we supplement the foregoing measures with a shift away from air transport for short haul traffic:

1. Increase load factor as above. The potential saving is $0.28 \times (2.91 + 4.66) + 0.08 \times (2.49 + 1.54) = 2.44$ quadrillion Btu's.

2. Reduce flight speed as above. The potential saving is $0.045 \times (11.6 - 2.4) = 0.41$ quadrillion Btu's.

3. Substitute high speed rail (150 mph at 1000 Btu's/pm[15]) for domestic air travel up to 400 miles. Such trips account for 10 percent of passenger miles and 13 percent of air transport energy. For such trips, only a minimal loss in door-to-door travel time would be involved. This modal shift involves decreasing air and increasing rail travel by 75 billion passenger miles in 2000. The energy changes (in quadrillion Btu's) would be:

Reduced energy for air travel	*Increased energy for rail*	*Savings*
.42	.08	.34

4. Shift all freight for trips less than 250 miles to truck (at 2800 Btu's/tm) and for trips between 250 and 400 miles to rail (670 Btu's/tm).[16] This would mean shifting 1.8 billion tm to truck and 2.8 billion tm to rail. The energy changes (in quadrillion Btu's) would be:

[15] Richard Rice, "Toward More Transportation with Less Energy," op. cit. (Note 8).

[16] Eric Hirst, "Energy Intensiveness of Passenger and Freight Transport Modes, 1950–1970," op. cit. (Note 11).

Reduced energy for air	*Increased energy for trucks*	*Increased energy for rail*	*Savings*
0.285	.005	.002	0.28

ZEG scenario

For the *ZEG* scenario we envision a slower overall rate of growth in energy demand. The Air Transport Association of America has recently forecast growth in turbine fuel requirements at only 4.5 percent up till 1982.[17] For *ZEG*, we adopt as a basis for energy conservation calculations this growth rate extended to the year 2000. From this basis, we calculate savings as in the *TF* scenario. Thus we obtain the following for air transport energy use in the three scenarios:

Year	*HG (quadrillion Btu's)*	*TF (quadrillion Btu's)*	*ZEG (quadrillion Btu's)*
1970	1.2	1.2	1.2
1975	1.9	1.9	1.9
1985	4.4	3.3	2.2
2000	11.6	8.2	4.1

Comparison of scenarios

The following is a comparison of air travel and air freight in the three scenarios:

Freight
(billions of ton miles)

	HG		*TF*		*ZEG*	
Year	*Domestic*	*International*	*Domestic*	*International*	*Domestic*	*International*
1970	3.3	2.4	3.3	2.4	3.3	2.4
1975	5.5	5.6	5.5	5.6	5.5	5.6
1985	14.2	16.8	14.2	16.8	9.5	11.2
2000	47.0	58.0	42.4	58.0	21.2	29.0

Passengers
(billions of passenger miles)

	HG		*TF*		*ZEG*	
Year	*Domestic*	*International*	*Domestic*	*International*	*Domestic*	*International*
1970	110	50	110	50	110	50
1975	163	72	163	72	163	72
1985	336	158	336	158	224	105
2000	752	358	679	358	339	179

[17] The Fuels Committee of the Air Transport Association of America, "United States Airline Industry Forecast of Turbine Fuel Demand," Washington, D.C., June 1973.

It is also of interest to express air travel in pm per capita as follows:

	HG		*TF*		*ZEG*	
Year	*Domes-tic*	*Inter-national*	*Domes-tic*	*Inter-national*	*Domes-tic*	*Inter-national*
1970	537	244	537	244	537	244
1975	758	335	758	335	758	335
1985	1420	669	1420	669	954	449
2000	2840	1350	2560	1350	1280	675

In addition to commercial air travel, a small part of air travel is done by general aviation. On the basis of data in the 1972 *National Transportation Report,* energy use in general aviation (in all scenarios) can be estimated as follows:

Year	*(quadrillion Btu's)*
1970	0.09
1975	0.12
1985	0.24
2000	0.30

We assume an average energy use rate per plane of 3.5×10^6 Btu's/hr.

Truck energy requirements

Historical Growth scenario

Total energy use for trucks is projected to grow as follows in the *HG* scenario:[18]

Year	*Energy (quadrillion Btu's)*
1970	3.5
1975	4.2
1985	5.3
2000	6.5

These projections can be disaggregated by assuming that the distribution of trucking requirements in the future is the same as in 1969, when 545 billion ton miles of freight were hauled by the following modes:[19]

7 percent by vehicles with 2 axles, 4 tires (about 30,200 Btu's/tm)
26 percent by other single unit vehicles (about 6300 Btu's/tm)
67 percent by combination vehicles (about 2980 Btu's/tm)

[18] Based on the projection in "Reference Energy Systems and Resource Data for Use in the Assessment of Energy Technologies," prepared by Associated Universities, Inc. for the Office of Science and Technology, April 1972.

[19] From A. French et al., "Highway Travel Forecasts Related to Energy Requirements," op. cit.

We obtain the following projections for truck transport growth:

	2 Axles, 4 tires		*Other single units*		*Combinations*	
Year	*Ton miles (billions)*	*Energy (quadrillion Btu's)*	*Ton miles (billions)*	*Energy (quadrillion Btu's)*	*Ton miles (billions)*	*Energy (quadrillion Btu's)*
1970	44	1.3	161	1.0	413	1.2
1975	51	1.6	193	1.1	497	1.5
1985	63	1.9	232	1.5	600	1.9
2000	79	2.4	292	1.9	750	2.2

Technical Fix scenario

We consider the following energy conservation measures for trucks:

1. switch gasoline fueled trucks to diesel.
2. shift long haul traffic to rail

Convert to diesel: According to a recent U. S. Department of Transportation report,[19] diesel powered trucks are about 30 percent more efficient than gasoline powered trucks under similar load and operating conditions. We assume that by 2000 all gasoline trucks are converted to diesel. Since all 2 axle, 4 tire vehicles, 77 percent of other single unit vehicles, and 32 percent of combination ton vehicles are assumed to be gasoline driven, the total potential saving in 2000 is 1.4 quadrillion Btu's.

Shift to rail: Trucks are about one-quarter as efficient in transporting freight as trains. Rather important energy savings could be achieved with a substantial modal shift of freight traffic from truck to rail. That the rails today do not command a larger share of freight traffic is due, in part, to the fact that government subsidizes the truck and the airplane, but not the railroad. The poor service and long delays that characterize rail transport, and the rail rate structures that inhibit the competitive role of the railroad are also to blame. There are indications that a significant fraction of intercity truck traffic can be shifted to rail if government policies would shift to neutral. About 40 percent of all freight tonnage in 1967 could have been hauled by either truck or rail. Actually, trucks carried over 80 percent of this "competitive" cargo[20].

Another fact that bears on the feasibility of switching freight traffic from truck to rail is that a substantial fraction of truck ton-mileage is long haul, for which the switch would be more attractive.

Finally the economics of a switch should be examined. There are indications that with even-handed government policies, some upgrading of rail service and the adoption of marginal cost pricing, the economics of a switch to rail would be favorable for over half the intercity freight now moving by truck[21]. Here we assume that 20 percent of combination truck traffic is shifted to rail by 1985 (corresponding to hauls longer than 500 miles) and 40 percent by 2000 (corresponding to hauls longer than 250

[20] A. L. Morton, *Competition in the Intercity Freight Market,* USDOT Office of Systems Analysis and Information.

[21] *1972 National Transportation Report,* op. cit., U.S. Department of Transportation.

miles). (Note that the shift of air freight shipments to trucks, for trips under 250 miles long, causes a negligible increase in the total truck ton-milage). Combination truck traffic switched to rail in the *TF* scenario amounts to:

$$1985: \quad 0.2 \times (600 \times 10^9) = 120 \times 10^9 \text{ tm}$$
$$2000: \quad 0.4 \times (750 \times 10^9) = 300 \times 10^9 \text{ tm}$$

The savings for this switch (assuming a rail energy intensivity of 670 Btu's/tm and a diesel truck energy intensivity of 2340 Btu's/tm) are:

$$1985: \quad (120 \times 10^9 \text{ tm}) \times (2340 - 670) \text{ Btu's/tm} = 0.2 \text{ quadrillion Btu's.}$$

$$2000: \quad (300 \times 10^9 \text{ tm}) \times (2340 - 670) \text{ Btu's/tm} = 0.5 \text{ quadrillion Btu's.}$$

Total truck energy requirements in the *TF* scenario amount to:

Year	*Energy (quadrillion Btu's)*
1970	3.5
1975	4.2
1985	5.0
2000	4.4

ZEG scenario

We assume that in 2000 *ZEG* truck energy requirements are down 15 percent, as a direct result of the decrease in manufacturing output in *ZEG* relative to other scenarios. Thus truck energy requirements are: .85 × (4.4 quadrillion Btu's) = 3.7 quadrillion Btu's.

Railroad energy use

Historical Growth scenario

The basic energy demand for railroad transportation is projected to grow at 3.7 percent per year.[22] for the *HG* scenario. From the 1970 railroad energy consumption level of 0.55 quadrillion Btu's[23] this gives the following total rail energy requirements for the HG scenario:

Year	*Energy (quadrillion Btu's)*
1970	0.55
1975	0.66
1985	0.96
2000	1.67

We assume that both freight and passenger transport grow at this rate of 3.7 percent per year, so that growth in freight and passenger modes is as follows:

[22] Based on the projection in "Reference Energy Systems and Resource Data for Use in the Assessment of Energy Technologies," by Associated Universities, Inc., op. cit.

[23] From Eric Hirst, "Energy Intensiveness of Passenger and Freight Transport Modes 1950–1970," op. cit.

Year	Freight		Passengers	
	Ton miles (billions)	Energy (quadrillion Btu's)	Passenger miles (billions)	Energy (quadrillion Btu's)
1970	770	0.52	11	0.03
1975	920	0.62	13	0.04
1985	1340	0.91	19	0.05
2000	2340	1.58	33	0.09

Here we have adopted the following values for specific energy requirements:[23]

670 Btu's/tm for freight
2900 Btu's/pm for passengers

Technical Fix scenario

In the *TF* scenario there is an increase in rail transportation, owing to shifts from air and truck transport modes, as discussed above. We assume that with the emphasis on rail passenger transport, the energy intensivity of rail passenger travel reduces to 1000 Btu's/pm by 2000.[24] Rail transportation is projected to be:

Year	Freight		Passengers	
	Ton miles (billions)	Energy (quadrillion Btu's)	Passenger miles (billions)	Energy (quadrillion Btu's)
1985	1460	0.98	19	0.05
2000	2640	1.77	108	0.11

ZEG scenario

In the ZEG scenario we assume that:

1. Freight hauling requirements drop 15 percent by 2000
2. Rail passenger travel increases at 10 percent per year after 1975.

Thus we obtain:

Year	Freight		Passengers	
	Ton miles (billions)	Energy (quadrillion Btu's)	Passenger miles (billions)	Energy (quadrillion Btu's)
1985	1460	0.98	35	0.10
2000	2250	1.50	157	0.16

[24] From Rice, "Toward More Transportation with Less Energy," op. cit.

Comparison of scenarios

	Total rail energy (quadrillion Btu's)		
Year	HG	TF	ZEG
1970	0.6	0.6	0.6
1975	0.7	0.7	0.7
1985	1.0	1.0	1.1
2000	1.7	1.9	1.7

	Personal rail travel (miles per capita per year)		
Year	HG	TF	ZEG
1970	54	54	54
1975	60	60	60
1985	80	80	150
2000	125	410	590

Farm machinery

In 1969, farm machinery consumed 7.6 billion gallons of fuel, amounting to 1.1 quadrillion Btu's.[25] About 290 million acres of cropland were harvested in 1969. It is expected[26] that this will increase to 350 million acres by 1985. We assume that this rate of expansion (0.8 percent per year) continues until 390 million acres are under cultivation in the year 2000. We assume that farm machinery energy requirements are proportional to the acreage, and adopt the same energy requirements for all three scenarios:

Year	Energy (quadrillion Btu's)
1970	1.1
1975	1.2
1985	1.3
2000	1.5

Miscellaneous (mostly ships)

Following the Associated Universities, Inc., study,[27] we assume growth in this category at 2 percent per year from a base of 0.93 quadrillion Btu's in 1970:

Year	Energy
1970	0.93
1975	1.03
1985	1.26
2000	1.69

[25] D. Pimentel et al., "Food Production and the Energy Crisis" *Science,* November 2, 1973.

[26] Lee Martin, "Agriculture as a Growth Sector—1985 and Beyond," draft report to the Energy Policy Project, April 1974.

[27] Associated Universities, Inc., "Reference Energy Systems," op. cit.

We adopt these values for the *HG* and *TF* scenarios. For *ZEG* we assume a 15 percent reduction (0.25 quadrillion Btu's energy savings) in 2000, corresponding to a cutback in industrial output.

Summary of transportation energy use

Transportation energy use and passenger travel data for the scenarios are summarized in Table A-8 and A-9.

Table A-8–**Transportation energy for scenarios** (quadrillion Btu's)

			HG		*TF*		*ZEG*	
	1970	*1975*	*1985*	*2000*	*1985*	*2000*	*1985*	*2000*
Auto	8.2	10.0	12.5	15.2	7.5	6.8	7.0	3.8
Bus	0.2	0.2	0.2	0.2	0.2	0.2	0.5	1.0
Air	1.2	1.9	4.4	11.6	3.3	8.2	2.2	4.1
Truck	3.5	4.2	5.3	6.5	5.0	4.4	5.0	3.7
Railroad	0.6	0.7	1.0	1.7	1.0	1.9	1.1	1.7
Farm machinery	1.1	1.1	1.3	1.5	1.3	1.5	1.3	1.5
Miscellaneous	0.9	1.0	1.3	1.7	1.3	1.7	1.3	1.4
Totals	15.7	19.1	26.0	38.4	19.6	24.7	18.4	17.2

Table A-9–**Travel in scenarios** (passenger miles per capita)

		HG		*TF*		*ZEG*	
	1970	*1985*	*2000*	*1985*	*2000*	*1985*	*2000*
Urban							
Auto	3,140	4,130	4,440	4,130	4,440	3,720	2,750
Bus	120	130	150	130	150	540	1,260
Total urban	3,260	4,260	4,590	4,260	4,590	4,260	4,010
Rural auto	5,070	5,020	4,970	5,020	4,970	5,020	4,820
Intercity bus	120	120	105	120	105	120	105
Air	780	2,090	4,190	2,090	3,905	1,400	1,955
Rail	60	80	125	80	410	150	590
Total intercity	6,030	7,310	9,390	7,310	9,390	6,690	7,470
Total travel	9,290	11,600	14,000	11,600	14,000	11,000	11,500

MANUFACTURING ENERGY

Historical Growth energy requirements

To estimate future energy needs in manufacturing for the *HG* scenario, we extrapolate energy consumption growth patterns, for various

Table A-10—**Energy growth in manufacturing, HG scenario**

	Direct fuel requirements (quadrillion Btu's)			
	Annual growth (percent)	*1968*	*1985*	*2000*
Process steam	3.6	10.1	18.6	30.8
On-site power	3.6	0.4	0.8	1.2
Direct heat		6.6	8.9	10.5
Feedstocks	6.1 (5.1 after 1985)	2.2	6.2	13.3
Subtotal	3.3 (for 1968–2000)	19.3	34.5	55.8

	Electricity requirements (quadrillion Btu's)			
	Annual growth (percent)	*1968*	*1985*	*2000*
Electric drive	5.3 (4.7 after 1985)	4.8	11.7	24.0
Electrolytic	4.7	0.7	1.6	3.1
Direct heat		0.3	4.5	13.5
Other	6.7	0.2	0.6	1.7
Less on-site generation	−3.6	−0.4	−0.8	−1.2
Subtotal Purchased electricity	6.3 (for 1968–2000)	5.6	17.6	41.1
Total energy requirements	4.2 (for 1968–2000)	24.9	52.1	96.9

Notes: Electrical energy requirements are expressed in terms of fuel inputs to electrical generation. Fuel processing activities such as petroleum refining are included as manufacturing operations.

On site power requirements are counted twice under direct fuel requirements—first under energy requirements for process steam, and second, explicitly under the category on-site power where it is measured as central station input energy displaced. The overcounting is then compensated for by subtracting the latter quantity from input energy requirements for electricity, thereby reducing purchased electricity requirements.

end uses of energy, from the period 1960–1968 to the year 2000.[28] (See Table A10.) Some variations of these historical growth rates are assumed:

- After 1985, growth rates in feedstocks and electric drive are

[28] The 1968 consumption numbers and 1960–1968 growth rates are taken from "Patterns of Energy Consumption in the United States," prepared for the Office of Science and Technology by Stanford Research Institute, January 1972.

reduced slightly, reflecting a trend toward maturation in the petrochemical industry and in automation.

- On-site electric power energy input maintains a constant relationship to process steam energy input.
- The end use demand for direct heat in industry is assumed to grow at the historical rate (slightly less than 3 percent per year), but a substantial shift to use of electric resistance devices is assumed in meeting this demand, so that the fuel input to direct heat grows much faster (about 4 percent per year). Such a shift appears necessary to account for the substantial growth in electric power requirements that characterizes most forecasts.

We note the following general features of this growth pattern:

- The direct use of fuels is projected to grow at about the rate of energy use in the economy as a whole (3.3 percent per year).
- There is an emphasis on electrification of industry. Purchased electric energy requirements are projected to grow at 6.3 percent per year, so that the electric energy share of industrial energy is expected to grow from 23 percent in 1968 to 43 percent in 2000.
- Total industrial energy requirements are projected to grow at 4.2 percent per year.

Energy requirements in the Technical Fix and ZEG scenarios

To identify savings for the *TF* and *ZEG* scenarios we proceed as follows:

1. Specify end uses of energy in five energy intensive industries (see Table A-11) where we can explicitly identify savings opportunities by way of process modification. Calculate for these areas the savings that could be realized in the *TF* and *ZEG* scenarios.
2. Calculate the quantities of energy associated with miscellaneous uses (i.e., those not explicitly identified in step (1) described above) in the areas of process steam, direct heat, and electricity. For these miscellaneous activities, estimate potential savings both from reduced energy processing requirements and from use of more efficient industrial processes.
3. Achieve further savings in *ZEG* by reducing the overall level of manufacturing output or shifting manufacturing output to less energy-intensive activities.

Potential savings in five energy intensive industries.

For certain energy intensive industries potential energy savings can be explicitly estimated. A study done for EPP by Thermo-Electron Corporation[29] indicates what savings can be achieved in the production of paper, aluminum, steel and cement. Another energy intensive item is plastics. We roughly estimate the energy conservation potential in plastics manufacturing by considering general energy savings schemes for steam

[29] E. P. Gyftopoulos, Lazaros J. Lazaridis, and Thomas F. Widmer, "Potential for Effective Use of Fuel in Industry," a Report to the Energy Policy Project, Cambridge, Mass.: Ballinger, 1974.

Table A-11–**Energy requirements for materials processing** (million Btu's per ton of product)

	Feedstocks	*Purchased Fuels*	*Purchased Electricity*	*Total*
Paper				
A	–	19.5	5.0	24.5
B[a]	–	15.9	2.0	17.9
C	–	11.7	−5.9	5.8
Primary Aluminum				
A	–	40	150	190
B	–	30	115	145
C	–	26	105	131
Steel				
A	–	23.6	2.9	26.5
B	–	15.3	1.9	17.2
C	–	12.5	1.5	14.0
Cement				
A	–	6.6	1.3	7.9
B[a]	–	5.3	0.9	6.2
C	–	3.6	0.9	4.5
Plastics				
A	45	106	6	157
B[a]	45	98	− 7	136
C	45	102	−29	118

A = present technology
B = improved technology, 1985
C = improved technology, 2000

Note: Purchased electricity requirements are expressed in terms of fuel input to power generation.

[a] Energy requirements for 1985 represent the average for a mix of some plants with present technology plus other plants using the improved technologies described in the Thermo-Electron study. For these industries, we assume that all new productive capacity after 1980 (including that which replaced old capacity retired at a rate of 2 percent per year) uses the improved technology. We assume that the energy efficiency of old capacity improves at a rate of 1.3 percent per year through various "leak plugging" measures.

production, and direct heat generation as discussed in the Thermo-Electron study.

The specific energy requirements for materials processing are shown in Table A-11. To calculate total energy requirements, we also need to know projections of future production for different scenarios. These projections, for the *HG* and *TF* scenarios, are based primarily on extrapolations beyond 1980 of projections that the Conference Board (CB) made in a report to the Project.[30]

For the *ZEG* scenario we assume slower growth than for *TF* and *HG* in steel, aluminum, and plastics production. Our projections for all three scenarios are shown in Table A-12. We use the production projections, along with the specific energy requirements in Table A-11, to obtain the projected energy requirements of these five key industries for all three scenarios. (See Table A-13).

From Table A-13 we obtain the following savings for these five industries:

[30] The Conference Board, *Energy Consumption in Manufacturing,* a Report to the Energy Policy Project, Cambridge, Mass.: Ballinger, 1974.

Table A-12–**Production in key industries**
(millions of tons per year)

	1967	1985			2000		
		HG	*TF*	*ZEG*	*HG*	*TF*	*ZEG*
Paper	(60 million tons in 1971)	104	104	104	170	170	170
Aluminum	4.2	10.4	10.4	8.4	21.5	21.5	17.3
Primary	3.3	7.7	6.4	4.3	15.9	12.4	8.0
Scrap	0.9	2.7	4.0	4.1	5.6	9.1	9.3
Steel	127	198	198	175	290	290	218
Plastics	6.8	35	35	25	75	75	37
Cement	71	129	129	129	192	192	192

Paper: For all scenarios extend Conference Board (CB) output projection for 1975–1980 (3.3 percent per year) to the year 2000.

Aluminum: Assume total production in *HG* and *TF* scenarios grows like
5.4 percent per year from 1971–1980 (following CB)
4.9 percent per year from 1980–2000 (following AUI)[a]
For *ZEG* subtract from this total the aluminum used for aluminum cans.
See section below on aluminum recycling potential, page 464.

Steel: For *HG* and *TF* scenarios assume future production grows at 2.5 percent per year, extending CB projection for 1975–1980.
For *ZEG* assume slower growth at 1.5 percent per year.

Plastics: Use CB growth rates till 1980 for *HG* and *TF* scenarios:
7.1 percent per year, 1967–1975
6.6 percent per year, 1975–1980
Assume a declining growth rate beyond 1980:
5.8 percent per year, 1980–1985
5.1 percent per year, 1985–2000
This reflects an assumption of the gradual maturing of the plastics industry. For *ZEG* we assume a slower growth schedule for production:
4.5 percent per year, 1975–1980
3.1 percent per year, 1980–1985
2.7 percent per year, 1985–2000
This slowed growth in part reflects a curtailment in growth of plastics packaging after 1980. Today such packaging accounts for about 24 percent of production. After curtailment of growth in plastics packaging, the level of plastics use for packaging would be twice as great as it is today.

Cement: Extent CB growth rate projection for 1971–1980 (2.7 percent per year) to the year 2000 for all scenarios.

[a] "Reference Energy Systems and Resource Data for Use in the Assessment of Energy Technologies," prepared for the Office of Science and Technology by Associated Universities, Inc., April, 1972.

	TF relative to HG (quadrillion Btu's)			*ZEG relative to TF (quadrillion Btu's)*		
	Fuels	*Electricity*	*Total*	*Fuels*	*Electricity*	*Total*
1985	2.5	1.5	4.0	1.7	0.3	2.0
2000	5.5	6.1	11.6	6.6	−0.5	6.1

Potential savings in miscellaneous categories

The five industries we have considered so far make up only about 30 percent of industrial energy requirements in the *HG* scenario. Table A-14 gives industrial energy use in various "miscellaneous" categories. We consider the following opportunities for reducing energy requirements in these miscellaneous categories:

Table A-13–**Energy Consumption in key industries**
(quadrillion Btu's)

	1985			*2000*		
	HG	*TF*	*ZEG*	*HG*	*TF*	*ZEG*
Paper						
Purchased fuels	2.0	1.6	1.6	3.3	2.0	2.0
Purchased electricity	0.5	0.2	0.2	0.9	−1.0	−1.0
Total	2.5	1.8	1.8	4.2	1.0	1.0
Aluminum						
Purchased fuels	0.3	0.2	0.2	0.7	0.4	0.3
Purchased electricity	1.2	0.7	0.5	2.4	1.3	0.8
Total	1.5	0.9	0.7	3.1	1.7	1.1
Steel						
Purchased fuels	4.5	3.0	2.7	6.5	3.6	2.7
Purchased electricity	0.6	0.4	0.3	0.8	0.4	0.3
Total	5.1	3.4	3.0	7.3	4.0	3.0
Plastics						
Purchased fuels	5.3	5.0	3.6	11.4	11.0	5.4
Purchased electricity	0.2	−0.2	−0.2	0.5	−2.2	−1.1
Total	5.5	4.8	3.4	11.9	8.8	4.3
Cement						
Purchased fuels	0.9	0.7	0.7	1.3	0.7	0.7
Purchased electricity	0.2	0.1	0.1	0.2	0.2	0.2
Total	1.1	0.8	0.8	1.5	0.9	0.9
Total purchased fuels	13.0	10.5	8.8	23.2	17.7	11.1
Total purchased electricity	2.7	1.2	0.9	4.8	−1.3	−0.8
Grand total	15.7	11.7	9.7	28.0	16.4	10.3

Note: Purchased electricity requirements are expressed in terms of fuel input to power generation.

- Reduced energy processing losses owing to lower energy needs in *TF* and *ZEG* scenarios.
- Cogeneration of electricity and steam.
- Direct use of fuels along with heat recuperators and regenerators in direct heat applications.

Table A-14–**Miscellaneous energy use in HG scenario**
(quadrillion Btu's)

	1985	*2000*
Misc. process steam	14.0	22.1
Misc. purchased electricity	10.4	22.8
Misc. feedstocks	4.6	9.9
Misc. direct heat		
Electric	4.5	12.4
Other	2.1	—

- "Belt tightening" plus use of other conservation technologies induced by rising energy prices.

Potential savings in energy processing: Extra energy is consumed in converting and delivering energy. The *TF* and *ZEG* scenarios save processing energy as well as saving energy at the point of end use. In fact, over half the total energy savings we project for *TF* over *HG* by 2000 arises from savings in energy processing. The estimates we adopt here for processing a unit of energy are given in the section on energy processing losses (page 467). Processing losses for electric power generation, petroleum refining, natural gas processing, uranium enrichment, and coal synthetics processing are summarized in Table A-15 for all three scenarios.

For petroleum refining we assume processing losses in the *HG* scenario are the same in the future as they are today. For the *TF* and *ZEG* scenario we assume for new refinery capacity the process improvements which the Thermo-Electron study suggested[31]. We assume that half of petroleum imports are refined domestically.

The losses given in Table A-15 for electric power generation and transmission for the different scenarios are separated out from other energy processing losses in the totals of Table A-15. The reason for this is

Table A-15–**Energy processing losses**
(quadrillion Btu's)

	HG		*TF*		*ZEG*	
	1985	*2000*	*1985*	*2000*	*1985*	*2000*
Electricity generation and transmission	24.5	48.5	15.2	19.3	15.2	18.7
Petroleum refining	5.0	7.5	3.6	4.3	3.6	3.2
Domestic gas processing & transport	2.8	3.3	2.6	3.1	2.4	2.4
Transport of synthetic and imported gas	0.1	0.1	0.0	0.1	0.0	0.1
Synthetics processing	0.8	3.2	0.0	1.2	0.0	0.0
Uranium enrichment	0.6	2.2	0.3	0.3	0.2	0.1
Totals	33.8	64.8	21.7	28.3	21.4	24.5
Totals excluding electricity losses	9.3	16.3	6.5	9.0	6.2	5.8

that electric losses are allocated to specific end uses in savings estimates for each sector of the energy economy. We obtain the following savings estimates for energy processing except electric losses:

	TF relative to HG (quadrillion Btu's)			*ZEG relative to TF (quadrillion Btu's)*		
	Fuels	*Electricity*	*Total*	*Fuels*	*Electricity*	*Total*
1985	2.1	0.7	2.8	0.2	0.1	0.3
2000	4.5	2.8	7.3	2.9	0.3	3.2

Cogeneration of electricity and process steam: In the combined generation of electricity and process steam, as described in the Thermo-Electron study to EPP,[31] 80 percent of the potential energy in the fuel can be used. Specifically, 1.55 units of fuel can be used to produce 1.0 units of steam plus 0.24 units of electricity. But this much electricity would displace 0.7 units of fuel input to a central station power plant. We assume that cogeneration can be accomplished for 10 percent of the remaining miscellaneous process steam by 1985, and for 50 percent by 2000. The following savings are achieved with cogeneration:

	TF relative to HG (quadrillion Btu's)			*ZEG relative to TF (quadrillion Btu's)*		
	Fuels	*Electricity*	*Total*	*Fuels*	*Electrictity*	*Total*
1985	−0.3	0.7	0.4	0	0	0
2000	−2.3	5.2	2.9	0	−0.3	−0.3

The total amount of on-site power generation arising from all sources (in the five key industries, in petroleum refining, and in miscellaneous applications) amounts to:

	Equivalent fuel input for central station electric power (quadrillion Btu's)			*Equivalent generating capacity*[a] *(thousands of megawatts)*		
	HG	*TF*	*ZEG*	*HG*	*TF*	*ZEG*
1985	0.8	2.4	2.0	18	55	45
2000	1.2	10.8	9.2	30	265	225

[a] Assuming a 50 percent capacity factor. The actual capacity factor for industrial power plants would likely be greater than this, so that the capacity estimates given here may be high.

Direct use of fuels and heat recuperators in direct heat applications: According to the Thermo-Electron study, fuel savings on the order of 23 percent can be achieved through the use of heat recuperators in direct heat applications. Use of recuperators should be a good investment even when applied to existing furnaces. We assume that in the *TF* and *ZEG* scenarios fuel is burned directly to provide direct heat instead of using electric resistive heat (as in the *HG* scenario) and that the use of recuperators reduces fuel requirements by 23 percent. The savings are:

[31] E. P. Gyftopoulos et al., *Potential Fuel Effectiveness in Industry,* op. cit.

	TF relative to HG (quadrillion Btu's)			*ZEG relative to TF (quadrillion Btu's)*		
	Fuels	*Electricity*	*Total*	*Fuels*	*Electricity*	*Total*
1985	−1.6	4.5	2.9	−0.1	0	−0.1
2000	−4.3	9.7	5.4	−0.1	0	−1.0

Belt-tightening and use of other conservation technologies: We assume that the remaining energy requirements in the area of miscellaneous process steam and miscellaneous electricity are reduced one percent per year below the HG levels by general belt-tightening and leak plugging efforts plus use of other conservation technologies, in response to higher energy prices, in the *TF* and *ZEG* scenarios. Other technological innovations include use of solar energy in steam production and use of bottoming cycles in power generation, as described in the Thermo-Electron study. The savings achieved amount to:

	TF relative to HG (quadrillion Btu's)			*ZEG relative to TF (quadrillion Btu's)*		
	Fuels	*Electricity*	*Total*	*Fuels*	*Electricity*	*Total*
1985	1.1	0.9	2.0	0	0	0
2000	2.2	4.4	6.6	−0.1	0	−0.1

Reaching ZEG

With the energy conservation measures considered thus far, industrial energy consumption in the *ZEG* scenario amounts to about 55 quadrillion Btu's in the year 2000. In order to have zero energy growth, it is necessary to further assume that industrial activity in the year 2000 requires no more than about 47 quadrillion Btu's. The additional 8 quadrillion Btu's reduction in industrial energy consumption requires either a slower growth in industrial output for *ZEG* (compared to the *TF* and *HG* scenarios) or it means that there must be a shift in the mix of industrial output to less energy intensive activities. We consider first the economic consequences of slowing down the growth in industrial output with no change in the mix of output.

Reducing energy consumption by 8 quadrillion Btu's in the year 2000 without changing the mix of industrial output corresponds to a reduction of industrial output of

$$\frac{8 \text{ quadrillion Btu's}}{78{,}000 \text{ Btu's/\$}} \sim \$100 \text{ billion},$$

where 78,000 Btu's/$ is the ratio of energy consumption to value added for the *ZEG* scenario.* This $100 billion plus the $30 billion arising from

* This is derived as follows:

$$78{,}000 \text{ Btu's/\$} = \frac{55 \text{ Quadrillion Btu's}}{\$710 \text{ billion}}$$

$710 billion = ($850 − $30 − $110) billion

$850 billion = value added for manufacturing and energy sectors of the economy for the year 2000 in the *HG* scenario (See Appendix F)

$30 billion = value of reduced production of plastics, aluminum, and steel considered so far.

the reduced growth in output of plastics, aluminum, and steel considered above, corresponds to a 20 percent reduction below the level of manufacturing value added for the *HG* scenario in the year 2000.** Gross national product would be affected much less than this. Because manufacturing accounts for only 20 percent of the GNP in 2000, the cumulative effect of this reduction in manufacturing output by 2000 would be to reduce *ZEG* GNP by 4 percent below the *HG* level.

This estimate is likely to overstate the economic impact of achieving *ZEG*, because in the real economic world there would almost certainly be a shift in the mix of manufacturing output to less energy intensive activities. As indicated in Appendix F, the DRI model shows only a 1.5 percent reduction in manufacturing value added in the *ZEG* scenario for the year 2000, compared to the 20 percent effect indicated above, where no shift in the mix of industrial output is taken into account.

Summary of industrial energy use

The energy savings calculated here are summarized in Tables A-16 and A-17. From Tables A-10, A-16 and A-17, we obtain the following summary of industrial energy use for the three scenarios:

Industrial energy use
(quadrillion Btu's)

		HG		*TF*		*ZEG*	
	1968	*1985*	*2000*	*1985*	*2000*	*1985*	*2000*
Fuels	19.3	34.5	55.8	30.7	50.2	28.9	33.0
Electricity	5.6	17.6	41.1	9.3	12.9	8.9	14.0
Total	24.9	52.1	96.9	40.0	63.1	37.8	47.0

Recycling potential for aluminum

Recycling aluminum would be pursued aggressively in the *TF* and *ZEG* scenarios, since recycled aluminum requires only 9.5 million Btu's per ton for processing, compared to 190 million for primary aluminum. Here we provide a basis for estimating scrap availability.

- Scrap production was estimated by the Conference Board to be 25 percent of total production in 1980. We assume this value persists to the turn of the century for the *HG* scenario.
- To determine scrap production in 1985 for *TF* and *ZEG* scenarios, we must estimate the potential for recycling. The CB assumes that growth in demand for various end uses grows as follows till 1980:

$110 billion = value added of energy production needed for the *HG* scenario but not for the *ZEG* scenario.

** One effect of this cutback would be to further reduce the output of plastics, aluminum, and steel to 31, 15, and 185 million tons per year respectively in the year 2000.

	Annual growth rate (percent)	*Average life (years)*
Building & construction	6.9	50
Transportation	5.3	5
Consumer durables	4.2	5
Electrical	5.1	25
Machinery & equipment	4.1	10
Containers & packaging		
Can stock	12	2
Other	11	2

Table A-16–**Summary of industrial energy savings, TF scenario relative to HG scenario** (quadrillion Btu's)

	Fuels		*Purchased electricity*		*Totals*	
	1985	*2000*	*1985*	*2000*	*1985*	*2000*
5 Key industries	2.5	5.5	1.5	6.1	4.0	11.6
Energy processing (except electricity)	2.1	4.5	0.7	2.8	2.8	7.3
Miscellaneous cogeneration of steam and electricity	−0.3	−2.3	0.7	5.2	0.4	2.9
Heat recuperation with direct use of fuels	−1.6	−4.3	4.5	9.7	2.9	5.4
Belt tightening	1.1	2.2	0.9	4.4	2.0	6.6
Totals	3.8	5.6	8.3	28.2	12.1	33.8

Table A-17–**Summary of industrial energy savings, ZEG scenario relative to HG scenario** (Quadrillion Btu's)

	Fuels		*Purchased Electricity*		*Totals*	
	1985	*2000*	*1985*	*2000*	*1985*	*2000*
5 Key industries	4.2	12.1	1.8	5.6	6.0	17.7
Energy processing (except electricity)	2.3	7.4	0.8	3.1	3.1	10.5
Miscellaneous cogeneration of steam and electricity	−0.3	−2.3	0.7	4.9	0.4	2.6
Heat recuperation with direct use of fuels	−1.7	−3.5	4.5	7.9	2.8	4.4
Belt tightening	1.1	2.1	0.9	4.4	2.0	6.5
Cutback in output	0.0	7.0	0.0	1.2	0.0	8.2
Totals	5.6	22.8	8.7	27.1	14.3	49.9

We assume these growth rates persist to 1985. To estimate aluminum available for recycling we use for each category

$$M(\text{available}) = M_{1985}e^{-kt}$$

where

M_{1985} = aluminum consumption in 1985
k = growth rate
t = average life

We thus obtain for 1985:

	M_{1985} *(million tons)*	*M(available) (million tons)*
Building & construction	3.46	0.11
Transportation	2.30	1.75
Consumer durables	1.02	.63
Electrical	1.59	.45
Machinery & equipment	.69	.45
Containers & packaging		
Can stock	2.02	1.59
Other	1.50	1.20
Other	0.85	–
Totals	13.43[a]	6.18

[a] About 22 percent of aluminum needs would be met with imports with *HG* and *TF* scenarios. About 26 percent would be imported with *ZEG*.

The difference between *TF* and *ZEG* is that we assume
50 percent of this old scrap (3.1 million tons) is recycled in *TF*
75 percent (3.4 million tons) is recycled in *ZEG**
Also, new scrap, according to the Thermo-Electron study, amounts to 14 percent of primary production.

• To determine scrap production for *TF* and *ZEG* in 2000, we assume that beyond 1985 all demand sectors grow at 4.9 percent per year, so that the M(available) is given by the following:

* With *ZEG*, the maximum recyclable old scrap material is diminished since there *are no* aluminum cans.

	M_{2000} *(million tons)*	*M(available)* *(million tons)*
Building & construction	7.23	0.64
Transportation	4.80	3.76
Consumer durables	2.14	1.80
Electrical	3.33	0.98
Machinery & equipment	1.44	0.88
Containers & packaging		
Can stock	4.20	3.80
Other	3.14	2.86
Other	0.85	–
Totals	27.13	14.72

As in 1985, we get old scrap available as follows:

7.4 million tons for *TF* scenario
8.2 million tons for *ZEG* scenario

Also we assume new scrap amounts to 14 percent of primary production.

Energy processing losses

Electric power generation (processing energy per unit of delivered electricity):

	1985	*2000*
Transmission losses	0.1	0.1
Generation losses	1.8	1.6
(Heat rate in Btu/kwh)	9560	8870
(Heat rate at point of use)	9900	9220

Uranium enrichment (processing electric energy per unit of generated nuclear electricity):

.048

Source: *Fuel Cycles for Electric Power Generation* by Thomas Pigford et al. EPA Rept. No. 68-01-0561, January 1973.

Petroleum refining (processing energy per unit of energy input):

Present practices		*With process improvements suggested by thermo electron*	
Fuels	0.1215	Fuels	0.106
Purchased Power[a]	0.0050	Purchased Power[a]	−0.011
Total	0.1265	Total	0.095

[a] Energy inputs to power generation.

Natural gas consumed in oil and gas processing (as fraction of gas consumption):

Oil and gas Fields	0.063
Pipelines	0.033
Total	0.096

Source: Minerals Yearbook, Vol. I, Metals, Minerals, and Fuels, U.S. Bureau of Mines, 1970, p. 740.

Synthetic fuels from coal (processing energy per unit of oil or gas produced):

0.4

Appendix B–Capital requirements for conservation: Technical Fix vs. Historical Growth

The conservation measures of the *Technical Fix* scenario generally involve the investment of capital. The result of such investments is to provide the same energy benefits as those available in the *Historical Growth* scenario. Table B-1 shows the capital requirements for fuel and electricity production in the *Historical Growth* scenario. All dollar figures are in constant 1970 dollars. Marketing capital costs have been excluded from the energy industry investment, since the capital costs for marketing fuels can be assumed to be replaced by marketing costs for energy conservation technologies. In any case, the capital involved is small, less than $100 billion, relative to the total in Table B-1.

These capital costs are only approximate; they could be changed significantly (for a given level of energy use) by the mix of fuels, the mix of technologies for using the fuels, environmental regulations, and the amounts of petroleum crude and refined product imports.

The estimates of the capital requirements for energy conservation are even more approximate than the energy industry capital costs. Some of them are based on EPP studies,[1] others are based on available publications, and still others on extrapolation or conversations with knowledgeable people (for example, in the transport sector); a few minor items are speculative. The capital costs for the steel plants, the plants to manufacture boilers or heat exchangers, etc., have not been taken into account since the basic materials and their processing are required to build refineries and pipelines as well. Moreover, these capital costs are implicit in the production rate of various materials, which are the same in the *HG* and *TF* scenarios.

Residential and commercial sectors

The capital costs for the residential sector are shown in Table B-2.

We assume commercial sector capital requirements per Btu saved are approximately equal to those in the residential sector, since the main savings in both sectors arise from space heating and cooling. Cumulative capital requirements in the commercial sector are thus:

$$\left(\frac{\text{Btu's saved in the commercial sector in the year 2000}}{\text{Btu's saved in the residential sector in the year 2000}}\right) \times \$122 \text{ billion} = \left(\frac{4.9}{11.7}\right) \times 122 = \$51 \text{ billion.}$$

Transportation

The main energy savings in the transportation sector occur from a shift to more efficient cars, or smaller cars, or both. In general, the capital requirements for smaller cars are lower than those for larger ones, but this may be offset by the necessity for more safety equipment in

[1] For example, E.P. Gyftopoulos, Lazaros J. Lazaridis, and Thomas F. Widmer, *Potential Fuel Effectiveness in Industry,* a Report to the Energy Policy Project, Cambridge, Mass.: Ballinger, 1974

Table B-1–**Cumulative capital requirements (1975–2000) for the energy industry, HG scenario**[a]

	billions of 1970 dollars
1. Domestic oil and gas[b, c]	750
2. Natural gas pipeline	150
3. Coal production and transport	70
4. Nuclear fuel cycle	30
5. Electric generation and transmission	750
Total	1750

Source: Jerome E. Hass et al., *Financing the Energy Industry,* a Report to the Energy Policy Project, Ballinger Publishing Company, Cambridge, Mass., 1974.

[a]The Hass study estimates capital needs to 1985. These have been approximately extrapolated to 2000. A linear extrapolation is used for the petroleum, natural gas, and coal sectors, and the extrapolation of electric generation is based on new capacity at $500/kw for generation, transmission, and distribution.

[b] Includes oil and natural gas, synthetic oil and gas from coal, and shale oil.

[c] Petrochemical plants, natural gas transmission, marketing, and exploration costs are excluded.

Table B-2–**Approximate cumulative capital requirements (1975–2000) for the residential sector, TF scenario**

	Dollars per house	*Number of homes (millions)*	*Cumulative capital (billions of dollars)*
1. Space heat			
Backfitting insulation & storm windows	500	50	25
Insulating new homes	700	60	42
Heat pumps instead of resistance heat and air conditioning	400	35	14
Solar space heating and water heating	1000	10	10
Heating efficiency (fossil fuel)	100	70	7
2. Air conditioning	100	150	15
3. Water heat (gas instead of electric)	100	40	4
4. Refrigerators	—	—	—
5. Miscellaneous	100	50	5
6. Total			122

Sources: Private communication with insulation manufacturer Owens Corning; "Comparison of Total Heating Costs with Heat Pumps vs. Alternate Heating Systems," Westinghouse Electric Corporation; John Moyers, "The Value of Thermal Insulation in Residential Construction," Oakridge National Laboratory, Oakridge, Tennessee, December 1971.

smaller cars, and by added capital requirements for large but efficient cars. We assume that the difference in capital outlay for the purchase of cars will not be substantial between the *TF* and *HG* scenarios. This assumption gains some credence from the current trend toward smaller but more "luxurious" cars, which don't cost much less than the large ones, but have significantly higher efficiencies.

Large savings also accrue in air transport and ground freight transport. There is a shift of 75 billion passenger miles of short haul passenger air traffic to intercity rapid rail. On long distance flights, the load factor in the *TF* scenarios is 67 percent, compared with 54 percent in the *HG* scenario. This means that besides the capital saving from lower investments in energy production, less capital will be needed in *TF* for airplanes, airports, etc. On the other hand, substantially larger investments will be needed for the intercity rapid rail systems.

Since the intercity rapid rail service is a short haul service, it will probably be confined to the major metropolitan corridors. Thus full development would mean about 1,500 to 2,000 miles of track along the Eastern Seaboard, 1,000 to 1,500 miles along the West Coast and 1,000 to 2,000 miles of miscellaneous runs between major cities—for example, Houston–New Orleans, St. Louis–Kansas City, Chicago–Milwaukee. The costs per mile of these systems are of course speculative, but we can arrive at an estimated order of magnitude by examining the costs of various rail systems that are planned or are being built. A recent National Academy of Sciences study[2] estimates the capital cost of high speed rail travel between Washington and Boston on new right-of-way at $2.6 billion. This works out to about $5 million per mile. If we use an estimate of $5 million per mile, the capital investment for a fairly comprehensive system of high speed short haul intercity rail comes to $25 billion. In addition, about $5–10 billion would be required for cars and power units.[3]

This capital required for railroads would be offset by reduced requirements for planes and trucks. Shifting about 75 billion passenger miles of air travel to rapid rail and improving the airplane load factor means that perhaps 1,000 fewer planes would be needed. At $20 million per plane, this means that on the order of $20 billion less would be committed to airplanes in the *TF* scenario as compared to the *HG* scenario. In addition, about 300 billion ton miles of freight traffic would be shifted from truck to rail, corresponding to about 800,000 fewer trucks. An investment of $20 thousand per truck means that $16 billion less is required for trucks in the *TF* scenario.

The industrial sector

The main energy savings in the industrial sector are in the areas of steam production and direct heat. The Thermo-Electron study investigates the capital costs of energy savings in these areas for the energy intensive industries. We assume that the capital costs for saving energy in other areas of industry will be similar.

---

[2] "Jamaica Bay and Kennedy Airport," a report of the Jamaica Bay Environmental Study Group, National Academy of Sciences–National Academy of Engineering, Washington, D.C., 1971.

[3] Richard Rice, "Toward More Transportation with Less Energy," *Technology Review,* February 1974.

The capital costs of installing recuperators designed to save a billion Btu's per year in a direct heat-using process are about $2,500. The additional capital cost for the solar-diesel steam raising (over a conventional boiler system) would be in the range of $5,000–$10,000 for saving a billion Btu's per year, depending on the cost of the solar collector ($1 to $3 per square foot). The additional costs of combined steam and electricity generators are about $5,000 for saving a billion Btu's per year. The total additional capital costs for the direct heat and steam production sectors are shown in Table B-3. We assume that the capital costs for saving energy in other industries and processes will be of the same order of magnitude.

Table B-3–**Capital requirements for industrial energy conservation, TF scenario**

	Capital ($ per billion Btu's annual saving)	*TF energy savings (quadrillion Btu's)*	*Total cumulative capital (1975–2000) required (billions of $)*
Direct heat recuperation	2,500	3	7.5
Fossil fuel—solar	5,000–10,000	2	15
Steam—electric	5,000	6	30
Other	5,000	23	115
Total		34	167.5

Summary

The total capital requirements for energy conservation in the *TF* scenario are shown in Table B-4. We have added 20 percent to this total

Table B-4–**Approximate cumulative capital (1975–2000) required for conservation measures, TF scenario**

	$ billions
1. Residential and commercial sectors	170
2. Transportation	—
3. Industrial	170
Subtotal	340
20% for infrastructure	70
Grand total	410

to account for the infrastructure capital requirements such as R&D, setting up new industries, modification of existing industries, and so forth. This has been done because energy conservation technology is not as well developed as energy production technology, so that some general additional capital requirements should be anticipated. Even so, the total capital required for energy conservation—$410 billion—is 40 percent less than the capital required to develop the corresponding energy facilities if

Table B-5–**Fuel capacity required in the year 2000 and capital available**

	HG (quadrillion Btu's per year)	*TF (quadrillion Btu's per year)*	*Difference in cumulative capital costs, 1975–2000 ($ billion)*
1. Domestic oil and gas [a]	95	75	150
2. Natural gas pipelines	37	32	25
3. Coal[b]	44	27	50
4. Nuclear fuel	40	11	20
5. Utility electric	74	31	450
6. Difference in cumulative capital requirements in energy supply (HG-TF)			695

[a] Excludes petrochemical plants, natural gas transmission, marketing and exploration costs; includes oil and natural gas, synthetic oil and gas from coal and shale oil.

[b] The difference in annual production between *HG* and *TF* in the year 2000 is 680 million tons per year. To build up this additional capacity (production and transportation) would require about $2 billion per year.

the energy were not conserved. Table B-5 shows the capital requirements for producing and processing the energy that is conserved in the *TF* scenario.

The Hass study (see Table B-1) estimates that the share of available capital that will be used by the energy industry would increase from the present 21 percent to 26.6 percent in 1985 in the *HG* scenario. A continuation of such a trend would lead to a 30 percent share by 2000 in the *HG* case. Under such conditions, capital could become scarce for other industries—which "wouldn't give it up without a fight"—and this could lead to rising interest rates. The total cumulative capital requirements to 2000 for the *TF* scenario energy sector (both production and conservation) would be about $1,450 billion, compared to $1,750 billion in the *HG* case. The $1,750 billion figure represents roughly 25 percent of the cumulative 1975–2000 capital requirements of all industries. The average share of investment in the energy sector in the *TF* scenario would be about 20 percent—or about what it has been in recent years.

Appendix C–Energy supply notes

Here we describe our future energy production estimates, particularly for the *Historical Growth* scenario, in the context of recent research on energy supplies. We concentrate on the *Historical Growth* scenario because the largest demands on energy resources are in that scenario.

The oil and gas resource base (excluding shale oil and synthetics from coal) will be the most strained in all scenarios. We discuss oil production in relation to reserves and resources below, and assume that, as in the past, associated natural gas will make up much of the gas supply. Natural gas liquids such as propane are included in the energy numbers for oil.

The domestic oil and gas supply case of the *Historical Growth* scenario (see Table 3, Chapter 2) pushes hardest on oil and gas resources. In that case, the cumulative use of domestic oil between 1973 and 2000 is 800 quadrillion Btu's.[1] Therefore, with an annual production of 40 quadrillion Btu's in 2000 and a 10 to 1 reserves to production ratio, the total requirements for proved reserves of petroleum in the 1973–2000 period are about 1200 quadrillion Btu's. The oil requirements for other supply cases and scenarios are also large, but they are 10 to 25 percent less than those for the *HG* domestic oil and gas case.

The oil production and proved reserves requirements for the *HG* domestic oil and gas case are well within the estimates of reserves and estimated additional recoverable resources (see Table 2, Chapter 2, and Table D-2 Appendix D). This also holds for natural gas. The estimates that we have used for oil production are compatible with estimates in the report of Resources for the Future (RFF) to the Project[2] and with those the National Petroleum Council (NPC) made in a 1972 study.[3]

NPC estimated that oil output could be raised to 13–15 million barrels per day[4] range by 1985 (up from 11 million last year) with prices

[1] Projections for liquid fuels consumption must be consistent with demand projections for end uses that can be met only with liquid fuels. The liquid fuels requirements for the domestic oil and gas supply option of the *HG* scenario in 2000 are derived on the basis of the following minimum liquid fuel requirements:

	Quadrillion Btu's
Fifty percent of fuel consumed directly in the residential/commercial sectors (see table 1, Chapter 2)	10
Fuel for transportation (see Table 1)	38
Eighty percent of total feedstocks (see Table A-10, Appendix A)	10
Residual fuel for boilers (10 percent of total petroleum)	6
Total	64

[2] Resources for the Future, "Toward Self Sufficiency in Energy Supply," a draft report to the Energy Policy Project, September 1973.

[3] National Petroleum Council, "U.S. Energy Outlook," Washington, D.C. 1972.

[4] Note that 1 million barrels per day is roughly equivalent to 2 quadrillion Btu's per year.

in the $7–8 range[5] and favorable government policies concerning development, particularly offshore. In its report to the Project, Resources for the Future estimated, on the basis of past relationships between increases in price and the discovery of new reserves, that oil output by 1985 could be as high as 16 million barrels per day at $6 per barrel. These price figures are, of course, "equilibrium" estimates in the sense that they are the minimum believed necessary to elicit the specified supply levels, given appropriate government policies.

Few forecasters are willing to project supply availabilities for 2000, but RFF did make some approximate estimates of the possibilities. Based on an examination of the resource base and likely costs of recovery, RFF concludes that domestic oil production could continue to grow at least until the turn of the century to anywhere from 20 to 29 million barrels per day. The domestic oil and gas supply case of the *Historical Growth* scenario corresponds approximately to the lower range of the above RFF estimate—i.e., about 20 million barrels per day in 2000. RFF estimates that prices would have to be $7 to $9 per barrel to support production in the range of 20 to 29 million barrels a day.

In the *HG* scenario, the large amount of nuclear power required to be produced by thermal reactors would necessitate some development of uranium resources in the $10–20 per pound range (see Table 2, Chapter 2). If the reserves to production ratio is assumed to be approximately 10 to 1, the total proved reserves requirements to the year 2000 would have to be 800–1,000 quadrillion Btu's. For the *Technical Fix* and *Zero Energy Growth* scenarios, the nuclear power requirements are much smaller and there would still be ample supplies of uranium at less than $10 per pound available through 2000. In all scenarios, the existing reserves of coal (5,000 quadrillion Btu's) are more than sufficient to supply the required energy.

The "Other" category in the supply tables is given particular emphasis in the *ZEG* scenario (see Table 24, Chapter 4). This category includes primarily energy from organic wastes (urban and agricultural) and solar energy. We expect that most of this energy would come from organic wastes, since the technologies for the conversion and use of organic wastes are, in general, closer to commercialization than are solar energy technologies.[6] Use of urban trash for electricity, pyrolysis of urban and rural wastes to liquid fuels, and the conversion of organic wastes (primarily agricultural wastes and urban sewage) to methane (pipeline quality gas) are some of the technologies we anticipate will be used.

---

[5] All prices in this Appendix are 1972 dollars.

[6] Alan Poole, "Potential for Energy Recovery from Organic Wastes," draft report to the Energy Policy Project, June 1974.

Appendix D–Major energy resources

Estimating potential energy resources is far from an exact science and is very much subject to the views and judgments of specialists interpreting geological, technological, and economic data. Over time, specialists may change the definitions and nomenclature of resource classification. More important, changes in exploration technology, mining and development technology, economic costs of production, and market price of each resource can lead to changes in estimates, both of overall resource quantities and of amounts in different resource categories.

Here we summarize some of the more up-to-date resource estimates for major energy sources, to provide readers with a quick overview of the terminology, data, and sources of energy resource estimates.

Terminology

Sources of energy—In some publications the term petroleum is used to refer to both liquid and gaseous hydrocarbons. It has become common practice, however, to use "petroleum" to refer to liquids alone, and "natural gas" to refer to gaseous hydrocarbons. Here the term petroleum will be used to include crude oil (as defined by the American Petroleum Institute)[1] and natural gas liquids. The term natural gas will exclude liquids recovered from natural gas.

The term coal refers to bituminous and anthracite coal and lignite. Raw shale oil is a black, viscous substance that is derived from the organic material kerogen found in the marlstone rock called oil shale. It can be extracted and refined into petroleum products.

Resources and reserves—The definitions presented here are those jointly used by the U.S. Department of the Interior's Geological Survey and Bureau of Mines.[2] Here resources refers to concentrations of naturally occuring solids, liquids, or gaseous material in or on the earth's crust, discovered or surmised to exist in such form that economic extraction of a commodity is currently or potentially feasible at higher future prices.

Resource availability is expressed in terms of (1) degree of certainty based on the extent of geologic knowledge about the existence and characteristics of the resource, and (2) feasibility of its economic recovery. The degree of certainty is classified into identified and undiscovered. The feasibility of economic recovery is distinguished by the terms recoverable, paramarginal, and submarginal. (See Fig. D-1).

The term identified resources refers to resources whose location, quality, and quantity are known from geologic evidence supported by appropriate engineering measurements. Materials classified as reserves

---

[1] American Gas Association, American Petroleum Institute, and Canadian Petroleum Association, "Reserves of Crude Oil, Natural Gas Liquids, and Natural Gas in the U.S. and Canada and U.S. Productive Capacity as of December 31, 1971," Vol. 27, p. 13, May 1973. "Crude Oil" is defined as including liquids and condensates.

[2] U.S. Department of the Interior, "New Mineral Resource Technology Adopted," April 15, 1974 (press release); and V.E. McKelvey (U.S. Geological Survey), "Hydrocarbon Reserves and Resources in the United States," testimony before the Special Subcommittee on Integrated Oil Operations, Senate Committee on Interior and Insular Affairs, February 20, 1974 (manuscript).

Figure D-1–**Classification of Energy Resources**

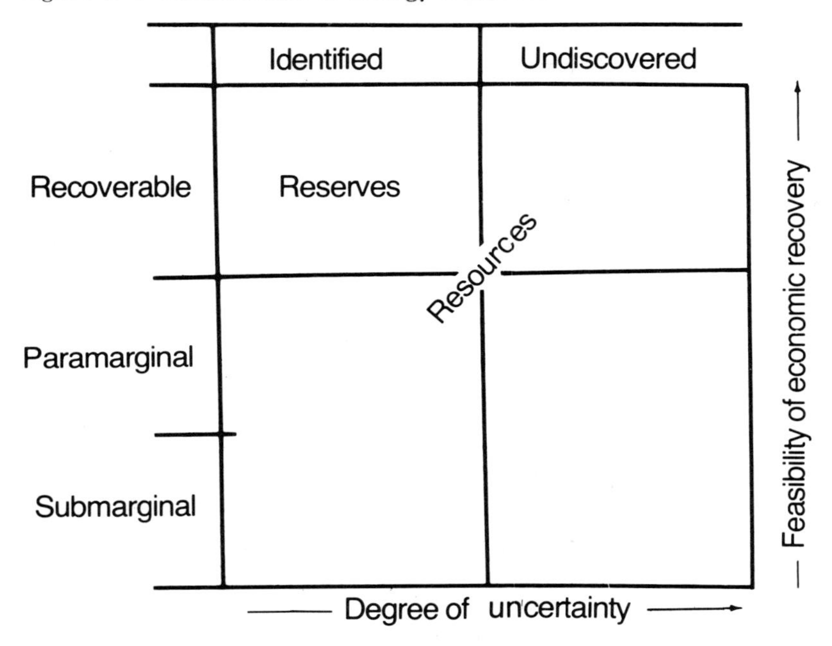

Source: U.S. Geological Survey

are those identified resources legally recoverable with existing technology under present economic conditions.

Resources which have been identified can further be subdivided into measured, indicated, and inferred categories. Measured resources are those whose quality and quantity have been estimated, within a margin of error of less than 20 percent, for geologically well known sample sites. The term indicated is applied to resources whose quantity and quality have been estimated partly from sample analyses and partly from "reasonable" geologic projections. The term inferred refers to materials in unexplored but identified deposits estimated on the basis of geologic evidence.

Undiscovered resources include those in areas which are surmised to exist on the basis of broad geological evidence and theoretical considerations.

The term recoverable or economically recoverable implies feasible exploitation of the resource under current technological and economic conditions. The term paramarginal refers to "that portion of sub-economic resources that (a) borders on being economically producible or (b) is not commercially available solely because of legal or political circumstances."[3] Resources that are identified as submarginal are those not recoverable under existing technological or economic conditions, because of size and location, or both. They are resources that would

[3] In an earlier paper, McKelvey and others included in the definition of "paramarginal" the condition of being "recoverable at prices as much as 1.5 times those prevailing now." See V.E. McKelvey and others, "Mineral Resource Estimates and Public Policy," *American Scientist,* 60: 1, pp. 32–40, 1972.

require prices substantially higher than current or foreseen prices ("more than 1.5 times the price at the time of determination"),[4] or advances in technology that would result in major reductions in the cost of production. These terms can apply to both identified and undiscovered resources that are considered subeconomic under current conditions.

Some of the quantities now classified as submarginal could become paramarginal if prices rise sufficiently and/or if new methods of discovery and recovery are developed to make development more economical. Also, resources that were considered paramarginal could become part of recoverable resources, as has happened with oil shale.[5]

According to the U.S. Geological Survey (USGS), "The estimates of identified recoverable resources in the upper left corner of the diagram (Fig. D-1) are within 20–50 percent of the correct value (but, as stated earlier, *measured* resources involve less than 20 percent error). The estimates of undiscovered submarginal resources in the lower right corner of the diagram reflect only an order of magnitude degree of accuracy. To be realized, these less known resources, which constitute the bulk of the resource base, will require great advances in the technologies of both search and extraction. Similarly, the submarginal identified resources require advances in extraction technology before utilization is feasible, whereas the undiscovered recoverable resources require a continuing effort in both exploration and exploration research as well as economic incentive."[5]

Choice of data

Most of the data presented here were generated by the main energy agencies—the Interior Department and the Atomic Energy Commission—but principally from the USGS, including estimates of others they incorporate into their own.

In many cases, "different" resource estimates are not independent. For example, two of the most widely cited estimates of coal resources in the United States are those of the Bureau of Mines (1971)[6] and of M. K. Hubbert (1969).[7] These are, however, both derived from the 1969 study by Paul Averitt of the U. S. Geological Survey.[8] The Bureau of Mines estimates of reserves are actually estimates of the USGS.

In the petroleum field, the only estimates of identified resources of petroleum liquids and natural gas for the United States are those published by the American Petroleum Institute and the American Gas Association; the USGS adopts these as part of their own estimates. The API-AGA estimates are prepared annually by local committees composed of oil and natural gas specialists from oil and gas companies and state

[4] U.S. Department of the Interior, "New Mineral Resource Technology Adopted," op. cit.

[5] P.K. Theobald, S.P. Schweinfurth, and D.C. Duncan, "Energy Resources of the United States," U.S. Geological Survey Circular 650, Washington, D.C., 1972.

[6] U.S. Bureau of Mines, *United States Coal Resources and Production—an Interim Report,* Washington, D.C., June 1971.

[7] M. K. Hubbert, "Energy Resources," in *Resources and Man,* National Academy of Science-National Research Council, Committee on Resources and Man, Washington, D.C., 1969, pp. 157–242.

[8] Paul Averitt, *Coal Resources of the United States, January 1, 1967,* U. S. Geological Survey Bulletin 1275, Washington, D.C., 1969.

governments. These local reports are aggregated by states and for the nation.

Estimates of undiscovered oil and gas estimates vary widely. The estimates of petroleum liquids and natural gas resources made by the U.S. Geological Survey tend to be higher than other reported estimates. This is, in general, because USGS estimates include a larger proportion of ground favorable for exploration. It is noteworthy that the most recent (1974) USGS resource estimates for petroleum, natural gas, and oil shale are lower than their 1972 estimates.[9]

The USGS oil shale resource estimates, although based mainly on work published in 1965, appear to be the only reliable estimates available. D. C. Duncan, whose 1965 paper is frequently cited,[10] worked closely with the National Petroleum Council in connection with the NPC's estimates for their recent reports. 1974 estimates to be published by the USGS refer to the earlier paper, but with the current estimates on the order of 50 percent of 1965 and 1972 estimates.[11]

Tabulations of resources

Resource estimates for coal, petroleum, natural gas, oil shale, and uranium are given in Tables D-1 through D-7. Summaries of these tables are contained in Table 2 of Chapter 2. A fuller review of resource estimates can be found elsewhere.[12]

---

[9] U.S. Department of Interior, "USGS Releases Revised U.S. Oil and Gas Resource Estimates," March 26, 1974 (press release).

[10] Donald C. Duncan and Vernon E. Swanson, *Organic-Rich Shale of the United States and World Land Areas,* U.S. Geological Survey Circular 523, Washington, D.C., 1965.

[11] P. K. Theobald, S.P. Schweinfurth, and D.C. Duncan, *Energy Resources of the United States,* U.S. Geological Survey Circular 650, Washington, D.C., 1972; D.A. Brobst and W.P. Pratt, eds., *United States Mineral Resources,* U.S. Geological Survey Professional Paper 820, Washington, D.C., 1973; and tables prepared for the World Energy Conference (1974) by the U.S.G.S. staff.

[12] "U.S. Energy Resources, a Review as of 1972," a background paper prepared for the Committee on Interior and Insular Affairs, U.S. Senate, Washington, D.C., 1974.

Table D-1–**U.S. coal resources, Jan. 1, 1972**
(trillions of short tons)

Estimated identified coal resources remaining in the ground, 0–3,000 feet[a]	
Bituminous coal and lignite	1.56
Anthracite coal	.02
Total	1.58
of which	
Subeconomic	1.18
Economic	
Recoverable	.20
Inaccessible[b]	.20
Estimated undiscovered coal resources geologically predictable as existing[c]	
Overburden, 0–3,000 feet	1.30
Overburden, 3,000–6,000 feet	.34
Total estimated undiscovered coal resources	1.64
Total coal resources, 0–6,000 feet	3.22

[a] Estimates are based on several studies made of different areas on various dates, minus depletion as a result of production and loss in mining from these dates to Jan. 1, 1972. Identified resources may or may not be profitably recoverable with existing technology or economic conditions.

[b] "Inaccessible" reserves include the amount lost in mining and those located under cities or gas wells and similar conditions.

[c] Such resources may or may not be economically recoverable.

Sources: D. Brobst and W. Pratt, eds., *United States Mineral Resources,* Geological Survey Paper 820, Washington, D.C., 1973, p.137, National Petroleum Council, *U.S. Energy Outlook, Coal Availability,* 1973; and conversation with Z. Murphy, Coal Specialist, Bureau of Mines, January 22, 1974; USGS paper presented at the World Energy Conference, Detroit, September 1974.

Table D-2–**Recoverable U.S. petroleum resources as of February 14, 1974**
(billions of barrels)

	Reserves		*Undiscovered*
	Measured	*Indicated-inferred*	*recoverable resources*
Onshore			
Coterminous U.S.	31	17–29	110–220
Alaska	10	5–10	25– 50
Total onshore	41	22–39	135–270
Offshore (to 200 meters deep)			
All, excluding Alaska	7	3– 6	35– 70
Alaska	1	—	30– 60
Total offshore (to 200 meters deep)	8	3– 6	65–130
Total	49	25–45	200–400

Note: This table lists only resources estimated to be recoverable under the economic and technological conditions at the time the estimates were made. Paramarginal and submarginal resources are not included. Table D-3 gives total petroleum resources, from earlier estimates.

Source: U.S. Department of Interior, "USGS Releases Revised U.S. Oil and Gas Resource Estimates," March 26, 1974 (press release).

Table D-3–**Total U.S. petroleum resources (including Alaska) as of December 31, 1970**
(billions of barrels)

	Identified (remaining)	*Undiscovered*
Recoverable	50	150– 450
Paramarginal and submarginal	290	280–2,100

Sources: P.K. Theobald et al., *Energy Resources of the United States,* USGS Circular 650, Washington, D.C., 1972. The lower estimates of undiscovered resources are derived from "Future Petroleum Provinces of the United States," prepared by the National Petroleum Council Committee on Possible Future Provinces of the U.S., 1970. Most other estimates fall between the USGS estimates and the NPC estimates.

Table D-4–**Recoverable U.S. natural gas resources as of February 14, 1974**
(trillions of cubic feet)

	Reserves		*Undiscovered recoverable resources*
	Measured	*Indicated-inferred*	
Onshore			
Coterminous U.S.	190	93–177	500–1000
Alaska	28	14– 28	105– 210
Total onshore	218	107–205	605–1210
Offshore (to 200 meters deep)			
All, excluding Alaska	46	22– 43	225– 450
Alaska	2	1– 2	170– 340
Total offshore (to 200 miles deep)	48	23– 45	395– 790
Total	266	130–250	1000–2000

Note: This table lists only resources estimated to be recoverable under the economic and technological conditions at the time the estimates were made. Paramarginal and submarginal resources are not included. Table D-5 gives total natural gas resources, from earlier estimates.

Source: U.S. Department of Interior, "USGS Releases Revised U.S. Oil and Gas Resource estimates," March 26, 1974 (press release).

Table D-5–**Total U.S. natural gas resources (including Alaska) as of December 31, 1970**
(trillions of cubic feet)

	Identified (remaining)	*Undiscovered*
Recoverable	290	1,200–2,100
Paramarginal and submarginal	170	4,000

Sources: P.K. Theobald et al., "Energy Resources of the United States," USGS Circular 650, Washington, D.C., 1972. The lower estimate for undiscovered recoverable resources is from "Potential Supply of Natural Gas in the United States as of December 31, 1970," prepared by the Potential Gas Committee, October 1971.

Table D-6–**U.S. oil shale resources, 1972**
(billions of barrels, by oil yield)

Oil Shale yield	*Identified deposits*	*Undiscovered resources*
25 to 100 gallons per ton	418	900
10 to 25 gallons per ton	1,600	25,000
5 to 10 gallons per ton	2,200	138,000

Note: These data are based mainly on 1965 estimates. More recent estimates use a different breakdown and tend to be more conservative. In the tables prepared for the World Energy Conference, the U.S. Geological Survey lists only known reserves with yields of fifteen gallons or more per ton. The Survey lists 209 billion barrels in deposits yielding 30 or more gallons per ton and 1,000 billion barrels in deposits yielding 15 or more gallons per ton.

Sources: D. Brobst and W. Pratt, eds., *United States Mineral Resources,* Geological Survey Professional Paper 820, Washington, D.C., 1973; C. Duncan and V. Swanson, *Organic-Rich Shale of the United States and World Land Areas,* Geological Circular 523, Washington, D.C., 1965.

Table D-7–**U.S. uranium resources at various costs**
(thousands of tons of U_3O_8)

Cost	*Reserves*	*Additional potential resources*
Up to $8 per pound	*277*	*450*
Up to $10 per pound	*340*	*700*
Up to $15 per pound	*520*	*1000*
$15 to $30 per pound	*140*	*630*

Note: The USGS estimate (see P.K. Theobald et al., *Energy Resources of the United States*) of total uranium resources (identified plus undiscovered) is much greater than the USAEC estimate, giving a total of nearly 9 million tons recoverable at up to $20 per pound.

Sources: USAEC news release, March 27, 1974, for uranium up to $15 per pound. J.A. Patterson, "Outlook for Nuclear Fuel," paper presented by the Assistant to the Director, Division of Raw Materials, USAEC, at the IEEE-ASME Joint Power Generation Conference, Pittsburgh, September 29, 1970, for uranium from $15 to $30 per pound.

Appendix E–Government organization and reorganization for energy

A great number of government organizations at the federal, state and local levels make decisions which affect energy policy. The many applicable laws, regulations, and government agencies came about, not in response to a need for energy policy, but in response to a variety of social goals and objectives which range from environmental protection to control over monopolistic corporate practices. The result is great fragmentation of energy policy decision-making. Consolidation and centralization of responsibilities via extensive government reorganization is neither desirable nor possible. In the concluding chapter of this report, we have suggested a means of providing common policy direction and coordination. The purpose of this Appendix is to examine in a nutshell the government energy structure, and explore areas where reorganization might be useful.

Five functional areas of policy responsibility will be treated separately:

- Policy development and program coordination;
- Regulation of the energy sector, including,
  - Economic controls,
  - Fuels allocation and import controls,
  - Facility siting and land use,
  - Environmental and safety regulations;
- Research and development;
- Energy resource development;
- Energy conservation.

For each functional area, this Appendix will describe the major energy agencies and legislative bodies, their authorities and their basic functions. Major problem areas will be briefly highlighted; and principal reorganization proposals will be mentioned.

A. Policy development and program policy coordination

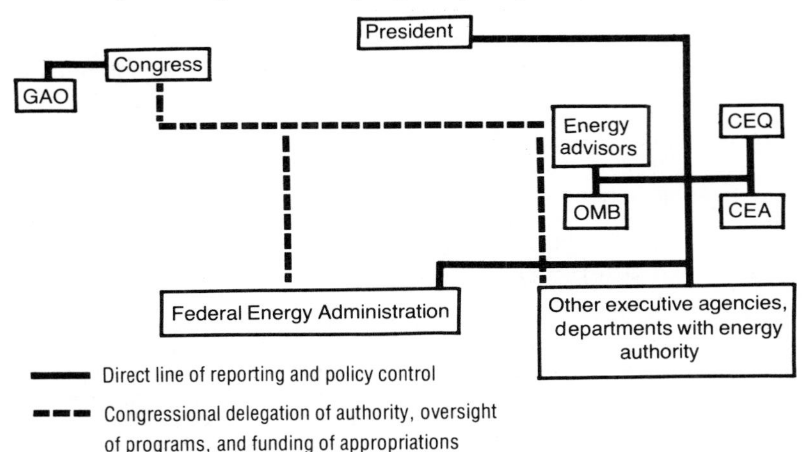

Presidential energy messages in 1971 and 1973 were recognition by the White House of the need for more comprehensive energy policy.

When energy was elevated to a major national issue in later 1973 and early 1974, a series of policy steps were taken, which culminated in the establishment of the Federal Energy Administration by Act of Congress. Its exact overall policy role is uncertain, however, and energy policy is still being made on an *ad hoc* basis.

As described in the following sections, each agency has a particular program responsibility. Agency programs are monitored, examined, and often determined in the Executive Office of the President. They are also subject to congressional oversight and funding, and ultimately, delegation or removal of program authority. A number of Congressional committees have responsibility as delineated in the following sections.

In the Executive Office, four identifiable groups are responsible for controlling energy policy and advising the President on energy matters.

Energy Advisors: This is not a permanently constituted group, and very often those performing this function are difficult to identify. In the past three years. this function has been performed, *ad seriatim,* by the Office of Science and Technology, the Domestic Council, a triumvirate of presidential advisors, the Secretary of the Treasury, several "energy czars," and an energy council.

Office of Management and Budget (OMB): It exercises powerful control over agency programs in developing the President's annual budget for the executive branch. Through this mechanism, it often has a substantial influence on program policy.

The Council of Economic Advisors (CEA): This is a permanent three-man body created by the Employment Act of 1946. It analyzes national economic policy, appraises the policies of the Federal Government, and advises the President on economic developments and new policies for economic growth and stability.

The Council on Environmental Quality (CEQ): This is also a three-man advisory body, created by the National Environmental Policy Act of 1969.* It is charged with developing and recommending national policies for the promotion of environmental quality and continually analyzing changes or trends in the national environment.

The *Federal Energy Administration,* established in 1974 by Act of Congress for an initial term of two years, is charged with developing near-term energy policies, collecting energy data, evaluating economic impacts of energy programs, and other functions, including fuels and energy price regulatory functions as discussed later.

On the Congressional side, the individual committees are assisted by the *General Accounting Office,* which has power to study, monitor and report on the energy activities of Executive Branch agencies.

At the state level, a number of states have taken actions to help governors and legislatures with energy policy problems. These vary

---

* The National Environmental Policy Act (NEPA) also establishes broad environmental goals for the nation. It has specific "action-forcing" provisions directed at federal agencies. Most important, to date, has been the requirement for preparation of environmental impact statements for all major federal actions affecting the quality of the environment. As pointed out in the text of this report, (see, e.g., the chapter on Federal Energy Resources) the impact statements often have been lacking as planning and public information documents. NEPA also requires federal agencies to develop program alternatives to better protect the environment, and undertake environmental studies.

widely state by state. Included are energy agencies, special energy task forces, standing policy planning and advisory groups to governors, and study efforts in universities and government agencies.

A major generic problem is that no permanent mechanism exists for developing energy policy options and presenting them to the President and the Congress for consideration. Instead, key decision-makers receive and review fragments of energy policy, never a comprehensive package of policy options. A second major gap is the lack of oversight and coordination of the major energy programs. Energy policy is merely the sum of the individual major energy programs. A system is needed to monitor all major energy program policy developments, evaluate the cross-effects of these developments with overall energy policy and other national policies, and provide guidance and recommendations to reconcile conflicts.

An Energy Policy Council, created by Act of Congress, is one viable concept for maintaining an effective ongoing program and insuring that assessment of the energy system is subject to Congressional and public scrutiny. Creation of such a council by legislative mandate, complete with a description of broad national energy goals, would provide a popular nationwide focus for energy policy. The Senate in 1973 passed a bill which would create a three-man Energy Policy Council, but it was not acted on by the House. An alternative is to have these functions performed by a line agency such as FEA. The Nixon Administration proposal is to create a single Department of Energy and Natural Resources, which would bring together many of the federal energy functions.

In Congress, conflicts and fragmentation in committee jurisdiction now parallel the conflicting maze in the Executive Branch. Some consolidation of energy jurisdiction is desirable to achieve effective oversight of Executive Branch activities and coordinate review of budgets and policy.

B. Regulation of the energy sector

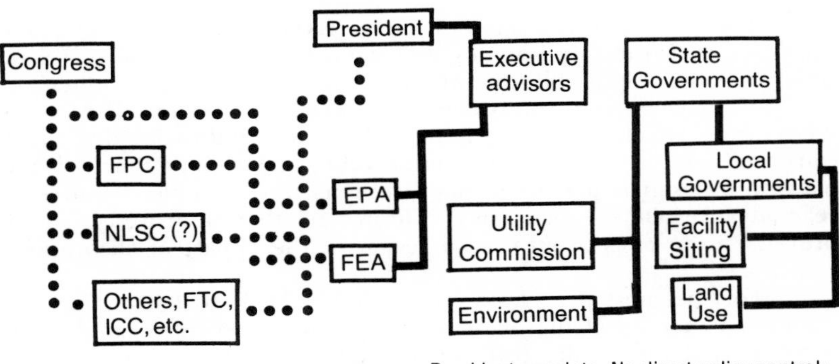

1. Economic regulation

The *Federal Power Commission,* the principal economic regulatory body in the Federal Government, is an independent agency with five commissioners appointed by the President. Its major function is regulation under the Natural Gas Act of the price of natural gas sold in interstate commerce. Under the Federal Power Act, the FPC regulates the rates and other aspects of wholesale electricity sold in interstate commerce. It also issues certificates for construction and operation of gas pipeline facilities.

The *Federal Energy Administration* has short-term authority to regulate the price of oil.

The enforcement of antitrust regulations and promotion of competition is primarily the purview of the *Federal Trade Commission,* which investigates and studies the status of competition in the energy sector. Enforcement of the antitrust laws is the responsibility of the *Department of Justice.* The *Securities and Exchange Commission* insures that any single holding company owns no more than one gas or electric company. The *Interstate Commerce Commission* regulates the rates for interstate transportation of fuels by railroad and highway.

State utility ratemaking commissions actually establish the prices which may be charged by electric and natural gas utilities. They operate independently, subject only to some broad coordination on the electric side through the Federal Power Commission.

2. Fuels allocation and controls

Under the Emergency Petroleum Allocation Act of 1973, the *Federal Energy Administration* has developed a comprehensive allocation program for petroleum and certain petroleum products. The intention is to allocate shortages in as equitable a manner as possible.

Energy imports are subject to varying controls. Petroleum imports were limited under the mandatory oil import control program until 1973, when the quota system, administered by the *Interior Department,* was lifted and a system of fees and tariffs substituted. Natural gas imports are licensed by the *Federal Power Commission.* And uranium imports are controlled by the *Atomic Energy Commission.*

3. Energy facility siting and land use

The control over location of energy facilities such as power plants, refineries, and transmission lines, and the use of land for energy-related activities is the responsibility primarily of state and local governments. A few states have enacted comprehensive siting laws or statewide planning mechanisms which provide generally that those industries desiring to construct a facility obtain state certification, and submit their advance plans to a designated state agency. Montana, for example, requires ten year advance plans. Individual certificates are then approved or disapproved on the basis of an overall review of the impact of all proposed projects.

For the most part, however, facility siting is based on the more traditional exercise of the planning and zoning power delegated by the states to local and county governments. Exercise of this power is usually not subject to overall review at the state level. Thus, for example, the siting of an oil refinery or a power plant is based on the decision of the particular company which then, having selected a site, seeks whatever zoning changes or variances are necessary to construct on that particular piece of land.

Another important land use control exercised by state governments is control over reclamation of strip-mined land. Again, there is a wide variation in the strength of the laws and efficacy of regulation. Under proposed federal legislation, this authority would remain in the states subject to uniform, federally-established standards.

The only Federal agency with significant facility siting authority is the *Federal Power Commission* which approves the siting of all hydroelectric facilities on interstate waters, and may, under recent court decisions, have some authority over fossil-fuel plants using interstate waters. The *Atomic Energy Commission* has some say over the location of nuclear power plants. The Bureau of Land Management exercises significant control over all facilities to be constructed on federal lands through the use of special land use permits. The Geological Survey is charged with regulating the operations of all federal energy mineral lessees, including oil and gas drilling and mining operations.

4. Environmental and safety regulation

This has been a field of growing federal preemption. The principal programs are those of the *Environmental Protection Agency.* Under the Clean Air Act, with principal amendments in 1970, and the Federal Water Pollution Control Act as amended in 1973, EPA is charged with overall responsibility for protecting the nation's environment. Both these laws establish national goals for air and water quality and EPA establishes standards to meet these goals. Actual enforcement of the standards, including development of plans to implement them, is left to the states; but if the states fail to meet tough EPA guidelines, that agency may then step in and take over regulation of the program.

Protecting the nation and its people from the hazards of civilian nuclear power, including nuclear power plants, uranium mills, and other facilities which process, store or transport nuclear materials, since 1954 has been the responsibility of the *Atomic Energy Commission,* operating under the Atomic Energy Act. The major thrust of this regulatory effort has been directed at construction and operation of nuclear power plants under a two-phase regulatory procedure of construction licenses and operating permits. A new, independent Nuclear Safety and Licensing Commission, splitting these functions from the AEC to eliminate the conflict of interest arising from its concurrent responsibility to promote nuclear power development, seemed likely to pass Congress in 1974.

A number of other federal agencies have less significant, but important, safety regulatory responsibility. Among these, the *Interior Department*'s Mine Enforcement Safety Administration (MESA) is charged with protecting the nation's miners under the Coal Mine Health and Safety Act.

On the Congressional side, regulatory oversight and authorization responsibility is vested primarily in the House and Senate Interstate and Foreign Commerce Committees. There are some divergences from this general rule. The Judiciary Committees are concerned with the antitrust and monopoly regulation of the FTC and Justice Department. The Senate Public Works Committee has been the prime mover in legislating both the Clean Air Act and the Federal Water Pollution Control Act, as well as maintaining close oversight of the Environmental Protection Agency. The Joint Committee on Atomic Energy oversees the nuclear regulatory functions.

The principal defect in the energy regulatory programs is their inability to respond to national goals and objectives. Regulatory decisions are bound by the objectives and constraints of each program. These tend to be parochial, not coordinated with other goals and objectives or with other units of government. The legislation of national energy conservation goals and preparation of guidelines suggested in the conclusion of this book should provide the basis of common energy objectives in the programs. (For environmental goals, this has been accomplished with some success at the federal level through the National Environmental Policy Act.) An additional reform with merit, suggested by a recent study,* is creation of offices to coordinate the federal licensing processes and to coordinate federal and state energy regulation.

C. Research and development

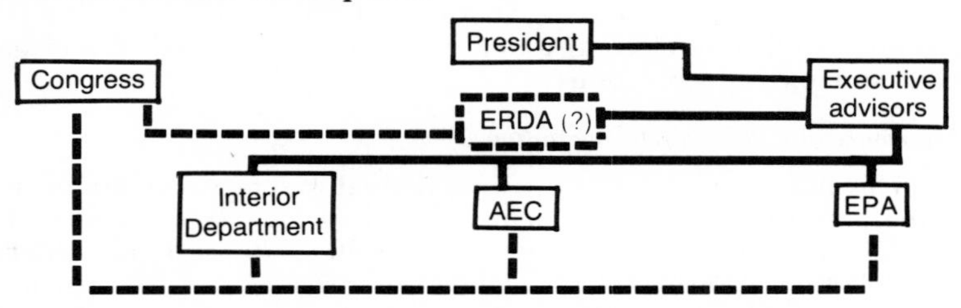

Since Congress passed the Atomic Energy Act of 1954, nuclear power has absorbed roughly three-fourths of the energy research and development budget of the federal Government. (The state efforts have been relatively minuscule.) With the exception of a limited coal research program, other new energy technologies have been ignored until recently. This particular allocation of energy research and development resources reflects the viewpoint and influence of certain institutions, primarily the Atomic Energy Commission and the Joint Committee on Atomic Energy which were expressly designed to promote nuclear energy.

On the other hand, fossil fuels research, traditionally, has been a function of the private sector. The first major attempt to involve the Federal Government in supporting coal research grew out of protracted congressional action marked by a Presidential veto. A weak compromise resulted in the creation in the Interior Department of the *Office of Coal Research* (OCR), which until 1971 did not receive any priority attention. Additional coal R&D and some oil shale research was performed by the Interior Department's Bureau of Mines. In addition, Congress placed pollution control research in the Environmental Protection Agency; this bifurcation of coal-related R&D did little to strengthen the Government's efforts.

Congress is attempting to change this imbalance and give some priority emphasis to new energy technologies by establishing a single Energy Research and Development Administration (ERDA). It will absorb the Interior Department and AEC energy research and development functions as well as some others, such as solar R&D, scattered in other agencies. Creation of ERDA combined with a consolidated Congressional appropriations system for energy R&D, instituted for the fiscal year 1975 budget, should bring better balance to the federal energy R&D effort.

D. Energy resource management

This function is described and analyzed in detail in Chapter 10, *Federal Energy Resources.* The resource management functions are

---

* *Federal Energy Regulation: An Organizational Study,* Federal Energy Regulation Study Team, Washington, April 1974.

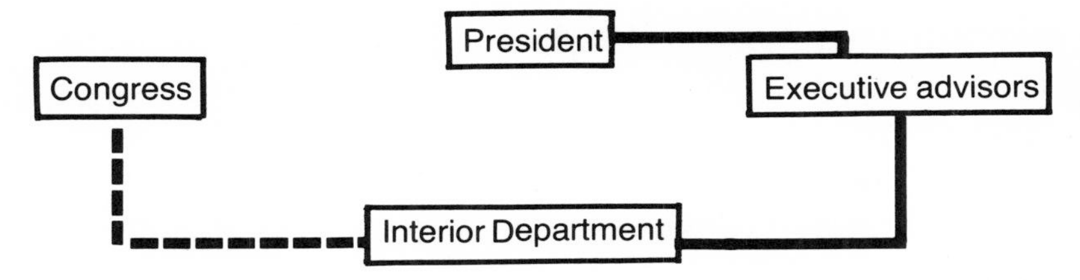

performed by the Department of the Interior, primarily through the United States Geological Survey and the Bureau of Land Management. The House and Senate Interior and Insular Affairs Committees are the principal Congressional bodies with authorization and oversight responsibility.

E. Energy conservation

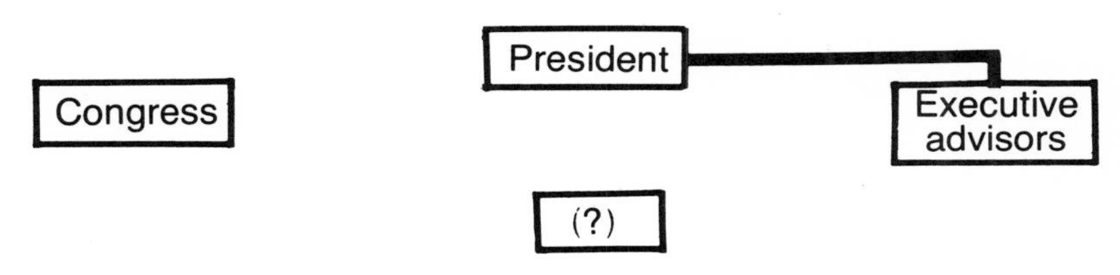

At this writing, energy conservation, although clearly a major functional area of federal energy responsibility, has not been institutionalized, and has no "home" in the federal bureaucracy. The Federal Energy Administration has some responsibility for managing demand in the context of near-term emergencies; and some conservation program initiatives have been taken by FEA along with the Environmental Protection Agency and the Council on Environmental Quality. Several federal agencies will have to play an important role in initiating and implementing conservation policies. The Department of Transportation and the Department of Housing and Urban Development are two obvious examples.

Several states have started to move toward establishing conservation policies. Minnesota, for example, in energy legislation passed in 1974, established an Energy Agency, which among other things is to promulgate energy conservation regulations, conduct energy conservation studies, and certify the need for large energy facilities.

Summary

Government organization per se is not a panacea for the nation's energy problems, but some reforms are necessary. During the time frame of the Energy Policy Project, many critical organizational issues have been identified and addressed. Assuming establishment of an Energy Research and Development Administration and independent Nuclear Licensing and Safety Commission, the most important remaining reform is to designate a single, permanently-constituted body with national energy policy oversight and coordination responsibility.

Appendix F–Economic Analysis of Alternative Energy Growth Patterns, 1975–2000

A report
to the Energy Policy Project
by Data Resources, Inc.
Edward A. Hudson
Dale W. Jorgenson

Summary

This study presents the results of simulations of U.S. economic growth over the 1975–2000 period under different energy supply and demand conditions. The economic impacts of moves from *Historical Growth* patterns to a *Technical Fix* growth path, and from this to a *Zero Energy Growth* path are examined. The main conclusions are:

- Substantial economies in U.S. energy input are possible within the existing structure of the economy and without having to sacrifice continued growth of real incomes.
- This energy conservation does have a non-trivial economic cost in terms of a reduction in real income levels vis-a-vis the *Historical Growth* position; in 2000, real income under *Technical Fix* and *Zero Energy Growth* are both about 4 percent below the *Historical Growth* figure.
- Adaptation to a less energy intensive economy will not have a cost in terms of reduced employment; in fact, it will result in a slight increase in demand for labor. This, with the reduced real output, means that labor productivity is reduced and, correspondingly, real wages are slightly lower in *Technical Fix* or *Zero Energy Growth* than in *Historical Growth*.
- Adaptation to a less energy intensive economy will not have a cost in terms of total capital requirements; in fact, *Technical Fix* or *Zero Energy Growth* should require slightly less total capital input than *Historical Growth*.
- The shift to reduced energy use will result in an increase in rates of inflation from a predicted 3.8 percent a year under *Historical Growth* to 4.1 percent under *Zero Energy Growth*.

The quantitative economic changes involved in the move to *Technical Fix* or *Zero Energy Growth* are summarized in Table 1 (p. 494).

Introduction

This report examines and compares the general economic environment corresponding to the three alternative energy growth patterns being studied by the Energy Policy Project. These growth patterns are: *Historical Growth* where past energy supply and demand patterns are assumed to continue into the future; *Technical Fix* growth, where energy conservation practices and known energy-saving technologies are incorporated into production and consumption patterns to the extent possible within existing life styles and economic organization; and *Zero Energy Growth* (ZEG) where, in addition to the technical fix measures, changes in life styles and economic structure are introduced in order to move towards a situation of constant per capita energy consumption. Economic growth paths under each of these three scenarios were simulated using

Table F-1–**Summary of differences between growth paths** (percentage difference in the level of each variable between growth paths)

	Historical Growth vs. Technical Fix		*Historical Growth vs. Zero Energy Growth*		*Technical Fix vs. Zero Energy Growth*	
	1985	*2000*	*1985*	*2000*	*1985*	*2000*
Real GNP	−1.64	−3.78	−1.61	−3.54	0.03	0.25
Price of GNP	2.00	4.81	2.26	6.03	0.25	1.17
Employment	0.90	1.52	1.25	3.32	0.35	1.77
Capital input	−1.02	−1.83	−0.88	−1.17	0.15	0.67
Energy input	−16.60	−37.70	−19.30	−46.10	−3.20	−13.40

the DRI energy model. The DRI energy model simulates production, transactions and consumption aspects of the economy to generate predictions of sectoral output levels, sectoral prices and patterns of energy use. These data can then be used to obtain a broad picture of the economic system along each of the alternative growth paths and, most importantly, to assess the differential impact of the two energy conservation programs vis-a-vis the *Historical Growth* path. Information on the differential impacts of *Technical Fix* and *ZEG* is extremely important as it provides the basis for ascertaining the nature and magnitudes of the economic costs of the two conservation programs so that, as a basis for energy policy decisions, costs can be compared to the benefits resulting from reduced energy consumption.

The conclusion of this study is that the transition to *Technical Fix Growth,* or even to *Zero Energy Growth,* can indeed be accomplished within the current economic structure without major economic upheaval or collapse.

A final purpose of this report is to complement other technically oriented studies of energy consumption being conducted by the Energy Policy Project. The present economic approach, conducted at an aggregate level and incorporating observed patterns of economic complementarity, substitutability and adjustment, provides a broad-based measure of the impact of reduced energy use on production and prices. The engineering approach examines the possibilities for energy conservation at a detailed, process level by ascertaining which conservation measures would be cost-effective, given current technology and given projected energy prices. Both approaches incorporate the same motivating force —cost minimization in production or consumption activities. Also, each approach is based on similar information concerning technically feasible adjustments in the economy with the difference being that the macro data are more aggregated and reflect adjustment patterns actually observed, whereas the micro data incorporate adjustment possibilities predicted by the process analyst. Thus, there is no inherent conflict between the two approaches; they just view the same problem from different perspectives. In fact, the two approaches do yield consistent results. This means that each reinforces the other, for it establishes first, that the economic adjustments predicted by the macro approach on the basis of observed behavioral responses do have a valid technical basis at the process level, and second, that the adjustments predicted by the process approach have a valid economic basis in the senses of being mutually consistent within the broad system of economic interdependence and of being consistent with observed patterns of business and consumer adjustments to changes in the availability and price of energy.

The Data Resources Inc. energy model

The Data Resources Inc. (DRI) energy model has already been presented in detail in the DRI report to the Energy Policy Project: "Energy Resources and Economic Growth," DRI, September 30, 1973. This section presents a brief outline of the model with the intention of illustrating the general derivation of the results presented below. The starting point in the projections is provided by a macro econometric model of U.S. economic growth. This model integrates both the demand and supply sides of macro economic activity into a single framework. This is used to project the general economic environment within which the energy simulations can be conducted. Specifically, the macro model is used to define the prices and availability of capital and labor inputs and the total levels of final expenditures, variables that are used as inputs in the detailed energy simulations. The energy analysis is then based on an interindustry model of the U.S. economy in which production and consumption are broken down in the following pattern:

- production is classified into nine sectors, each of which is represented by a production submodel. These nine sectors are agriculture (together with nonfuel mining and construction), manufacturing, transport, services (together with trade and communication), coal mining, crude petroleum and natural gas extraction, petroleum refining, electric utilities, and gas utilities;
- the nine producing sectors purchase inputs of primary factors imports, capital services and labor services;
- the nine producing sectors must also purchase inputs from each other; for example, manufacturing makes purchases from transport and the transport sector makes purchases of manufacturing output;
- the nine producing sectors then sell their net output to final users—personal consumption, investment, government and exports.

These components are then integrated within an interindustry, or input-output, model. The feature of input-output analysis is that transaction flows are brought into consistency so that each sector produces exactly that amount needed to meet final demands as well as the intermediate demands from other producing sectors. The critical feature of the DRI energy model is that the patterns of input into the producing sectors, as well as the final demand levels, are functions of, *inter alia*, prices. This means that the model allows for production to substitute, within the bounds of given technical parameters, relatively less expensive for relatively more expensive inputs. This feature is of central importance in energy analysis for it captures the fact that producers and consumers react to higher energy prices by economizing on energy use by substitutions between different fuels, and by substitutions between fuel and non-fuel purchases, as well as by cutting back on "non-essential" energy input without accompanying substitutions.

The actual solution of the model moves through the following steps:

1. prices are determined endogenously in terms of production coefficients, efficiency levels, primary input prices and other information,
2. these prices are then used to solve for the pattern of inputs

into each producing sector that is most economical in terms of these prices,

3. these prices are also fed into final demand submodels to obtain final demands for each type of output,

4. the input-output system is then solved to find the primary inputs and the interindustry transactions that are required to satisfy these final demands. Thus the model simulates, on the base of exogenous parameters characterizing the general economic environment, the entire flow of transactions in the economy—transactions from factors to producers, producers to producers, and producers to consumers. Specifically, the model generates transactions flows and totals in current dollars and real terms (constant dollars) together with the corresponding sectoral price levels and energy usages.

The parameters of the production models were obtained by econometric estimation of the models on the basis of U.S. interindustry transactions over the 1947–71 period (these data were prepared by the Energy Policy Project).

The approach uses information about production relationships that have actually occurred in the past as the basis for predicting future production responses to price changes. In particular, the projected reaction to energy price increases is based upon the observed patterns of past production responses. This requires the assumption of reversibility in the sense that producers' reactions to the very substantial declines in real energy prices in the past will apply, but in reverse, to the adjustment to increasing real energy prices in the future. In fact, this assumption is likely to be rather conservative in estimating the scope for energy conservation. The predicted responses are based on behavioral adjustments within existing technical knowledge and within advances in this knowledge along past trends. In fact, much future technical knowledge is likely to be of an energy-conserving nature which would permit even greater conservation than that predicted from historical relationship. Therefore, the projections presented below probably err on the side of underestimating the potential for future energy conservation.

Methodology

The simulations of the alternative energy scenarios were made in two steps. First, the DRI energy model was calibrated so as to produce the Energy Policy Project *Historical Growth* path of economic development. This involved selecting and inserting into the model initial assumptions covering productivity advance, fuel imports, income growth, primary input prices, energy supply conditions and so on, in such a way that the predicted energy demand growth path exhibited the same general characteristics and trends as observed in the *Historical Energy* growth patterns. Once the model was calibrated in this way, the exogenous assumptions were fixed and only those parameters corresponding to a move from *Historical Energy* use patterns to a *Technical Fix* situation and then from *Technical Fix* to a *Zero Energy Growth* situation were varied. In other words, the general specification of the model was held unchanged in the three different energy scenarios; only energy-specific parameters were varied to secure the move between the three alternative growth paths.

The simulations focused on three years:–1975, which was used as the common starting point for all three alternative growth paths; 1985, when the three growth paths had clearly diverged; and 2000, by which time the full effects of each energy conservation program had been felt and the differential impacts of the energy conservation programs were most clearly visible. Thus three economic growth paths, starting from the same initial position in 1975, are examined at two points in time—1985 and 2000. The differences between the growth paths can still be examined in detail; limiting the comparison to two years has no cost but saves the complexity involved in simulating every year from 1975 to 2000. The solution presupposes that the economy has had time to make the adjustment from its initial to its equilibrium configuration. The use of 1985 and 2000 as comparison years is entirely consistent with this assumption since the time lags to these years are more than sufficient to cover the transition period needed for the economy to adapt to policies and conditions implemented in the near future.

The predictions are based upon an economic model which simulates aggregate production, expenditure and consumption relationships. Since the model is a simplified and idealized representation of actual processes, its forecasts cannot be considered as pin-point accurate predictions of future economic events. Actual future developments will vary from those predicted in this study because the assumptions made about future exogenous developments may not be completely accurate, and also because the model does not replicate economic processes with perfect accuracy. However, the focus of this report is on the differences in economic performance under different energy conditions rather than on future levels of economic indicators. The model does give meaningful estimates of these: first, because differencing itself eliminates any systematic bias introduced into the forecasts through incorrect assumptions and through biases in the model itself; second, because extensive testing of the model suggests that it does produce reasonable estimates of the changes in aggregate economic behavior produced by changes in exogenous parameters.

Historical Growth

The pattern of economic growth and energy consumption corresponding to the *Historical Growth* scenario is summarized in Table 2. This growth pattern is, by design, essentially a continuation of recent trends so that, even in 2000, the forecast composition of the economy is similar to that of the 1975 starting point.

Production increases at rates similar to, although slightly below, recent growth rates. The decline in growth rates is expected to become significant only in the 1980s in response to the low fertility rates currently being experienced. The assumption is made that the fertility rates experienced over the 1970–73 period, rates which imply an eventually constant population size, will continue so that when today's infants begin to enter the labor force in the late 1980s, the rate of labor force expansion slows, leading to a general reduction in the rate of increase of real GNP. Per capita income and output is not reduced, but a smaller labor force means a smaller total output.

The composition of production does change somewhat over the forecast period; in terms of gross output—that is, total sales of each sector—transport expands the most rapidly, followed by the energy industries, then manufacturing, then services. These trends in composition reflect developments that can be discerned today:

Table F-2–**Historical Growth path**

	1975	1985	2000	Growth Rates (% per annum) 1975–85	1985–2000
Output (gross) (billion 1971 $)					
agriculture	306.8	387.9	532.2	2.4	2.1
manufacturing	848.6	1228.4	1966.1	3.8	3.2
transport	94.4	140.2	244.8	4.0	3.8
services	976.6	1364.7	2109.5	3.4	2.9
energy	97.8	144.3	249.6	4.0	3.7
Demand (billion 1971 $)					
consumption	838.3	1211.8	1990.9	3.8	3.4
investment	309.7	430.5	670.4	3.3	3.0
government	275.0	388.4	604.9	3.5	3.0
net exports	19.2	33.2	78.8	5.6	5.9
GNP (billion 1971 $)	1442.2	2064.0	3345.0	3.6	3.3
Output (value added, billion 1971 $)					
agriculture	135.8	186.6	290.1	3.2	3.0
manufacturing	345.1	459.2	662.5	2.9	2.5
transport	52.3	64.3	82.7	2.1	1.7
services	703.3	1011.2	1669.6	3.7	3.4
energy	63.1	101.3	190.4	4.9	4.3
services of durables	142.6	226.6	446.8	4.7	4.6
Employment (billion manhours)					
agriculture	16.478	19.696	26.006	1.80	1.87
manufacturing	41.689	48.049	59.807	1.43	1.47
transport	6.927	7.524	8.683	0.83	0.96
services and government	105.452	129.834	168.061	2.10	1.74
total	173.115	205.103	262.557	1.71	1.66
Energy (quadrillion Btu's)					
coal	13.15	18.54	34.40	3.49	4.21
petroleum	34.87	39.48	58.91	0.36	2.70
electricity	6.81	13.16	27.37	6.81	5.00
gas	24.47	34.51	42.23	3.50	1.35
nuclear, other	5.55	22.50	51.25	15.02	5.64
Total energy input	78.03	115.03	184.71	3.96	3.21
Energy consumption (quadrillion Btu's)					
personal consumption	23.165	31.606	48.359	3.2	2.9
services and government	10.936	15.480	26.743	3.5	3.7
electricity generation	21.080	40.739	84.716	6.8	5.0
industry	26.900	36.900	49.476	3.2	2.0
transport	2.672	3.469	4.867	2.6	2.3
total input	78.032	115.031	184.706	4.0	3.2
Prices					
agriculture				4.70	5.09
manufacturing				3.04	3.68
transport				2.03	2.56
services				3.88	4.23
coal				1.78	5.72
crude petroleum				4.76	4.44
refined petroleum				5.74	4.54
electricity				−0.90	2.53
gas				5.96	4.94
consumption				3.63	3.98
investment				3.58	4.08
government				3.79	4.16
GNP				3.61	3.98

• increased demand for transport for business and vacation travel, and to service increasingly dispersed economic activity, together with some increase in the relative importance of public transport, result in a continued rapid increase in transport activity;

• energy output also grows rapidly in large part because of the rapid growth of electricity usage which, since electricity is a secondary energy form suffering large energy conversion losses, places great demands on the primary energy sources;

• manufacturing output grows in line with total production, driven both by demand for manufactured goods as an input into the other producing sectors as well as by continuing growth in final use demands for manufacturing output;

• services grow less rapidly than manufacturing in terms of total output for, although final demand for services in current dollars is rising more rapidly, the faster rate of price increase for services converts this to a slower rate of increase in real output. The historical forecast implies a continuation of the relative increase in the importance of service activities, but services prices increase more rapidly than those of manufacturing, leading to real service output growing less rapidly than real manufacturing output;

• agriculture and construction real output grows at the slowest rate, primarily because demand for these types of output is linked to population more than income so that increasing consumption demand flows more to the other producing sectors.

The value added in each production sector moves a little differently from the growth pattern of real output. Services of consumer durables show the fastest increase; that is, the imputed flow of services to consumers from owner-occupied housing, automobiles and other home and personal appliances increases more than the market-transacted output. The greatest rate of increase in value added in marketed output occurs in energy production; the rapidly growing demand for energy sources along with the increasingly difficult supply conditions in fuel production result in inputs being drawn into these sectors relative to other production. Services show the next most rapid increase and are predicted to continue to increase relative to real GNP. This increase is due to the continuing rapid growth of final demand for services, along with the very low rate of productivity advance expected in service activities, drawing capital and labor services into service occupations faster than the general rate of increase in the supply of these inputs. This process is reflected also in the increasing share of services in total employment. Agriculture and construction value added increases less rapidly than GNP, mainly because total demand for output from these activities is not growing as rapidly as GNP. Manufacturing and transport value added increases least rapidly of all sectors. The reason for this is the continued high rate of productivity advance expected in these activities since this allows their output to increase without a correspondingly rapid increase in primary inputs.

The employment pattern changes in a similar way to changes in the pattern of value added, with services, agriculture and construction increasing, and manufacturing and transport declining in relative importance. Services and government increase their share of total employment from 60 percent to 64 percent over the forecast period. Total employment (which includes a labor quality improvement index) increases at around 1.7 percent a year although this rate of increase declines over time due to the effect of low fertility rates in slowing labor force growth.

Prices are projected to increase at around 3.75 percent a year which is, although not as rapid as the inflation currently being experienced, still substantially faster than average inflation rates of the last 10 or 15 years. On the demand side, consumption, investment and government purchase price indices all rise at about the same pace. On the production side, however, there is more substantial variation in rates of price increase. Fuel prices, apart from electricity, rise the fastest of any prices as it becomes increasingly difficult to produce the fuel to meet the rapidly growing demand. Electricity prices show much less increase. The reason for this lies in the productivity assumptions upon which the *Historical Growth* forecasts are based. The past rapid growth in electricity use has been, in large part, due to the past steadiness, and even decline, in electricity prices which, in turn, have been made possible by a very rapid rate of productivity increase in the electricity generation sector. This productivity advance has moderated in the past four years due, apparently, to short run influences; but, in line with the historical conditions objective of the *Historical Growth* forecast, this slowdown is assumed to be temporary, with productivity advance in electricity generation returning to typical past rates. This efficiency permits fuel, capital and labor price increases to the electricity generation sector to be absorbed without comparable increases in electricity sales prices.

Nonfuel prices also show differences in their growth rates. Productivity advance in manufacturing and transport allows these sectors to absorb some input price increases with the result that their output prices increase a little less rapidly than the general rate of inflation. Service, agricultural and construction activity, however, does not exhibit such rapid productivity growth and this, together with their relative intensity of use of an input—labor—whose price is rapidly increasing, causes their prices to rise more rapidly than general inflation.

Energy use continues broadly along past trends. The dominant feature in energy is the rapid increase in the consumption of electricity. This increase is partially due to the productivity and price behavior of electricity generation already discussed. The growth in electricity production leads to rapid growth in the use of primary fuels used in the generation of electricity, with this growth being evidenced primarily in nuclear generation, but also in the demand for coal. Petroleum and gas consumption, on the other hand, increases more slowly, for here the price increases resulting from demand facing a restricted supply lead to some moderation in the demand for these fuels. Total U.S. energy input increases by around 3.5 percent a year which is close to past average rates of increase.

This *Historical Growth* projection approximates a continuation of the conditions, especially those relating to energy supply, existing in the 1960s. Developments of the recent past, such as limitations on fuel imports, restrictions on construction of nuclear electricity plants, slower productivity growth in electricity generation, restrictions on oil and gas exploration and production, major increases in fuel prices and so on are not incorporated in the *Historical Growth* projections. In other words, these projections assume no significant price or regulatory pressure to alter energy demand and no serious problems in obtaining the fuel resources to satisfy these demands.

Recent events have shown the set of assumptions underlying the *Historical Growth* forecast to be unrealistic. Thus, although this forecast is extremely useful as an analytical reference point, we need to supplement it by alternative forecasts which incorporate the recent energy developments. Therefore, we proceed to examine the *Technical Fix* and *ZEG* alternative growth paths, both of which incorporate less favorable conditions concerning the availability of energy or which, alternatively, could be viewed as projections of economic growth under policies designed to restrict energy demand.

Technical Fix growth

The growth path of the economy under *Technical Fix* conditions is summarized in Table 3. Also, this table shows the difference between the *Historical* and the *Technical Fix* growth paths. The summary information is that in 2000, a reduction of 38 percent in total energy input can be accommodated with only a 3.8 percent decrease in real GNP, a small increase in the rate of inflation and no increase in unemployment. That is, the economy can adjust to a substantial decline in energy use without major dislocation. The differences between the *Historical Growth* and the *Technical Fix* growth paths are now considered in more detail.

The motivating forces introduced into the energy model to secure the move from the *Historical Growth* path to the *Technical Fix* growth path were increases in petroleum products prices and in electricity prices. These price increases, when their impact on other prices, on input patterns and on demand levels has been solved through in the model, lead to a new economic configuration requiring a reduced energy input. The critical output from this analysis is the economic changes that are produced by these price increases; the underlying cause of the price increases is not directly relevant. In fact, the initial price increases in the model were secured by assuming unfavorable domestic petroleum supply conditions and restrictions on imports of petroleum, which served to produce a dramatic increase in petroleum product prices, and by assuming a continuation of recent slow productivity advance in electricity generation, which served to increase electricity prices. (The corresponding *Historical Growth* assumptions were that domestic oil production and/or imports could expand to accommodate petroleum demand growing at historical rates, and that electricity generation productivity advance returned to its rapid, historical trends after the slowdown of the past four years). Alternatively, the price increases might be viewed as being produced by taxes on petroleum and electricity sales with the revenue being returned to the private sector by decreases in income taxes; or the results might just be viewed as showing the effect of petroleum and electricity prices on the rest of the economy, without specifying the cause of the price rises. The main results concern the economic differences between the *Historical* and the *Technical Fix* growth paths and it is these differences which we now examine.

The *Technical Fix* growth path involves an increase in energy input at a little less than half the rate associated with *Historical Growth*, specifically at 1.6 percent a year instead of at 3.5 percent. The comparative reduction in energy use is 17 percent in 1985 and 38 percent in 2000. This reduction is concentrated in electricity and petroleum use. In 1985, electricity and petroleum consumption are each reduced by over 20 percent while the reduced electricity output leads to a reduced level of coal use and to a substantial reduction in nuclear input. But, higher petroleum and electricity prices lead to an increase, due to inter-fuel competition and substitution, in the price of gas. This produces a decline in use of all fuels, although the gas and coal use reduction is of a smaller

Table F-3–**Technical Fix growth**

				Growth Rates (% per annum)		Difference from Historical Growth level (%)	
	1975	*1985*	*2000*	*1975–85*	*1985–2000*	*1985*	*2000*
Output (gross) (billion 1971 $)							
agriculture	306.8	381.3	512.3	2.2	2.0	−1.70	−3.74
manufacturing	848.6	1214.3	1906.1	3.6	3.1	−1.15	−3.05
transport	94.4	138.3	236.6	3.9	3.6	−1.36	−3.35
services	976.6	1347.8	2045.5	3.3	2.8	−1.24	−3.03
energy	97.8	115.4	144.0	1.7	1.5	−20.03	−24.31
Demand (billion 1971 $)							
consumption	838.3	1188.2	1904.5	3.5	3.2	−1.96	−4.35
investment	309.7	425.9	652.2	3.2	2.9	−1.07	−2.71
government	275.0	383.1	585.5	3.4	2.9	−1.36	−3.21
net exports	19.2	33.0	76.7	5.6	5.8	−0.60	−2.66
GNP (billion 1971 $)	1442.2	2030.2	3218.5	3.5	3.1	−1.64	−3.78
Output (value added, billion 1971 $)							
agriculture	135.8	185.5	285.1	3.2	2.9	−0.59	−1.72
manufacturing	345.1	456.3	650.8	2.8	2.4	−0.63	−1.77
transport	52.3	63.5	80.8	2.0	1.6	−1.24	−2.30
services	703.3	1010.6	1658.8	3.7	3.4	−0.06	−0.65
energy	63.1	76.5	96.2	1.9	1.5	−24.50	−49.50
services of durables	142.6	226.6	446.8	4.7	4.6	0.0	0.0
Employment (billion manhours)							
agriculture	16.478	19.696	25.962	1.80	1.86	0.00	−0.17
manufacturing	41.689	47.914	59.454	1.40	1.44	−0.28	−0.59
transport	6.927	7.452	8.488	0.74	0.86	−0.96	−2.25
services and government	105.452	130.262	168.532	2.13	1.74	0.33	0.28
total	173.115	206.949	266.548	1.80	1.70	0.90	1.52
Energy (quadrillion Btu's)							
coal	13.15	17.37	25.13	2.82	2.49	−6.31	−26.95
petroleum	34.87	31.58	37.30	−0.99	1.12	−20.01	−36.68
electricity	6.81	9.43	13.51	3.31	2.43	−28.34	−50.64
gas	24.47	32.36	32.04	2.83	−0.07	−6.23	−24.13
nuclear, other	5.55	14.62	22.57	10.17	2.94	−35.02	−55.96
Total energy input	78.03	95.92	115.00	2.09	1.22	−16.61	−37.74
Energy consumption (quadrillion Btu's)							
personal consumption	23.165	26.085	27.264	1.2	0.3	−17.5	−43.6
services and government	10.936	13.548	17.836	2.2	1.9	−12.5	−33.3
electricity generation	21.080	29.198	41.506	3.3	2.4	−28.3	−51.0
industry	26.990	33.295	39.787	2.1	1.2	−9.8	−19.6
transport	2.672	3.232	4.161	1.9	1.7	−6.8	−14.5
total input	78.032	95.924	115.005	2.1	1.2	−16.6	−37.7
Prices							
agriculture				4.85	5.20	1.48	3.20
manufacture				3.13	3.81	0.89	2.67

Table F-3–**Technical Fix growth (continued)**

	1975	*1985*	*2000*	*Growth Rates (% per annum)* *1975–85*	*1985–2000*	*Difference from Historical Growth level (%)* *1985*	*2000*
transport				2.13	2.63	1.00	2.05
services				3.98	4.31	1.00	2.31
coal				2.63	7.52	8.59	40.30
crude petroleum				3.85	4.60	−8.37	−6.15
refined petroleum				8.38	6.11	27.98	59.98
electricity				4.33	5.66	67.27	162.56
gas				6.12	6.41	1.50	25.05
consumption				3.88	4.22	2.46	5.95
investment				3.69	4.19	1.07	2.80
government				3.94	4.30	1.39	3.31
GNP				3.82	4.17	2.00	4.81

order of magnitude than the reduction in petroleum and electricity use. Similarly, in 2000, electricity consumption (and nuclear input) are reduced by 50 percent, with petroleum use down by 37 percent and coal and gas use down by 25 percent.

Higher petroleum and electricity prices lead to a general upward pressure on prices due both to the consequent increase in production costs and to the redirection of demand and input patterns which places more demand pressure on other production. Thus, in 2000 for example, the electricity price more than doubles and the petroleum products price goes up by 60 percent; this leads to smaller, but still substantial, increases in coal and gas prices, as well as to increases in prices of nonfuel products, by about 2 to 3 percent over the period 1975 to 2000.

On the demand side, the higher energy prices have a substantial and immediate impact on the price index of consumption goods and services, and this increase is further boosted by the rise in prices of nonfuel goods and services. Thus, the rise in consumption prices is double the rise in prices of investment and government purchases. However, the overall impact on prices is not catastrophic; the GNP price deflator is increased by about 4 percent which corresponds to a 0.2 percentage point higher rate of inflation under *Technical Fix* than under *Historical Growth*—that is, inflation increases from 3.8 percent to 4.0 percent.

Output and real incomes are reduced slightly by the reduction in energy use but here, too, the reduction, although significant, is not catastrophic: real GNP under *Technical Fix* is 1.6 percent lower in 1985 and 3.8 percent lower in 2000 than the corresponding *Historical Growth* path levels. Energy output suffers the greatest reduction, a fall of 42 percent in constant dollar terms in 2000, for example. But other output is not drastically affected. Services output is reduced the least with agriculture output reduced the most; but the reductions, even in 2000, are only of the order of 3 percent. In terms of value added, service output is hardly affected while other output is reduced by about 2 percent in 2000. On the final use side, personal consumption suffers the greatest reduction, but even in 2000, real consumption is only 4.4 percent below the *Historical Growth* level. Total output, as measured by real GNP, is reduced by 3.8 percent in 2000 which corresponds to a reduction in real growth rates of 0.15 percentage points, from 3.42 to 3.26 percent a year.

The relatively small impact of such a large reduction in energy use on real output is a striking and important result. Its economic explanation lies in the following considerations:

- Final demand energy use is curtailed as a result of higher energy prices. This may take the form of turning down thermostats, switching to smaller cars, installing home insulation and so on. (These avenues for energy conservation mean that, after a transition period, lower energy input is consistent with the original level of effective energy-based personal and household services.) This reduction has very little impact on the rest of the economy, for the demand reduction corresponds to only a part of the output of what is, in economic terms, a relatively small sector of the economy. Even in the 2000 *Historical Growth* projection, the energy-producing sectors represent only 4.2 percent of the entire economy in terms of gross output and 5.7 percent in terms of value added. Since, in turn, personal consumption of energy absorbs only about one third of total fuel output, it can be seen that the direct impact of a reduction of personal energy consumption on the total output of the economy is not very large.
- Use of energy in the producing sectors can be reduced somewhat without reducing output merely by reducing waste and by adopting more energy-efficient techniques. Further, there exists significant scope for substitutions between inputs into production, and the emergence of higher fuel prices stimulates use of nonenergy-intensive inputs. One area where this is important concerns capital input; capital and energy are complementary; so higher energy prices lead to reduced use of capital services and to the substitution of other inputs, particularly labor, for these services. The results of this substitution process are illustrated by the behavior of capital and labor inputs under *Technical Fix* growth: in 2000 for example, capital input is reduced by 1.8 percent from the *Historical Growth* level, whereas labor input increases by 1.5 percent. Also, substitutions between capital and materials, between capital and services and between other inputs are possible. The net result is that producing sectors can achieve substantial economies in energy use at the expense of comparatively small reductions in output.
- Any saving in the use of electricity by final consumers or by producers, even if offset by increased use of other energy services, leads, due to the conversion losses in electricity generation, to approximately three times the reduction in primary energy input. Further, to the extent that the input of uranium into electricity generation is reduced, the energy saving is even greater since the enrichment of uranium by present technologies is a heavy user of energy. Thus, increases in electricity prices, and the consequent reduction in electricity use, are a powerful instrument in reducing total energy input.

The relative magnitudes of each of these forms of energy saving are shown in Table 3. In 2000, for example, the total reduction in energy input between *Historical Growth* and *Technical Fix* is 38 percent (69.7 QBtu). Energy use in electricity generation is reduced by the largest proportion, 51 percent (43.2 QBtu), while personal consumption use is reduced by 44 percent (21.1 QBtu), service and government use by 33 percent (8.9 QBtu), industrial use by 20 percent (9.7 QBtu) and transport (which excludes private automobiles) use by 15 percent (0.7 QBtu). This indicates that significant economies in energy use are possible in all forms of energy consumption, with personal consumption, service and government economies particularly significant. The greatest Btu savings are achieved through a reduction in the inputs absorbed in electricity generation. Electricity use is reduced due to economizing in fuel use in general as well as by the partial substitution of other fuels for electricity. The net result is that electricity conservation releases 62 percent of the total energy savings achieved in the move to *Technical Fix* growth.

The share of energy in total real personal consumption expenditure is shown in Table 4. In *Historical Growth* conditions, energy purchases constitute an increasingly important component of consumption purchases, increasing from 5.54 percent in 1975 to 6.99 percent in 2000. The economies in personal energy use achieved under *Technical Fix* conditions are, however, sufficiently large to reverse this upward trend so that energy purchases in 2000 represent only 3.75 percent of real consumption expenditure. This is a significant reduction in the energy share but nonetheless, energy remains an important component in consumption spending and per capita personal consumption of energy is still higher than in 1975. The composition of personal energy use is also changed in response to the relative price changes. Electricity is clearly the major energy source in both *Historical Growth* and *Technical Fix* conditions, but the increase in the relative price of electricity under *Technical Fix* results in the partial substitution of both petroleum products and gas for electricity use.

Table F-4–**Energy use in consumption, manufacturing and services**

Real expenditure on energy in proportion to total real expenditure (percent)

	1961	*1971*	*1975*	*1985*	*2000*
Personal Consumption					
Historical Growth	4.74	5.53	5.54	5.96	6.99
Technical Fix			5.54	4.59	3.75
ZEG			5.54	4.43	3.23
Manufacturing					
Historical Growth	1.88	2.14	2.07	2.16	2.08
Technical Fix			2.07	2.08	1.68
ZEG			2.07	2.05	1.54
Services					
Historical Growth	1.30	1.76	1.85	2.14	2.58
Technical Fix			1.85	1.61	1.40
ZEG			1.85	1.55	1.21

Composition of energy input in 2000

	Personal Consumption		*Manufacturing*		*Services*	
	HG	*TF*	*HG*	*TF*	*HG*	*TF*
Coal	—	—	11.8	13.2	—	—
Petroleum	16.9	21.0	23.4	27.8	41.1	39.9
Electricity	73.6	62.5	42.0	40.8	48.4	48.8
Gas	9.4	16.4	22.8	18.2	10.4	11.3
Total energy use	100.0	100.0	100.0	100.0	100.0	100.0

(Note: *HG* = Historical growth path energy use pattern.
TF = Technical fix growth path energy use pattern.)

Manufacturing and services also redirect their input patterns to economize on energy in response to the increase in energy prices under *Technical Fix* growth. These input patterns are shown in Table 4. Energy input into manufacturing remains stable in *Historical Growth* but, under *Technical Fix,* the input proportion is reduced in 2000, from 2.08 to 1.68 percent. The overall reduction in energy use is accompanied by a redirection of energy purchases towards the relatively inexpensive fuels, particularly petroleum. In services, the trend to the increasing relative importance of energy input under *Historical Growth* is reversed under *Technical Fix* so that, in 2000, energy forms 1.40 percent of total real inputs compared to 2.58 percent.

The composition of energy input in *Technical Fix* is a little different from that in *Historical Growth;* energy conservation in services takes the form of general reduction in fuel use rather than substitutions between fuels. Technical considerations in services use of energy, and to a lesser extent in manufacturing, constrain the possibilities for substitution between energy forms, but those substitution possibilities that do exist, together with economy in energy input in general, permit significant reduction in service and manufacturing energy use.

The substitution between inputs and adjustment in input patterns that result from higher energy prices is shown for the manufacturing and service sectors in the following table. The forces at work are initially illustrated by the input patterns along the *Historical Growth* path for the increasing relative use of capital and decreasing use of labor resulting from the increasing relative price of labor. This induces producers to substitute, within technical limits, capital for labor. Also, the relatively inexpensive energy available in *Historical Growth* leads to the continuing increase in the share of energy input. The move from *Historical Growth* to *Technical Fix* or *ZEG* paths with their causal and induced price changes leads to a further set of adjustments being superimposed on these. The price increases primarily relate to energy but these cause, in turn, a smaller change in the structure of other prices. The induced changes in input proportions in manufacturing and services can be followed from the input proportions given in the following table. The reduction in energy input has already been outlined. But, all inputs are affected by the change in prices. In manufacturing, capital-energy complementarity leads to capital input being reduced, although not to the same extent as energy. The small degree of complementarity between energy and inputs of materials leads to the material input proportion being reduced. The reduction in capital, energy and materials input into manufacturing is offset by increased use of the nonenergy-intensive and now relatively less expensive, input—labor services. Thus in 2000, for example, labor input which is already 26 percent of total input under *Historical Growth* increases to 27 percent of input under *Zero Energy Growth*. Similar forces are at work in the service sector although with slightly different results. In services, capital and energy are substitutes rather than complements, so increased energy prices lead to a slight increase in capital input (for example, capital might be absorbed in energy saving uses such as increased insulation, installation of more efficient heating and air conditioning equipment, and so on). Some complementarity exists between materials and energy, so the rise in energy prices leads to a reduction in the proportion of materials inputs. Use of labor, the nonenergy-intensive input, increases to replace the reduction in energy and materials inputs and to permit service production to absorb these reductions without a comparable reduction in output.

Table F-5–**Composition of inputs into manufacturing and services** (percentage that specified input represents in total input, based on constant dollar purchases)

	1961	*1971*	*1975*	*1985*	*2000*
(a) Manufacturing					
Capital input					
Historical Growth	10.2	10.6	11.6	12.4	13.6
Technical Fix			11.6	12.4	13.5
ZEG			11.6	12.4	13.4
Labor input					
Historical Growth	33.4	28.2	30.0	28.1	26.0
Technical Fix			30.0	28.3	26.6
ZEG			30.0	28.4	26.9
Energy input					
Historical Growth	1.9	2.1	2.1	2.2	2.1
Technical Fix			2.1	2.1	1.7
ZEG			2.1	2.1	1.5
Materials input					
Historical Growth	54.5	59.1	56.3	57.3	58.3
Technical Fix			56.3	57.2	58.2
ZEG			56.3	57.1	58.2
(b) Services					
Capital input					
Historical Growth	26.5	29.6	32.7	35.6	41.4
Technical Fix			32.7	35.9	42.1
ZEG			32.7	35.9	42.3
Labor input					
Historical Growth	47.3	42.6	39.0	35.9	30.4
Technical Fix			39.0	36.5	31.4
ZEG			39.0	36.6	31.6
Energy input					
Historical Growth	1.3	1.8	1.8	2.1	2.6
Technical Fix			1.8	1.6	1.4
ZEG			1.8	1.6	1.2
Materials input					
Historical Growth	24.9	26.0	26.5	26.4	25.6
Technical Fix			26.5	26.0	25.1
ZEG			26.5	25.9	24.9

Note: Materials are all nonfuel inputs that are purchased from other intermediate sectors and from imports.

The changes in input proportions in manufacturing and services involved in the shift from *Historical Growth* to *Zero Energy Growth* conditions are significant. But these shifts are well within the range of recent experience. Thus, the largest changes in input proportions involve energy input, but even these changes correspond only to reversing the *Historical Growth* trend to increasing energy inputs so that energy input proportions in 2000 are in the region of the actual 1961 proportions.

Zero Energy Growth

The economic and energy information describing *Zero Energy Growth* is presented in Table 6. The move from *Technical Fix* growth to *ZEG* was simulated by imposing an energy sales tax (a uniform tax rate

Table F-6–**Zero Energy Growth**

				Growth Rates (% per annum)		*% difference from level of*			
						Historical Growth		*Technical Fix*	
	1975	*1985*	*2000*	*1975–1985*	*1985–2000*	*1985*	*2000*	*1985*	*2000*
Output (gross) (billion 1971 $)									
agriculture	306.8	380.5	507.8	2.2	1.9	−1.91	−4.58	−0.21	−0.88
manufacturing	848.6	1213.2	1898.2	3.6	3.0	−1.24	−3.45	−0.09	−0.41
transport	94.4	138.4	237.9	3.9	3.7	−1.28	−2.82	0.07	0.55
services	976.6	1350.1	2066.8	3.3	2.9	−1.07	−2.02	0.17	1.04
energy	97.8	111.7	124.9	1.3	0.7	−22.6	−49.9	−3.21	−13.3
Demand (billion 1971 $)									
consumption	838.3	1185.3	1885.4	3.5	3.1	−2.19	−5.30	−0.24	−0.99
investment	309.7	424.9	643.8	3.2	2.8	−1.30	−3.97	−0.23	−1.29
government	275.0	387.8	623.3	3.5	3.2	−0.15	3.04	1.23	6.46
net exports	19.2	32.8	74.3	5.5	5.6	−1.20	−5.71	−0.61	−3.13
GNP (billion 1971 $)	1442.2	2030.8	3226.7	3.5	3.1	−1.61	−3.54	0.03	0.25
Output (value added, billion 1971 $)									
agriculture	135.8	185.4	284.7	3.2	2.9	−0.64	−1.86	−0.05	−0.14
manufacturing	354.1	456.5	652.8	2.8	2.4	−0.59	−1.46	0.04	0.31
transport	52.3	63.5	80.4	2.0	1.6	−1.24	−2.78	0.00	−0.50
services	703.3	1013.2	1682.7	3.7	3.4	0.20	0.78	0.26	1.44
energy	63.1	74.7	79.3	1.7	0.4	−26.3	−58.4	−2.35	−17.6
services of durables	142.6	226.6	446.8	4.7	4.6	0.0	0.0	0.0	0.0
Employment (billion manhours)									
agriculture	16.478	19.706	26.063	1.81	1.89	0.05	0.22	0.05	0.39
manufacturing	41.689	47.982	60.028	1.42	1.50	−0.14	0.37	0.14	0.97
transport	6.927	7.452	8.562	0.74	0.93	−0.96	−1.39	0.00	0.87
services and government	105.452	130.652	179.691	2.17	2.15	0.63	6.92	0.30	6.62
total	173.115	207.667	271.274	1.84	1.80	1.25	3.32	0.35	1.77
Energy (quadrillion Btu's)									
coal	13.15	16.90	22.01	2.54	1.78	−8.45	−36.0	−2.71	−12.4
petroleum	34.87	30.64	32.59	−1.28	0.41	−22.4	−44.7	−2.98	−12.6
electricity	6.81	9.15	11.73	3.00	1.67	−30.5	−57.1	−2.97	−13.2
gas	24.47	31.07	27.04	2.42	−0.92	−9.97	−36.0	−3.99	−15.6
nuclear, other	5.55	14.25	20.00	9.89	2.29	−36.7	−61.0	−2.53	−11.4
Total energy input	78.03	92.87	99.60	1.76	0.47	−19.3	−46.1	−3.18	−13.4
Energy consumption (quadrillion Btu's)									
personal consumption	23.165	25.170	22.340	0.8	−0.8	−20.4	−53.8	−3.5	−18.1
services and government	10.936	13.104	16.441	1.8	1.5	−15.3	−38.5	−3.3	−7.8
electricity generation	21.080	28.319	36.298	3.0	1.7	−30.5	−57.2	−3.0	−12.5
industry	26.990	32.245	34.448	1.8	0.4	−12.6	−30.4	−3.2	−13.4
transport	2.672	3.177	3.844	1.8	1.3	−8.4	−21.0	−1.7	−7.6
total input	78.032	92.865	99.600	1.8	0.5	−19.3	−46.1	−3.2	−13.4
Prices									
agriculture				4.85	5.27	1.69	4.36	0.22	1.13
manufacturing				3.16	3.89	1.13	4.10	0.23	1.39
transport				2.15	2.68	1.21	3.10	0.21	1.03
services				3.99	4.34	1.09	2.76	0.09	0.44
coal				3.06	8.56	13.22	68.50	4.26	20.33
crude petroleum				3.86	4.62	−8.31	−5.87	0.06	0.30
refined petroleum				8.74	6.92	32.25	85.28	3.33	15.81

Table F-6–**Zero Energy Growth (continued)**

						% difference from level of			
				Growth Rates (% per annum)		*Historical Growth*		*Technical Fix*	
	1975	*1985*	*2000*	*1975–1985*	*1985–2000*	*1985*	*2000*	*1985*	*2000*
electricity				4.67	6.54	72.71	207.0	3.25	16.93
gas				6.57	7.39	5.95	49.68	4.39	19.70
consumption				3.91	4.29	2.77	7.33	0.30	1.31
investment				3.72	4.27	1.30	4.13	0.23	1.30
government				3.95	4.32	1.51	3.75	0.12	0.43
GNP				3.85	4.24	2.26	6.03	0.25	1.17

applied to each dollar of sales from the energy sector) with the tax revenue then being spent by the government on health, education and transport services (the revenue was allocated as follows: 75 percent to purchases of labor and services, 20 percent to purchases of manufactures and 5 percent to purchases from the transport sector). This is a dual mechanism: energy use is directly discouraged by taxes, and demand is further redirected by a change in spending patterns towards nonenergy-intensive production, which is superimposed on an economy which already has adapted to the energy-efficient *Technical Fix* position.

The move from *Technical Fix* to *ZEG* involves a reduction in energy input of 3 percent in 1985, and of 13 percent in 2000. The uniform energy tax discourages all energy use with the result that consumption of each energy source is reduced by comparable proportions; in 2000, ZEG consumption of nuclear power is reduced by 11 percent from the *Technical Fix* position, consumption of coal is down 12 percent, that of petroleum and electricity 13 percent, and of gas 16 percent. When compared to the *Historical Growth* energy consumption pattern, the *ZEG* energy consumption in 2000 is reduced by 46 percent with electricity and nuclear down by around 60 percent, and other fuels down by around 40 percent. The reduction in energy consumption varies between uses. The move from *Technical Fix* to *ZEG* in 2000 involves a 13 percent (15.4 QBtu) reduction in total energy input with final demand use reduced by 15 percent (4.2 QBtu), electricity generation use down by 13 percent (5.5 QBtu), and industrial use, including use of electricity, down by 12 percent (7.3 QBtu).

The tax rate required to produce the move between *Technical Fix* and *ZEG* is 3.3 percent in 1985 and 15 percent in 2000. The 1985 shift is comparatively small and the tax revenue is similarly small, but the 2000 shift is more substantial and the revenue raised by the energy sales tax is $131 billion ($50 billion in today's prices). This substantial revenue affords the opportunity to divert a significant amount of final demand from energy-intensive to nonenergy-intensive types of expenditure. (In fact, revenues of this size are of the order of magnitude required to sustain currently mooted national health insurance programs). The energy tax does result in substantial increases in energy prices; fuel prices in 2000 under *ZEG* are about 18 percent higher than under *Technical Fix*. Nonfuel product prices also increase, but by much smaller proportions, generally of the order of 1 percent. In total, therefore, *ZEG* involves only small increases in prices above those forecast for *Technical Fix* growth —the increase in the rate of inflation (of the GNP price deflator) is only 0.05 percentage points, from 4.03 to 4.08 percent a year.

Real incomes and real output are not reduced by the move from *Technical Fix* to *ZEG*, despite the reduction in energy consumption. The reason for this lies in the redirection of final demand caused by governmental purchases in services financed by the energy tax revenues. Reduced energy use without an exogenous change in spending patterns would lead to a reduction in real incomes and real output, as in the move from *Historical Growth* to *Technical Fix* growth, but the increase in demand for services caused by increasing government purchases creates sufficient new demand to offset the reduction in real output and, as the new demand is relatively energy-nonintensive, the restoration of output and incomes can be sustained at the new lower level of energy consumption. The net effect is that, in 2000 for example, real output rises by 0.25 percent in *ZEG* compared to the *Technical Fix* position, despite the 13 percent reduction in energy use. The gain in real output is, in itself, trivial, but the critical result is that energy consumption can be reduced without any cost in terms of total real output and total real income. The mechanism that secures this result is differential government policy—specific discouragement of energy use by means of taxes, and specific encouragement of nonenergy-intensive production and consumption by means of increased governmental provision of service activities.

The composition of production differs in *ZEG* from the *Technical Fix* pattern, due primarily to the impact of the new government expenditure. Agricultural and manufacturing output is reduced, transport and service output is increased. On the final use side, the net result of the energy taxation and higher government expenditure is a relative increase in the proportion of government purchases in real GNP with an equal decrease in the share of personal consumption expenditure; investment and net exports are not affected. Real output and real income growth rates remain almost identical in *ZEG* and in *Technical Fix* growth. The composition of primary inputs does change, however. The energy tax and increased service purchases lead to an increase in labor input relative to capital input, although both inputs show an increase in *ZEG* compared to *Technical Fix* growth.

The increase in labor input associated with *ZEG* is the result of energy-capital complementarity. Higher fuel prices lead to the substitution of labor for capital. Increased purchases of services leads to an increase in primary inputs, again with emphasis on labor input. Labor input in all nonfuel sectors increases, reflecting labor-capital substitution, while employment in service and government sectors rises substantially since increased activity in labor intensive sectors is superimposed on labor-capital substitution. Thus, total employment (labor input in man-hours) is 1.8 percent higher in 2000 under *ZEG* than under *Technical Fix* growth. If all the increase in labor input were supplied by those previously unemployed, the unemployment rate would fall to 1.4 percent. But the decrease in unemployment would probably be less as the additional labor would be supplied partly from longer work-weeks, partly from higher participation rates and partly from decreased unemployment.

Conclusions

The basic result of these economic analyses is the qualitative finding that substantial reduction in U.S. energy input, compared to the

Historical Growth energy demand patterns, can be secured without major economic cost in terms of reduced total real output or reduced real incomes or increased inflation or reduced employment. The scope for inter-input substitution, for economizing on energy use and for redirection of demand patterns is such that the rate of growth of energy input over the remainder of this century can be more than halved without requiring fundamental changes in the structure of the economy and without requiring major sacrifices in real income growth.

Energy conservation, as represented by *Technical Fix* and *ZEG* conditions, will have an economic cost that is non-trivial. At the aggregate level the costs are that total real incomes and output are reduced; thus the level of real GNP, in 2000, for example, is 3.5 percent lower under *ZEG* than under *Historical Growth;* and that the rate of inflation is increased. The real GNP deflator increases at 3.8 percent a year under *Historical Growth* but at 4.1 percent a year under *ZEG*. However, energy conservation leads to increased employment, so fears of widespread unemployment due to energy shortages are unfounded. Once the economy has had time to adjust to more expensive and less plentiful energy, employment will actually increase as labor is substituted for capital and material inputs. There are also costs of energy conservation at the microeconomic level; new input patterns in production will require a relocation of some people and jobs in both geographical and occupational terms, and people will have to adapt to new ways of doing things. The model does not spell out these very detailed effects, but it does show that, on the basis of economic responses observed in the past, such adaption is well within the bounds of practicability within the economic system as it is presently constituted.

The opposite side of these economic costs is the marked reduction in energy usage that is possible over the remainder of the century. The benefits from reduced energy usage are reduced environmental degradation, reduced pollution, reduced dependence on foreign sources for a critical economic input, reduced need for nuclear and other energy sources whose full implications are, as yet, incompletely known, slowing the rate of depletion of U.S. fuel resources and so on. These benefits are fully explored in other Energy Policy Project studies. The present study demonstrates that these benefits can be obtained, admittedly at a cost, but not at the cost of major economic dislocation. In fact, the present projections indicate that economic activity can grow along a broadly similar pattern to that experienced in the past while simultaneously achieving major economies in energy consumption.

We conclude this study by pointing out that:

- energy conservation along the lines of *Technical Fix* or *ZEG* ideas is possible within the existing structure of the economy;
- the cost of reduced energy use in terms of higher inflation and reduced real incomes and output are significant but not catastrophic;
- these costs have been quantified above so that the costs and benefits of energy conservation can be explicitly faced and compared. This information can provide the basis for a rational choice regarding energy policy in the United States.